Abraham Colles, Robert McDonnell

Selections from the Works of Abraham Colles

Consisting chiefly of his practical observations on the venereal disease, and on the

use of mercury. Ed., with annotations

Abraham Colles, Robert McDonnell

Selections from the Works of Abraham Colles
Consisting chiefly of his practical observations on the venereal disease, and on the use of mercury. Ed., with annotations

ISBN/EAN: 9783337424626

Printed in Europe, USA, Canada, Australia, Japan

Cover: Foto ©berggeist007 / pixelio.de

More available books at **www.hansebooks.com**

SELECTIONS FROM

THE WORKS

OF

ABRAHAM COLLES,

CONSISTING CHIEFLY OF HIS

PRACTICAL OBSERVATIONS

ON THE VENEREAL DISEASE, AND ON THE

USE OF MERCURY.

————————

EDITED, WITH ANNOTATIONS,

BY

ROBERT McDONNELL, M.D., F.R.S.,

EX-PRESIDENT OF THE ROYAL COLLEGE OF SURGEONS IN IRELAND, AND OF THE PATHOLOGICAL
SOCIETY; MEMBER OF THE COUNCIL OF THE UNIVERSITY OF DUBLIN; ONE OF THE
SURGEONS TO DR. STEEVENS'S HOSPITAL, DUBLIN; ETC., ETC., ETC.

CONTENTS.

PART I.

ON THE USE OF MERCURY IN VENEREAL COMPLAINTS.

CHAPTER I.

ON THE NATURAL HISTORY OF THE VENEREAL DISEASE.

A knowledge of the course of a Disease leads to fixed principles for its treatment, page 25—Mr. Hunter's Treatise contains the first systematic description of the Venereal Disease, 28—His History imperfect, as it did not trace the Symptoms through their various gradations, 28—Opportunities for further investigating the Natural History of this Disease are rare, 28—The use of Mercury a great obstacle to prosecuting this inquiry, 29—It alters the character of the Symptoms, and changes the order of their succession, 30—This information may confidently be expected from those who have devoted themselves to the Non-Mercurial plan, 31—The lengthened duration of the Disease another obstacle, 31 —A knowledge of the course of this Disease may be acquired by each practitioner contributing his observations, 31—Course, as occasionally observed, 32—Sometimes, instead of acute Eruptive Fever preceding the Secondary Symptoms, we find marks of wasting, and of Hectical state, 33—Eruptive Fever no longer ushers in fresh Symptoms, 33— From ignorance of these natural changes in the Disease practitioners cannot with justice estimate the value of different plans of treatment, and hence the contradictory reports of different remedies; How the patients are afflicted and carried off in the last stages of this Disease, 35 — Secondary Symptoms capable of infecting, 36 — Case, 37— Dr. ——'s case and its consequences, 37 to 40—Dr. E. Barry's statement, 40.

CHAPTER II.

ON THE ADMINISTRATION OF MERCURY.

CHAPTER III.

ON THE ADMINISTRATION OF MERCURY.

CHAPTER IV.

ON CHANCRE.

CHAPTER V.
ON BUBO.

CHAPTER VI.
ON A DISEASE OF THE LYMPHATIC GLANDS OF THE GROIN ATTENDED WITH PECULIAR SYMPTOMS.

CHAPTER VII.
ON SECONDARY SYMPTOMS.

CHAPTER VIII.

OF VENEREAL DISEASES OF THE MOUTH, &C.

CHAPTER IX.

VENEREAL ERUPTIONS, NODES, AND DISEASES OF THE TESTICLE.

CHAPTER X.

ON THE USE OF MERCURY IN VENEREAL HECTIC FEVER.

CHAPTER X.*

MERCURY DURING VENEREAL ERUPTIONS.

CHAPTER XI.

ON THE TREATMENT OF SYPHILIS IN SCROPHULOUS PATIENTS.

CHAPTER XII.

MINUTE DOSES OF MERCURY IN LATE AND CHRONIC BUBO.

CHAPTER XIII.

SYPHILIS IN INFANTS.

CHAPTER XIV.

PSEUDO-SYPHILIS.

CHAPTER XV.

ON THE NON-MERCURIAL TREATMENT OF SYPHILIS.

PART II.

ON THE USE OF MERCURY IN AFFECTIONS OF THE NERVOUS SYSTEM.

PART III.ᵃ

SELECTIONS FROM MINOR WORKS AND ESSAYS OF ABRAHAM COLLES.

ᵃ By mistake printed "PART II." at p. 328.

EDITOR'S PREFACE.

—◆—

WHEN I was invited, on the part of the Council of the New Sydenham Society, to edit selections from the works of Abraham Colles, I gladly undertook the task. I felt that it would not be labour thrown away to collect into one volume the scattered fragments of work accomplished by one whom I have always admired as having done much for the School of Surgery in his own country, and who is an example to all of what may be achieved by accurate observations, and truthful record of those observations.

I will confess, however, that I had little idea of how difficult a task I had undertaken, in proposing to introduce such comments and annotations as I might think proper. The truth is, I found that it was not possible to annotate the greater part of Colles's writings, with a view to bring them up to the present day. Such writings form a part of the classical literature of the subject of which they treat; they must be read as such, the reader ever bearing in mind the period at which they were written. Thus and thus only can we form a fair idea of the part such works take in the gradual advancement of knowledge. We must compare them with the views and doctrines put forth before they appeared. We must see how far they aided in accomplishing the further steps of progress made in later times.

Abraham Colles prided himself chiefly upon his introduction of the treatment of syphilis by small doses of mercury.

With evident satisfaction, he writes such passages as this—
"Now this is precisely one of those cases in which *very small*
doses of mercury will be found of the most signal service;
although I make no doubt that many surgeons will be startled
by the mere proposal of using mercury under such discouraging circumstances."

His solution of the question as to whether secondary
venereal symptoms are capable of infecting and of communicating the disease, but still more notably his observation that
syphilitic children nursed at the breast often infect wet-
nurses, but never their own mothers, are great fundamental
propositions. The latter observation is, in fact, the basis of
what Mr. Hutchinson has aptly called " Colles's law."

Many of the shorter memoirs are now difficult, if not
impossible, to find, save in public libraries.

Although feeling the difficulty, as well as conscious of my
incapacity of properly annotating the writings of such a man
as Abraham Colles, I confess that I have found even my
endeavour to do so not an uncongenial task. I esteem it an
honour that my name should be associated in any way with
one whom I have learned so greatly to admire.

MEMOIR

OF

ABRAHAM COLLES.

BY THE EDITOR.*

———◆———

THIRTY-SIX YEARS have now elapsed since the death of the eminent Surgeon whose works are, in great part at least, reproduced in this volume. He was at that time in his seventy-first year.

To estimate with justice what he achieved during his lifetime, we should endeavour to carry ourselves back to the opening years of the present century. The University of Dublin, it is true, long before this had possessed a Faculty of Medicine which enjoyed a *bonâ fide* existence. From the time of the Restoration a Regius Professorship of Physic had existed as a separate institution. As early as the year 1711 chairs of Anatomy, Chemistry and Botany had been established. Surgery was combined with Anatomy, and students who sought a purely medical education were permitted to attend lectures along with students in Arts whose names were on the books of the University. The Royal College of Surgeons in Ireland was not founded until the year 1784. The increased demand for naval and military surgeons on the breaking out of the French war gave a stimulus to surgical education, but certainly not of the highest kind.

---

* The materials out of which this sketch has been composed were placed at my disposal by my friend and colleague Mr. William Colles, Regius Professor of Surgery in the University of Dublin,—a worthy son of a distinguished father. The sources from which I have compiled it are manuscripts and notes, books left by Abraham Colles and in his son's possession, and also in great part from a memoir, published at the time of his death, in the 'Dublin University Magazine.'

The hospitals of Dublin, and the rich supply of subjects for dissection obtained by the lawless methods practised before the passing of the Anatomy Act, attracted many young men anxious to qualify themselves to enter the public service with the least possible expenditure of time, labour, or money.

The profession of Surgery at this time did not stand high in the estimation of the public. It was, as yet, not very far in advance of the trade with which it had been once so intimately associated. The Surgeon did not, as an accomplished, highly educated professional gentleman, occupy a social position alongside the Physician, as he does in the present day. The literature and science of Surgery had been raised by the labours and writings of such men as Percival Pott and John Hunter; yet they were but pioneers, and the errors into which Hunter had fallen (as we now know), in perhaps the most important as certainly the most classical of his works, gave ample scope to those who followed to investigate and correct them. Mr. Hunter's erroneous doctrines naturally struck deep root. They had been set forth by a man of earnestness and genius, and they took hold upon his disciples, as teaching always does when accompanied by a sincere conviction of its truth.

Such of our readers, therefore, as desire to form a just appreciation of the merits of the subject of the present memoir, should strive to picture to themselves what the teachings and the science of Surgery were about the year 1800, when Mr. Colles became a member of the Royal College of Surgeons in Ireland. They should compare this with the state of things when Mr. Colles resigned the Professorship of Surgery in the College in the year 1836. It will be still in the memory of many, that on this occasion the assembled College voted an address expressive of their feelings, and assuring him, amongst other things, that "it was the unanimous feeling of the College that the exemplary and efficient manner in which he had filled the chair of the Theory and Practice of Surgery, for thirty-two years, had been the principal cause of the success and consequent high character of the school of Surgery in this country."

It is only to the members of the medical profession that

medical biography forms a very interesting portion of that branch of literature which records the lives and preserves the intellectual portraits of distinguished medical men. Even a rapid and imperfect sketch of one who, while living, was the object in a high degree of their respect and admiration, cannot fail to be interesting to the members of his own profession. To young men about to start in the same arduous course, an outline of a course so honourable is peculiarly instructive. Mr. Colles won his way to eminence by means which are, to some extent, within the reach of all. He may be said to have wrested success from the hands of adverse fortune by his energy and indomitable perseverance. A finer model of every attribute and moral quality, essential to the achievement of high success in the practice of Surgery and Medicine, can hardly be offered for imitation.

Mr. Colles was born in the city of Kilkenny, of a highly respectable family long settled in that county. His family was of the class of gentry, and previous to their immigration into this country occupied an honourable position in England, where some members of it had represented the city of Worcester in Parliament. The first settler in Kilkenny was himself a Physician who made a considerable fortune. Mr. William Colles, father of Abraham Colles, was educated at Ballytore by Mr. Ebenezer Shackleton, who some time before had been the preceptor of Edmund Burke, with whom William contracted a friendship, and at whose house in England he was subsequently a respected guest.

Abraham Colles was the second son of this gentleman. He was born in 1773 at Milmount, on the banks of the River Nore, about two miles from Kilkenny. His early education was received in the endowed school of Kilkenny, under the auspices of Dr. Elliotson, formerly Fellow of Trinity College, Dublin. An incident is mentioned as having occurred whilst he was under the care of this gentleman, which had, perhaps, its influence in deciding his professional destiny. The flank of a house on the river's side near St. John's Bridge, inhabited by a Dr. Butler, then a Physician in Kilkenny, happening to be swept away by a sudden flood in the River Nore, a treatise upon Anatomy was carried down by the stream and picked

up by the schoolboy in a meadow near his father's house.
He restored the volume to Dr. Butler, but putting some
pointed questions respecting the contents, satisfied that
gentleman that his young friend had looked into its pages.
The book was probably of no great value, for the Doctor
made it a present to Abraham Colles, who carried it home in
triumph and soon preferred it to his Horace or Lucian. Not
long after, during a visit to Dublin, at the house of an uncle
who was a member of the Bar, he discovered to that relative
the passion he had conceived for medical studies, and his
mother was advised to permit him to follow the bent of his
talents.

In the year 1790 he entered the University of Dublin. His
progress at school had been such as to promise the highest
collegiate success. But he aimed at success in life rather
than success in college. He left his brother to gather
academic laurels, and to obtain premium and scholarship.
He limited his own studies to the object of passing through
the course respectably, and at once, after matriculation,
commenced his medical studies. He became apprentice to
Dr. Woodroffe then resident Surgeon in Steeven's Hospital.
From the first he devoted himself with extraordinary earnestness to the profession of his choice. He embraced it with
ardour and paid to it no divided allegiance. There are still
in existence case-books filled with notes of cases taken during
his apprenticeship.

An anecdote connecting his name with that of Edmund
Burke ought not to be omitted. His uncle, Mr. Richard
Colles, at whose house he was now a frequent visitor, had
some dispute with a London bookseller concerning the
publication and copyright of a satirical poem. He had also
a correspondence with Mr. Burke on the subject. Edmund
Burke's letter suggested to the young Surgeon some remarks
"On the condition of political satire," which he committed to
paper and showed to his uncle, who privately sent them to
his illustrious friend in England. Edmund Burke returned
them with encomiums on their spirit and good sense, even
recommending their publication. The author, however, when
requested by his gratified relative to prepare them for the

press, thrust the papers into the fire; and when his uncle talked of the name he was sacrificing, replied "that he had deliberately determined not to allow himself to be drawn aside from the career he had laid down for himself."

Mr. Colles obtained his diploma from the Royal College of Surgeons in Ireland in the year 1795, and then repaired to Edinburgh. He resided during two sessions in the Scottish metropolis, where he obtained his M.D. degree. His letters at this time, some of which are preserved, show the steady earnestness with which he worked. They give proof also, even at this early age, of that strong common sense which formed so important an element in his character, and of his desire for early improvement. He says, in a letter dated 1796: "Since my arrival here I have not made the acquaintance of any Irishman, or one of any other nation. Indeed, I am afraid to get acquainted with men to whose character I am an utter stranger. On Tuesday I took my seat at the Speculative Society, which consists almost entirely of Scotchmen." February, '96: "You may have a specimen of my oratory, for I am resolved to attempt speaking, and have actually begun to manufacture a speech for next night's meeting. As it is of home-spun material, I fear the texture will be but rude and coarse; but you may depend on having a true account of it." In April, 1796, he writes: "I must begin by telling you that I am much improved in point of personal appearance and accomplishments; for, having this day cast aside my winter clothes and put on my summer dress, my landlady could not be restrained, even by the dictates ef female modesty, from telling me that I was a very 'spry buck,' and recommending me very strongly to wear hair-powder. I need scarcely tell you that, strong as my pride is, it could not persuade me to go to the expense of six guineas a year." . . . . "I neither pay nor receive a visit till after eight o'clock in the evening, when I and a friend take a walk together; and we have entered into a most extravagant plan, but I must not be the one to rail against it as it took its rise from me. It is neither more nor less than dining together and spending the evening together every Sunday. As you may wish for a pattern of a nice entertainment, I give you, as an example, my first dinner:—a hind quarter of lamb, two shillings; salad, one penny; but as this lasted me three days, perhaps my party was not very expensive. Ten pence for a cod-fish, which also served for dinner for three days. As for my landlady, Mrs. Williams, she is positively afraid I shall read myself into a coffin, and actually comes to keep me idle at different times, out of her desire to keep me in health."

After residing in Edinburgh for two winter seasons, he obtained his M.D. degree, and with this honour he started on foot for London, a tour of some difficulty, in the year 1795. Mr. Colles has left behind him an interesting diary of this expedition, giving proofs in every page of his professional ardor, and his manly, independent character. In London he attended assiduously the principal Hospitals, and devoted himself with great earnestness to the study of Anatomy. At this period of his life his acquaintance with Sir Astley Cooper commenced, one which afterwards ripened into intimacy. He entertained for this distinguished Surgeon the sincerest admiration and respect, as is seen by his dedication to him of the most important of his works, in which he speaks of his "having the vanity to wish it to be known that he is honoured with his friendship"; and he adds: "I cannot neglect the only opportunity that may ever present itself of paying such a tribute as I have to offer, not only to the highest professional station and celebrity, but to a juster object of admiration— that zeal in the cultivation of Science which has outlived in you all the ordinary incentives to human industry."

Mr. Cooper (as he then was) was a young man engaged in the investigation of the anatomy of Hernia, and preparing for publication the results of his labours. Mr. Colles rendered him assistance in making the dissections from which were executed the engravings illustrating this work. It was probably at this time that he himself conceived the idea of publishing his 'Treatise on Surgical Anatomy,' which, although never completed, forms a fragment of an important work, and one which in fact marks an era in the teaching of Anatomy. Although now surpassed by others of the same kind, it was one of the first in which Anatomy was taught strictly in its bearings on practice and in its relations to Surgery. It was probably while studying Anatomy with Sir Astley Cooper that Mr. Colles first became impressed with the inefficiency of the older modes of teaching Anatomy, and the desire to instruct himself and others in this all-important subject after better methods. He constantly dwells on the fact that the language in which anatomical descriptions are conveyed is no less tedious than trifling, by striving after minute and unattainable accuracy, which serves only to impress the idea of difficulty when no difficulty really exists. In the introductory 'Address on the preparatory Education necessary for the Surgical Student,' which forms the opening pages of his 'Treatise on Surgical Anatomy,' he particularly dwells on this subject :—

"The profound and comprehensive mind of the philosophic Bacon, having discovered and demonstrated the necessity of methodical arrangement in the cultivation of the Sciences, Anatomists hastened to avail themselves of its advantages. They accordingly divided this Science into several distinct branches, as Osteology, Myology, Neurology, corresponding to the different distinct parts of the animal frame. These divisions they termed 'Systems.' Each system they described separately, without taking any notice in this description of its connexion with other systems, unless where it happened that that which was the immediate subject of examination should have remained absolutely unintelligible without such a reference. And succeeding Anatomists have ever since continued to tread implicitly in the footsteps of their predecessors. By these means we are constantly enabled to examine the several parts with an accuracy, and to describe them with a precision, before unknown. But though the description of each particular part be now perfect, yet the plan is still so far defective that the description of any one part seldom reminds the Student of any other, the examination of any one system seldom leads him to trace its connexions and relations with the other systems, nor do so many detached views of the several parts enable him to take any general and connected view of the whole. Thus the Student, who has been shown the distribution of the venous, arterial, and nervous systems of the arm, does not know how each of them lies with respect to the other at the bend of the elbow, and therefore he does not know how he should attempt, in cases of Aneurism, to pass a ligature round the artery without at the same time including its accompanying nerve which communicates sensation to the principal part of the limb. Nor can he, in the common operation of blood-letting, account for that sharp pain of which the patient particularly complains when the basilic vein is opened, because these detached descriptions of the different systems did not lead him to observe that some considerable branches of the nerves run down along the face of this vein. In short, an attempt to explain the nature and structure of the animal machine, by dividing the several parts of which it is composed into distinct classes, and then giving only a detached and unconnected description of each class, without even considering them as the component parts of one organized whole, is, in my mind, as preposterous and unavailing as would be an attempt to explain the mechanism of a watch by taking it to pieces, and giving a separate description of each particular wheel and spring without afterwards attempting to show by what contrivance the one moves the other, or how each wheel contributes, by its particular motion, to regulate the general movement of the whole machine. Is it then to be wondered at, that a plan so little calculated to excite industry or stimulate curiosity—a plan which, so far from showing the subservience of Anatomy to Surgery, does not even teach Anatomy itself as a distinct Science—a plan which leaves the whole weight to press on the memory, and that, too, in the most unfavourable manner, should have but few attractions for the youthful student? Is it surprising that he should consider the study of the Science a drudgery rather than a pleasure?—that he should take it up with disinclination and turn from it with disgust? In fact, the Student who has been employed in acquiring an anatomical knowledge of the different divisions or systems of the human body, has but encountered all the difficulties without securing any of the benefits. For such a plan of study can neither enable him to form a perfect idea of the structure of any part of the body, nor can these partial and detached views of the Anatomy in any degree qualify him to perform a Surgical operation. The study of Anatomy too generally ends at that point where it begins to be useful. . . . . If the present work shall enable you to trace for yourselves in the dissecting-room those parts which are most necessary to be known; if when you retire to your closet it shall assist to imprint on your memory a knowledge of those parts which you had previously dissected; if it shall explain to you the different operations in Surgery, and demonstrate to you the anatomical principles on which each step of every Surgical operation is founded, the views of its author will be fully accomplished."

In 1797 Mr. Colles returned to the Irish metropolis. His connexions were not numerous, but he determined to try his fortune. His industry was astonishing. He was blessed with excellent health and the most amiable of tempers. The frankness and vigour of his character were written on his countenance, while his robust and well-made frame showed that he was well suited for physical exertion,

It needed some courage at this time, without friends or interest, and we may add, without much money, to plunge into such a career as he had before him. Even with all the fortitude which he possessed, his purpose was sometimes shaken, and he thought of seeking a regimental Surgeoncy, or sinking into the retirement of a provincial Dispensary. Having resolved, however, to enter on practice in Dublin, his determination was not easily shaken. He became attached at first to the Sick Dispensary in Meath Street, a Charity which had been first established by the Society of Friends. He also became an active District Visitor to the Sick and Indigent Roomkeeper's Society, where it was his duty to visit applicants for relief and ascertain the truth of their condition. This self-imposed duty, however, was not professional : it was a work of benevolence which brought him in contact with the poor, and gave him unbounded opportunities of exercising the kindly skill of his profession. The humanity and zeal with which he performed the duties thus confided to him attracted the notice of his professional friends, and amongst others, of Mr. Stuart, at the time Surgeon General, who, perceiving the peculiar bent of his talents, urged him to give up Medicine and devote himself to the Surgical branch of his profession. At his solicitation he became a candidate for the post of Resident Surgeon in Steeven's Hospital, which he obtained in the year 1799. At this early period in his career he commenced Clinical teaching : he inaugurated the system of giving regular Clinical Lectures and also lectured on Surgery in his own private rooms.

This marked the first great step in Mr. Colles's professional life. He has left behind him a small book containing notes of the commencement of his career, in which he gives an interesting record of his small beginnings as regards fees. In 1798 we find the following entry :—" Sum total of fees received, £8 10s. 7½d.," and the accompanying remark : " Apparently a trivial sum, yet considering the time I was sick and in the country, and that this is the first year after my return from Scotland, I do not look upon it as a dispiriting circumstance that my fees have been so small." In the following year he indicates a marked improvement :—" Total of fees and salaries received from January 3rd, 1799, to January, 1800, £178 4s. 4½d.," and the observation—" This is a very great sum of money for my second year's practice : compare it with the previous year's total, and the comparison is very flattering." The third year is a still more remarkable advance :—" My fees, £133 17s. 5d. ; salary, £60 ; two apprentice-fees, £227 10s. ;—total received, £421 7s. 5d." The fact of his already at this early period of his career, receiving apprentice-fees, shows that he had commenced to enjoy some reputation as a teacher. Some of his entries show the peculiar humour which those of his pupils who still remember him describe as characteristic of Mr. Colles. " For giving ineffectual advice for deafness, £1 2s. 9d. ; for attempting to draw a stump of a tooth, ——. Another fee for I know not what service, unless he may have thought the last fee too small." April, 1799, is marked " fee-less." " For telling him that he was not dyspeptic, one guinea." Such were the early beginnings on an annual professional income which attained its maximum in the year 1826. During this year Mr. Colles received in fees £6128 ; from 1823 to 1834, his income was nearly although not quite so much, being always over £5000, and sometimes closely approximating £6000 a year.

From the very commencement he kept notes of his cases. We perceive a constant tendency to make memoranda of his own shortcomings, and the points in diagnosis as well as in operation which were difficult or puzzling.—

" September 8th, 1799. A man came into the long ward with his arm badly wounded by glass. I was very cool, seeing at one view what actually might be necessary. So far I was pleased, but when the nurses and other attendants seemed to be too slow or to commit any error, then I was entirely too hasty and too anxious for them to hurry."

September 17th, 1799, we find the following memorandum : —

" This day I removed a cancer. My anxiety for my own character was the predominant sensation at the commencement of the operation ; but this gradually wore off, and I soon felt for the success of the operation as if it had been a child of my own, or rather I felt as if I had performed some piece of mechanical work and was anxious for its success. My anxiety at the beginning of the operation was greater than I wish it to be on any future occasion, but on the whole I was well pleased that my state of mind had been such as it was."

In the year 1802 Mr. Colles had already attained sufficient eminence as a lecturer to justify him in seeking for the chair of Anatomy and Surgery in Trinity College. He was not successful on this occasion, having been beaten by Dr. Hartigan. As to the validity of the election some doubts existed. With his wonted energy Mr. Colles took legal proceedings to defeat it. The election, however, stood, and Mr. Colles had to turn his energies into other channels. He became an active member of the Medico-Chirurgical Society, where he read communications on Medical and Anatomical subjects. He threw himself with great earnestness into the proceedings of this body. He attended its meetings with regularity, and exercised his mental powers in its discussions. He became President of this Society, and in this office he had to deliver an address at the close of each session. The ability with which he discharged this duty raised him still higher in the esteem of his professional brethren.

In the year 1804 the Royal College of Surgeons in Ireland secured Mr. Colles's services as a lecturer. In that year the chair of Anatomy and Surgery became vacant, and Mr. Colles was appointed to it in conjunction with Mr. Dease. This was the second great step in his public life. This important and honourable Professorship he held for thirty-two years, discharging its duties with assiduity, zeal, and ability. It was not long after his appointment that the present edifice in Stephen's Green (the College of Surgeons) was erected. Every year added to its reputation. The classes increased rapidly in number, and at one period it counted more pupils than any such Institution in Great Britain. It success was in a great degree due to Mr. Colles. Like that of all such men, his energy was contagious. It not only infected his pupils: it infected his collaborateurs. The taste for study became diffused: the spirit of enquiry was excited, and a vast general impulse was given to professional education. Some idea of the advance made in Surgical education in Dublin may be gathered from the fact that, in the year 1800, when Mr. Colles himself first became a Member of the College of Surgeons, the number of Students of Medicine in Dublin did not exceed 60, while in the year 1836 they were computed to be about 1000, and were certainly not less than 800.

Abraham Colles in the year 1807 married Sophia, the daughter of the Revd. Jonathan Cope, Rector of Ahascragh, in the County Galway, by whom he had a large family. He at the same time took a moderate-sized house in Stephen's Green, in which he continued to live until he removed to his larger residence in the same square which he continued to occupy for the rest of his life. At the beginning of the year 1835 his health began to decline. He began to find the fatigue of lecturing too much for him, and even his Clinical work overtaxed his strength. It was not, however, until the following year that he resigned his Professorship in the College of Surgeons. He was induced to take this step at the urgent solicitation of his friends, who thought it absolutely indispensable to his health that he should do so. He took it, however, with great reluctance, and it deeply affected his spirits. He cared nothing for the emolument, but it was painful to him to relinquish its duties. He did not, however, resign his appointment of Steeven's Hospital until the year 1841, when he was obliged also to restrict his hours of practice. His practice continued until about the close of his career. He died on the 16th November, 1843, in the 71st year of his age.

The best evidence of the excellence of his life was the manner in which the news of his death was received by all who knew him. All the Medical Schools in the Irish metropolis suspended their proceedings. The College of Surgeons met and resolved to pay to his memory the tribute of a public funeral. The same step was taken by the College of Physicians and the Apothecaries' Company. His remains were interred in the cemetery of Mount Jerome, and the entire body of the Medical profession in Dublin followed them to their last resting-place. A vast concourse of private friends took a part in the procession, so universally beloved was he both within and without the circle of his profession.

The popular notion which finds expression in the saying that few men are found to be heroes in the eyes of those who see behind the scenes of their private life, gets but little support from the consideration of Mr. Colles's character. There were perhaps none who formed so high an estimate of his merits — so true a judgment of the genuine heroism of such a life as his — as those who enjoyed the closest intimacy with him. He held in contempt alike hypocrisy and fanaticism in religion, as he did any semblance of artifice or charlatanism in his profession. The principal features of his character were little calculated to catch the public eye or attract idle curiosity. He was an accomplished gentleman in the truest sense of the word; amiable, upright, modest, intellectual, seemingly unconscious of the high estimation in which he was held by those around him; simple in manners, cheerful in disposition, of the most kindly good nature, endowed with the charm of natural courtesy which made him the delight of his intimate friends and absolutely the idol of his own family circle. An acute observer, eminently sagacious, he knew the world well, but was not corrupted by it. Having a healthy and well-balanced mind, he possessed solid judgment, sound understanding, and the kindest of hearts. He was a man of rare benevolence, a generous giver, ever forestalling solicitation and sparing the pain of an appeal.

His likeness is well preserved both in marble and on canvas. The halls of the College of Surgeons in Ireland are ornamented with his full-length portrait, and a marble bust stands in their museum. Both are striking likenesses and good specimens of art. He had a shrewd, clear, good eye, a fine forehead, and a character of marked decision around the mouth. He was about the middle size, his figure well-proportioned, and his manner unaffectedly dignified. He was an early riser from the beginning to the end of his career. He liked rural retirement, and preferred the society of his own house and intercourse with his intimate friends to all other pleasures.

As an author his language is correct, strong, and explicit. He strove rather to express his ideas simply than to aim at eloquence or show. The simplicity of his language gives a kind of classical elegance united with a profound acuteness.

He studiously avoided reference to obscure and general principles. He reasoned by analogy and induction from established facts. His mind appears to have been somewhat deficient in speculative power, a pardonable fault when we remember what a deceptive guide analogy often is in scientific reasoning and argument. He never confused his reader with uncertain hypotheses founded on shifting principles.

In early life it is evident that he took great pains to acquire the faculty of speaking readily. He undoubtedly trained himself to do so with great point and energy, enforcing what he said with an expressive but simple elocution. Those who recollect him as a lecturer speak with genuine enthusiasm of the vivid pictures of disease which he sought to paint in words for his pupils. His lectures, which were delivered *extempore*, but from copious manuscript notes which, however, he rarely used in the lecture-theatre, bear comparison with those of Sir Astley Cooper. He possessed the art of touching briefly on the salient points of his cases, and was gifted with what Dove has called "the incomprehensible talent" of separating the essential from the immaterial in complicated phenomena. Though lacking that boldness of invention possessed by more imaginative minds, yet he controlled the observations and discoveries of others by the most strict and cautious criticism. He possessed that unremitting attention which allows few circumstances of real moment to pass unnoticed. With so much acuteness, such keenness of observation combined with depth of thought, he had the hand of a skilled workman. In short, as a practitioner in Surgery he possessed all the essential qualifications, sound judgment, cool determination, and great manual dexterity. He shunned affectation and singularity. He held in extreme contempt those who aimed at gaining notoriety by assumed eccentricity or anything resembling the tricks of a mountebank. He thus acquired the universal confidence of his own professional brethren, and without any accidental or external help raised himself to one of the highest positions which men can attain—the foremost rank in a liberal profession.

It is obvious from his manuscript notes (which are immensely voluminous, and from the very commencement of his professional career) that from his earliest days he possessed the character of a really strong man. He did not fear to speak to his pupils of his own errors. He sought to improve them as he did himself, by pointing out his shortcomings and aiming at avoiding similar errors in future. "Having shown to you," he says, in some memoranda which we find as early as the year 1800, "the manner in which amputation of the thigh should be performed, it will be in conformity with the plan I have laid down to disclose to you those steps of the operation where I have erred from confusion, hurry, ignorance, or inadvertency. The first, and indeed a very capital error, occurred in my application of the tourniquet, for the tourniquet compress had not been applied with sufficient firmness, so that, when the patient brought her body more erect the compress was thrown off the artery": and so he proceeds, pointing out the errors to be avoided.

Throughout his entire career as a clinical teacher he lost no opportunity of frankly admitting his blunders, and making them the means of instructing his pupils. The following anecdote is characteristic:—A patient in Steeven's Hospital died after having been treated by Mr. Colles for stricture of the rectum by the introduction of a bougie. The death had been unexpected. At the *post mortem* examination a malignant stricture was found, and below it a ragged opening through the coats of the bowel into the peritoneum. The resident Surgeon, when showing the specimen to Mr. Colles, took some pains to explain that it was probably by extension of the ulceration that the opening had been formed. "It is below the stricture," said Mr. Colles: "give me the bougie." The bougie having been produced, he passed its tip into the hole. "It fits exactly," he said, and then added, turning to his class, "Gentlemen, it is no use mincing the matter: I caused the patient's death;" and then, without another word, and amid the absolute silence of his class, he walked out of the theatre.

No man less than Mr. Colles took a tradesman's view of his profession. He nevertheless always spoke of it to his pupils as of a calling wherein honourable men, striving to gain a livelihood, confer advantages upon themselves by their

service to the public. We frequently find passages similar to
the following in the manuscript notes of his lectures :—

" Let the history of this establishment (the College of Surgeons)
serve to animate your exertions. See what progress the science
and practice of Surgery have made within the few years since the
foundation of this institution ; a progress which must in justice be
ascribed to the enthusiastic zeal, profound good sense, and incor-
ruptible integrity of a few individuals. See how these men, by
promoting the progress of science, have at the same time raised
the profession and its members to a rank in society far beyond
what even their fondest wishes could have anticipated. Be assured
that in this, more than in any other walk of life, public benefit
and private advantage are so blended together that the most certain
means of advancing your private interest is to promote the public
good."

" Be assured, by manly fortitude and determined perseverance,
you will sooner or later overcome every obstacle, and thus render
your future career of practice not only lucrative but honourable to
yourselves and most useful to the community."

In politics Mr. Colles was a liberal. He held his political
opinions, as he did all others, with modest firmness, but
without the faintest tinge of violence or party fanaticism.
He was not one of those who sue for favours, and when, in
the year 1839, his claims as undisputed head of his profession
in Ireland were overlooked, he was neither offended nor
disturbed. The body of the profession, however, not only in
Ireland, but, to their honour be it said, in England also,
expressed their surpise and dissatisfaction. Sir Astley Cooper
was one of the foremost in doing so. He bore witness to the
general opinion of the profession in the British empire, and
declared that although Mr. Colles had achieved honour for
himself far beyond what any Government could bestow, that
nevertheless he was the legitimate channel through which
to decorate in Ireland the profession to which he belonged.
This opinion was so generally entertained and so strongly
expressed as to reach the ear of the dispensors of patronage.
They acknowledged their mistake, and strove to repair it.
A baronetcy was offered and its acceptance more than once
pressed on Mr. Colles. He, however, firmly but modestly,
declined the proffered distinction. If, he said, it had been
offered to him in the first instance, he should have considered

it due to his profession to accept it; but that, for himself personally, such distinctions had no attraction, and that in consequence of the distribution he intended to make of his property amongst his children, a hereditary title would be an inconvenient honour.

Mr. Colles twice filled the Presidential chair of the Royal College of Surgeons in Ireland (in 1802 and in 1830). His great reputation as a skilful practitioner, his works as reproduced in this volume, established for him a world-wide fame beyond what any title could bestow. His family and descendants may point with just and honourable pride to the unadorned name of Abraham Colles, and boast that they are sprung from one who merited a title but declined it.

SIR ASTLEY P COOPER, BART.,

ETC., ETC.

My DEAR SIR ASTLEY,

Two motives influence me in placing your distinguished name in the front of the book which I now publish. I have the vanity to wish it to be known that I am honoured with your friendship; and I cannot neglect the only opportunity that may ever present itself of paying such tribute as I have to offer, not merely to the highest professional station and celebrity, but to a juster object of admiration—that zeal in the cultivation of science which has outlived, in you, all the ordinary incentives to human industry. Enjoying a splendid fortune, and an undisputed precedence, you have nothing further in wealth or eminence to desire. In such circumstances the world can ascribe your unabated ardour and energy in those fields of inquiry, where you have already had such brilliant success, to nothing but the purest love of truth, and the virtuous ambition of promoting the interests of humanity. An example such as you set will always have more admirers than imitators: amongst the former permit me to claim a high place; and believe me,

My dear Sir Astley,

With these and every other sentiment,

of regard and respect,

Your devoted friend and servant,

ABRAHAM COLLES.

Stephen's Green, Dublin,
20th Jan. 1837.

AUTHOR'S PREFACE.

I HAVE long been of opinion that, in the whole range of surgical diseases, there is not one which may with more justness be styled the opprobrium of Surgery, than the Venereal Disease.

When we recollect that, for three centuries and upwards, medical men throughout all Europe have been employed, most extensively and uninterruptedly in the treatment of this disease; that during almost the entire of that period a remedy has been known which possessed the power of curing this disease; that this remedy, since its first discovery, has undergone a variety of modifications and combinations supposed to render it more suited to particular states or forms of the disease; that, in addition to this, some other remedies, and other plans of treatment, have been found to effect its cure: when, I say, we take all these plans into account, surely it might be reasonably expected that, with all these advantages, we should by this time have arrived at a knowledge of fixed rules for its treatment, and have acquired such a command over it, as to be able to say that we can control its course and check its ravages.

Yet I think it is not an exaggeration to assert that at no period, since the disease was first subdued by mercury, has the opinion of practitioners been more divided and unsettled, or their treatment more wavering and unsuccessful.

Anxious to have this reproach wiped out, not merely because it involved the character of the Profession, but rather because it too clearly showed that the human race was suffering severely from our imperfect knowledge of the treatment of this disease, I sought to discover the cause of o· r backwardness. After mature consideration, I was led to attribute it principally to two causes :—

First. The imperfect knowledge we possess of the natural course, or natural history, of the Venereal Disease.

Second. The very imperfect knowledge we possess of the means of directing the operation of mercury, so as to make it act in a salutary manner; and the equally imperfect knowledge we have of the earliest phenomena which would indicate that it is beginning to act as a poison rather than as a remedy.

If this opinion as to the causes which have so long prevented us from acquiring the command over this disease be well founded, it will then be seen how vain must be the attempt of any single individual, however extensive his opportunities, to supply all these material defects: this can only be hoped for from the co-operation of very many.

In conformity with these sentiments, I have stated in detail such observations as I have made as to the progress or natural history of the Venereal Disease; and I have offered some remarks on the mode of administering mercury, so as to induce the salutary action of this medicine. I have also attempted to point out a few early indications, which denote that its action will become poisonous if its use be persevered in; and I have detailed many of the symptoms of Syphilis, and stated my opinion as to their appropriate treatment.

Although I fear it will be found that I have not shed much light on this obscurity, yet I trust that the attempt will be favourably received by the Profession; and that it will have the effect of stimulating others, more capable, to pursue the same line of investigation. It is only from the contribution of many inquirers that a complete development of a subject of such a nature is to be looked for; and I trust that no one will withhold the result of his observation merely because he has but little to offer; for the delineation of one new feature in the character of the disease, and even a single additional rule to guide us in the management of mercury, must prove a valuable acquisition to our present stock of knowledge.

From this statement, then, the reader will not expect to find in the following pages a systematic treatise on the Venereal Disease; on the contrary, some symptoms have not even been mentioned, and others have been noticed only in a cursory manner. The remarks which I have offered are merely the result of my own observations and reflections. Of what value these may prove I leave to my professional brethren to decide, only assuring them that the facts are stated with the most scrupulous fidelity.

To the observations on the use of mercury in Syphilis I have added some observations on the use of this medicine in the treatment of diseases not venereal. In this second part of the work I have studiously avoided any mention of its use in the diseases in which its efficacy is generally known and long established; and I have designedly and strictly confined myself to those in which it has been only very rarely employed, but in which I have had the good fortune to have found it a most active medicine, and a most speedy and invaluable remedy.

In conclusion, I beg to remark that I feel fully sensible of the many inaccuracies, both in style and composition, which occur in the following pages. My great object has been to convey my own meaning in a clear and perspicuous manner; if I have frequently expressed myself in common-place and too familiar language, it must be attributed to my being but little accustomed to literary composition, and long in the habit of delivering lectures extemporaneously. I trust, however, that these deficiencies will be excused, or, at least, not regarded as of any material consequence in a work the sole object of which is practical utility.

PRACTICAL OBSERVATIONS.

PART I.

ON THE USE OF MERCURY IN VENEREAL COMPLAINTS.

CHAPTER I.

ON THE NATURAL HISTORY OF THE VENEREAL DISEASE.

Value of a knowledge of the Natural History of this Disease—Obstacles
to our acquiring this knowledge—Order of Symptoms—Mode in
which this disease ultimately proves fatal--Secondary Symptoms
are capable of communicating this disease.

THAT morbid like healthy action has its laws, and that
disease in many of its apparently fickle changes and Protean
forms, is but proceeding in obedience to certain rules, are
positions established by innumerable examples and by
convincing proofs; witness the progress of many fevers; the
whole history of small pox; most of the eruptive or exan-
thematous diseases; the influence of vaccination, &c. &c.

These, and many other similar facts, have taught us that
certain diseases appear to be under a certain influence as to
the course they take, or as to the order in which their
several symptoms appear, and that morbid poisons, in
particular, excite a peculiar train of action, each producing
one particular set of symptoms, and none other, and that
these occur at regular periods, and proceed in a certain
regular order, in a sort of fixed succession; and that in the
constitution they infect, as well as in the part to which they
are applied, they never fail to give rise to certain morbid
processes, and to effect certain changes, which finally either

lead to their own extinction, or to the destruction of the system or of the part in which they have resided. When we reflect on these facts, and on the almost unerring certainty and undeviating regularity with which these occurrences take place, we must at once come to the conclusion, that to trace these laws—that is to expose the course and progress of disease, so as to be enabled, even beforehand, to predict the next symptom it will produce—must be a task not only interesting, but one of essential service in a practical point of view, for without this knowledge, at least to some extent, the practitioner must be in frequent uncertainty and doubt, and consequently liable to commit many errors and to meet many disappointments in his practice.

Accordingly in those diseases in which the attention of the medical profession has been most successfully exerted in developing the particular course they are to run, and in tracing the order in which their several phenomena are to occur, we find that the plan of treatment to be adopted for their cure has been proportionately well established on principles simple and intelligible: let us hope that, in the treatment of that disease to which these pages more particularly apply, similar exertions may produce similar effects, for I cannot refrain from expressing my strong conviction that one very principal cause of those unsteady rules of practice which exist even in the present day in respect to the treatment of Syphilis, is to be found in the want of a real knowledge of the natural progress of this disease; and that it is to an ignorance of the laws under which it proceeds, or to a want of knowledge of its natural history, that we must attribute the vacillation of some, and the contradictory assertions of others, who have expressly treated of this affection, its consequences, and cure.

[There is still ample room for enquiry before it can be said that we have gained much knowledge of the natural progress of this disease. It is indeed surprising how little has yet been learned as to its natural history. Messrs. Lane and Gascoyne, in their valuable memoir on Syphilization, published in the Medico-chirurgical Transactions, have not been able to assign, with anything approaching to definiteness, the true therapeutic action to syphilization, simply because so little is still known as to the cause which simple, uncomplicated syphilis runs. Thus, Mr. Lane believes that it does exercise some beneficial specific influence over the progress of the disease. Mr. Gascoyne, on the other hand, thinks that the natural tendency to recovery which an uncomplicated constitutional syphilis exhibits with the lapse of time and under circumstances favourable to the general health, is sufficient to account for the subsidence of the secondary symptoms during syphilization. It is obvious that until something more positive is known as to the natural progress of the malady, – until, in fact, there is some definite standard of comparison,—it is not only difficult but impossible to assign its true therapeutic value to any remedy. Carmichael did much in this direction. Mr. Rose, of the Coldstream Guards, followed up what he had begun. They proved what had been, in fact, unknown up to that time, that all primary syphilitic affections could be cured without mercury, and that the secondary lesions which follow, also in time give way to mild and simple treatment. "In my opinion," says Sir William Lawrence, "this is the most important step that has been made towards understanding the nature and treatment of the venereal disease, and I should place the truth thus established far beyond any of the speculations even in the work of John Hunter."]

We meet with nothing like a scientific description of the venereal disease in any of the earlier authors; in their writings we find only a mere enumeration of various local symptoms, without any arrangement or reference to their order of succession, and without any account of the course which any one symptom pursues through its different stages. Strange as it must appear, yet it is most true, and has proved most unfortunate for our science, that, prior to Mr. Hunter's treatise on the venereal disease, the profession possessed no systematic account, no accurate description either of its primary symptoms, or of the period or the order in which its secondary effects usually occur. This invaluable treatise poured a flood of light, not only on the natural history of this disease, but also on its pathology and treatment. It was Hunter who first demonstrated with clearness and precision that secondary symptoms succeed the primary at an interval of six or seven weeks, and that they are generally preceded by an eruptive fever, which is ordinarily of an inflammatory type; he also described the third order of symptoms; and all this he has done with an accuracy and fidelity which the subsequent experience of the profession has amply confirmed.

It is much to be lamented that Mr. Hunter did not prosecute
his enquiries still further—that he did not, for example, trace the
full progress of secondary symptoms, or mark their natural tendency to amend, and then again to relapse, each unfavourable
change being preceded by constitutional disturbance,—and that
he has omitted to notice the probable period of time which
these symptoms occupy from their origin to their acmé, and
thence to their decline. Whether these omissions proceeded
from want of time, or from his having adopted the opinion
that the venereal disease " had no tendency to cure itself, or
that the constitution was unequal to the cure of this disease,"
is now of little moment. But does it not appear strange that
subsequent writers have not made some effort to supply these
deficiencies? Every person, even of those superficially
acquainted with the medical literature of this disease, who
has perused the brief history and the clear classification of its
symptoms which this distinguished surgeon has given, must
have observed with satisfaction the great improvements which
have been effected thereby in its pathology and treatment;
and it was reasonable to expect that a sense of those benefits
would have impelled some later writer to investigate with zeal
the most minute particulars in the progress of its symptoms.
No doubt, the opportunities for prosecuting these enquiries
must have been but rare, as a little reflection will show; for
after the first series of secondary symptoms have appeared, it
must be extremely difficult to trace the further steps of the
disease, because every practitioner is so impressed with the
sense of duty to do all in his power for the relief of his
patient, that he cannot, consistently with that feeling,
withhold such medicine, or refrain from such means as may
be best adapted to the case: hence the further course of the
present, as well as of any future symptoms, must be more or
less altered from that which they would have pursued if
totally uninterfered with.

Another obstacle which, till within these few years,
obstructed all attempts to trace the natural progress of the
venereal disease, was the uniform employment of mercury in
every form and symptom of Syphilis. If that medicine had
been administered in such a manner as to have had merely

that effect upon the symptoms which it had in Mr. Hunter's experiments, namely, removing one set of symptoms, without curing the disease, and allowing the next series to appear in their natural time and order, we could then with ease and certainty trace the progress of the disease through all its stages; but this is by no means the case, for we find in the majority of instances, where secondary symptoms have occurred after the use of mercury, that this medicine has been administered very freely, and that therefore their occurrence is not to be attributed to the too sparing use of mercury; on the contrary, in the majority of patients who have been thus unsuccessfully treated, mercury has been carelessly or injudiciously employed, and has been pushed so far as to injure the general health materially, while at the same time it has so changed the local symptoms that we cannot without great difficulty recognise their true pathognomonic characters: for example, a venereal ulcer in the throat will no longer present the true specific appearance, if the patient has been overdosed with mercury, but will rather exhibit the features of a scrophulous ulceration, and will ultimately heal with that silvery cicatrix which is the natural termination of scrophulous ulceration in the throat; here then the treatment has caused such a change in the local character that we have been unable to recognise its true syphilitic nature. Another very important change which is sometimes effected in this disease by this injudicious or overpowering use of mercury, is that the order of succession of the symptoms is completely altered; thus, for example, if it have been administered in this way for venereal sore throat, we shall probably find, that the patient will be attacked, even during its exhibition, with nodes, or venereal swelled testicle; and that after this third order of parts has been thus affected, a venereal eruption (a symptom of the second series) will make its appearance: hence any history of the course of the venereal disease drawn from the description of its symptoms, during its advanced stages, in cases where mercury has been administered freely and injudiciously, cannot be relied on; for in such, a new set of symptoms will often appear in a quick and in an irregular succession, even while the system is under the operation of the mercury: one order will be as it were, anticipated, while another, such as the eruption, will be postponed (contrary to the usual course), until after the third order of symptoms shall have appeared.

Too frequently are the powers of the constitution so lowered by this indiscriminate use of mercury, that it is no longer able, as it were, to exhibit the eruptive fever, but it seems to be prematurely sunk into that weakened state which attends the latest stages of Syphilis. We can only hope to arrive at an approximation to the true history of the course of the venereal disease, by paying close attention to those cases which have been treated on the non-mercurial plan ; in such, although the natural duration of each symptom may be disturbed by whatever treatment is pursued (as the eruption for instance may be made to disappear even quickly by the use of tartar emetic), yet the order in which the subsequent symptoms will occur, and the individual features of each, will be preserved as faithfully as if no medicine whatever had been prescribed. I make no doubt but that many surgeons, who have adopted the non-mercurial plan of treatment, will be ready to favour the profession with most valuable information on this all-important point : for by referring to their daily notes of venereal cases, and collating a number of these, they can trace with unerring accuracy the march of symptoms, their order, their periods of appearing ; and with equal certainty can describe the states of the system which preceded and accompanied the various changes in the local symptoms.

Another obstacle to our arriving at a correct and full knowledge of all the phenomena of the venereal disease, is the lengthened duration of it, when it is not controlled or cured by medical treatment ; for we meet with various instances in which we can distinctly trace back the history of this disease to a period of five or six years. The unfortunate sufferer cannot be supposed to have remained during all this time under the care of the same surgeon, however strongly he may at its commencement have been impressed with a high opinion of his talents and skill ; still the melancholy experience that he has not been cured by him, in a period of twelve or eighteen months, must shake his confidence, and

create a desire for change, in the hope that another practitioner may prove either more skilful or more fortunate in
the management of his case. The friends of this unhappy
patient will, at all events, insist upon his applying to some
other surgeon, in whom they have more confidence : nor is it
to be wondered at, that these changes should take place three
or four times during the continuance of such a protracted
state of suffering. If we wish to watch the course of this
disease in an hospital patient, we are prevented not only by
the fact that he also loses confidence in the skill of his
surgeon, and becomes anxious to put himself under the care
of some other, but also frequently by the rules of the institution, which will not admit of a patient being retained beyond
a limited period, when his case is not in a certain train to be
cured. During the long period in which I have been engaged
in practice I cannot recollect to have retained, during its
entire course, the confidence and treatment of a protracted
venereal case except one, and that I was fortunate enough to
cure after a period of six years !

These obstacles, however formidable they appear, should
not prevent us from attempting to gain a perfect knowledge of
the course of this disease ; it cannot be expected to be
attained by individual observation ; it will require the labour
and attention of very many, each of whom however, by
contributing some new fact, or confirming one already known,
may furnish the materials to supply this desideratum, and so
complete the natural history of this disease. Should medical
men generally look on this matter in the light that I do, I am
confident that in a very few years we shall see this very desirable
object accomplished, especially as so much valuable information
must have been collected by those practitioners who have of
late years devoted themselves to the non-mercurial treatment.

Notwithstanding all these very serious obstacles, still we
occasionally enjoy an opportunity of tracing some parts of the
progress of this disease ; thus we sometimes meet with an
individual who either from absolute ignorance of the nature of
the disease with which he is troubled, or from an indifference
and recklessness of consequences, has totally neglected his
disease from its very commencement, and only seeks for

medical assistance after the secondary symptoms have been
established. If such a person be, from his situation in life,
unable for a time to submit to active treatment, or if the state
of his health render him unfit for it, we are afforded an
opportunity of studying in part the uninterrupted progress of
the disease. Under such circumstances I have observed, that
when the local symptoms have become fully established,
which is probably in two or three weeks after they make their
appearance, they then become stationary, and the constitution
is relieved from the febrile disturbance. In this quiescent
state the symptoms may remain for about three weeks; they
will then show a strong disposition to amend, and sometimes
they will proceed so far as to impress the surgeon with a
hope, and the patient with a firm conviction, that he is about
to get perfectly well. But this illusion, having lasted for two
or three weeks, is dispelled in general by a fresh attack of
eruptive fever, and by an eruption or sore throat, although
sometimes other symptoms (iritis, for example) be added to
those under which the patient had been previously suffering.
How long the disease might continue in this condition of
alternate improvement and deterioration, I cannot pretend to
say, as I have not had an opportunity of witnessing in the
same individual more than two, or at the utmost three, of
those revolutions: from what I have seen of such cases I am
led to say we may calculate upon each relapse as likely to
recur every third month. These remarks apply to the
secondary symptoms as observed for two, three, or four of the
first revolutions.

At a later, and sometimes, though rarely, in an early stage
of the disease, we find a somewhat different process ushering
in a new attack; thus we sometimes observe that the patient
who during the four or five preceding weeks had been
improving in flesh, colour, appetite, and strength, now begins
to exhibit a different aspect; his countenance alters considerably, it becomes sickly, his complexion assumes a waxen
hue, and he evidently loses flesh from day to day, and all this
takes place while the patient himself is not at all aware of the
change, and is still less suspicious as to its cause. In some
time, however, the patient complains of night sweats, want of

sleep, loss of strength, and declining appetite, so that in the course of two or three weeks he is reduced to a state of great weakness and emaciation: this downward course proceeds steadily until a fresh order of symptoms appears and becomes established.

I must here remark that I have repeatedly observed these newly appearing symptoms to be, as it were, a repetition of those that preceded them: thus, for example, a patient may have had a papular eruption which remained stationary for two or three weeks, then began gradually to decline, so as to lead him to hope that he was about to become totally free from it; but this eruption having faded to a certain extent, has then become stationary for two or three weeks, and the general health has improved; at the end of this time, however, the health again begins to suffer in the manner above described, nor does it cease to decline until this new train of symptoms has become for some time established. More commonly, however, this long-continued wasting is followed by some new symptoms, often by those of the third order, succeeding the repeated renewals of secondary symptoms.

After an attack of this lengthened nature we seldom, if ever, observe a distinct eruptive or premonitory fever appear in this patient; and I may also add that the local syphilitic symptoms cease to present their characteristic signs as strongly marked as heretofore. Thus it would seem as if each eruptive fever, or rather each succeeding attack, brings the constitution into a weaker and weaker state. In the very advanced stages of the venereal disease we do not see those periodical changes; the constitution then appears unequal to any struggle, so that one continued and increasing state of debility, with slow fever and great emaciation, are conjoined to the local symptoms, while the latter also are but little disposed to undergo any change, except a slow and gradual deterioration: thus, I have sometimes, though rarely, seen a tubercular eruption, combined with pains of bones and joints, continuing during four or five years, and undergoing no material change, being at one time a little better, and at another a little worse.

D

It must be obvious that, without a knowledge of the various changes above alluded to, the practitioner will be very liable to err in forming a judgment of the different plans of treatment which have been pursued; for were he to administer mercury, sarsaparilla, acids, or any medicine at the particular time when the symptoms were spontaneously declining, he would be disposed to attribute the improvement to his prescriptions: and on the other hand, if he had commenced the treatment when the health was about to decline (as it always does previous to the appearance of a new set of symptoms), he would be equally disposed to condemn whatever he had prescribed as worse than useless; I consider, therefore, that inattention to these circumstances has been one and the great source of those various contradictory statements which have been made even by men of sound judgment and strict veracity, as to the value of the different modes of treatment, and of the different remedies which have been recommended at various periods for the cure of the venereal disease.

When the disease has arrived at that advanced stage in which the general health has been broken down, and the local symptoms are so much changed as scarcely to be recognised as venereal, it is still a matter of uncertainty how long the patient (if not relieved by art) is to drag on a miserable existence, or in what manner a termination is to be put to all his sufferings. Many of its unfortunate victims are destroyed by what may be considered a continuation of the disease; but by far the greater number appear to be carried off by other complaints to which we may presume they were naturally disposed, or to which they were rendered liable by the very weakened and reduced state of their general health.

Among the former we may mention an ulceration of the throat, which, creeping downwards, at length seizes upon the larynx, and causes a destruction of a greater or lesser portion of this organ. The lungs quickly suffer from this affection of the larynx, and the patient dies, seemingly worn out by difficult respiration and by severe cough, with profuse expectoration and hectic fever.

Others again are destroyed by the repeated exfoliation of the bones, frequently of those of the cranium, and sometimes of the long bones. Moreover, there are others who, having been forced by severe nocturnal pains to have recourse to opium, cannot afterwards be prevailed on to relinquish this medicine, but are rather disposed to increase the quantity of each dose. Although now almost free from the pains which had driven them to adopt this medicine in the first instance, such persons, we remark, cannot be induced to rise until late in the day, they have no appetite for breakfast, but will probably take some highly-seasoned food for dinner; in the early part of the day they are quite torpid, averse to every kind of exercise; in the afternoon they seem to revive, become more animated in the early part of the evening, and are unwilling to retire to rest until a very late hour. One or two years may pass on in this manner with little apparent change; at length they sensibly decline in flesh, a diarrhœa supervenes, which is sometimes attended with sickness of stomach and vomiting, and by this they are ultimately carried off. It is remarkable that from the time these patients give themselves up to the inordinate use of opium their symptoms exhibit but slight traces of a venereal character.

Many others are carried off by diseases which have no other connexion with the original venereal affection except that by it, or rather by the treatment adopted for its cure, their system has been rendered more susceptible to the attacks of disease in general. Perhaps the greater number of such sink under attacks on the chest, such as pleuritis or pneumonia, terminating in serous effusion. Many of those who are strongly predisposed to phthisis fall victims to that disease; not a few are found to have the liver enormously enlarged, extending across the abdomen, and even below the umbilicus; this is followed by slow emaciation, and after some months by ascites and anasarca. Dysentery also at times seems to destroy some of those who have very long suffered under Syphilis, and, as might be expected, ulceration of the intestines is discovered after death.

From this account then it is very evident that there is no certain period after the appearance of the primary disease at which death terminates such protracted cases of Syphilis.

We may say that in general this event takes place between the second and fifth or sixth year. *

A most important point for our consideration in the history of the venereal disease is the question, whether secondary symptoms are capable of infecting and of communicating this disease. Contrary to the opinion of Mr. Hunter, and perhaps of most surgeons of the present day, I am convinced that they do propagate this disease, and this opinion I shall endeavour to support by adducing the following facts and arguments.

[We have here, without doubt, one of the most important of Colles's observations. Up to this time the opinion that secondary symptoms were not capable of communicating disease was firmly believed in. John Hunter speaks very positively about it. "We may observe," he says, "that even the blood of a pocky person has no power of communicating and is not capable of giving the disease to another, even by inoculation; for if it were capable of irritating a sound sore into a venereal inflammation, no person that had this matter circulating, or had lues venerea, could escape having a venereal sore whenever he is bled or receives a scratch with a pin, the part so wounded turning into a chancre." Mr. Guthrie, long afterwards, is no less distinct upon this point. "It is allowed," he says, "by all that the secondary ulcers of syphilis, as they are called, cannot produce primary ones; a proof that the nature of the disease is changed, and that the primary and secondary are two distinct things."]

Many years ago a young surgeon of my acquaintance, paying his addresses to a young lady, had unfortunately at the time a secondary venereal ulcer on the lower lip. The lady contracted an ulcer on her lip, which was soon followed by an enlarged lymphatic gland under the lower jaw; the ulcer had the character of venereal so strongly marked that the case was pronounced to be syphilitic, and she was directed to use mercury; unfortunately this medicine was employed in

---

* This account of venereal patients who, after long suffering, have been carried off by diseases not at all akin to, or naturally attendant on, Syphilis, is partly derived from, and is materially confirmed by, the concurrent testimony of Dr. Thos. Ferrar, now Professor of Surgery in the Belfast Institution, whose zeal led him during four years, while a pupil at Stevens's Hospital, to follow many of these unhappy sufferers through the various stages of their misery, to watch over them in their own wretched private abodes, and to trace them through the different hospitals of this city.

doses too powerful for her delicate constitution to bear with impunity, and this joined to deep mental distress seemed to induce phthisis, to which her system was strongly predisposed, and she shortly died of what is vulgarly termed a galloping consumption. This case, which came under my observation when I was but a very young practitioner, made a deep impression on my mind.

The next fact which I shall adduce is one that occurred in this city twenty-five years ago; and as the circumstances have been recorded by my friend Dr. Healy in Tilloch's 'Philosophical Magazine,' vol. xxxix., page 90, I shall extract his account, and annex to it a few further particulars which came within my own observation.

"On the 23rd July, 1810, I was requested by a respectable friend to see his wife, who complained of great and general debility, loss of appetite, with violent pain of the head. She was nursing, and the child seemed very healthy. She said she had been attacked with hæmorrhoids about three weeks after her lying-in, which was on the 11th May, 1810. She had had small glandular swellings in her groin, which had subsided; had taken no medicine. About the 28th, reddish spots appeared on her arms, and small tumours scattered over her thighs, which disabled her from walking. She complained of profuse perspiration on her breasts, particularly at night. I directed the warm bath with alteratives.

"August 7th. Notwithstanding the treatment, the symptoms became much aggravated, and the reddish spots had spread upon her face. I requested my friend to confess whether he had not contracted the venereal disease; to which he answered in the most solemn manner in the negative.

"On August 14th I required a consultation, and met one of the most eminent physicians in Dublin. After examining the patient, he mentioned our suspicion to the mother of the lady, that the disease was venereal. It was deemed advisable to have a surgeon in consultation, and that we should meet the following day.

"On 15th we met, and that consultation removed every doubt of the nature of the disease; and as she did not receive it from her husband, I suggested to the medical gentlemen that the accoucheur might have conveyed the infection by his hand. That mode was deemed possible, though not very probable; and our opinion to that effect was communicated to the accoucheur.

All that train of misery incident to supposed connubial infidelity, aggravated by the sufferings of a loathsome disease, must have been the fate of my patient if she and her husband had not proper mutual confidence, and a friendly reliance on my further investigation.

"On the evening of the 14th August, I visited a patient convalescent from fever, from whom I heard that a Mrs. M— was dangerously ill, not only from a disease which she had contracted at the time of her lying-in, but also from a very sore mouth. On enquiry, it appeared that she was attended in her accouchement by the same gentleman who attended my patient.

"August 18th. I met Mrs. M—'s brother, with whom I was acquainted; and anxious to vindicate the character of my patient, I told him my suspicions of his sister's disease, and asked him whether I could with propriety mention my suspicion of the disease to Mr. M—. He answered in the affirmative, and introduced me to Mr. M—. I related to that gentleman the situation of my patient, and requested to know the name of the disease his wife was labouring under, and the time of her delivery. He said she lay in on 22nd May, 1810, that the accoucheur was treating her at present for cancer in the womb, or for a liver complaint. I submitted my opinion that it was the venereal disease, and also that she might have been infected at the time of her delivery, the accoucheur conveying the infection by his hand. He added, he suspected her disease to be venereal, and had mentioned that suspicion to his wife frequently. I was introduced to the lady, and after examining her I became more confident in my opinion (as there were buboes in her groins). I advised Mr. M— to have a conversation with the accoucheur as to the nature of the disease. In consequence of this, Mr. M— called a consultation of the same medical gentlemen with the accoucheur, who met on the 19th.

"The surgeon, before the consultation, called me out of the room, and communicated what the accoucheur had informed him of that morning; namely, that he had contracted the venereal disease in his finger, in the course of his practice, and had conveyed this disease in that manner. It was deemed advisable and even indispensable for the accoucheur to confess that he was the cause of this severe ailment to these ladies; which he complied with by letter, not only to these ladies, but to others who had been diseased in like manner. Meaning to view the subject merely as an instructive medical report, suffice it to say, that upon a full and legal investigation, it was deposed on oath by a medical gentleman on behalf of the accoucheur, that he had contracted the disease in the course of his practice, about two years previous to the preceding unfortunate event; that he had undergone a complete course of mercury, and used even a larger quantity than is usual, and that he conceived himself incapable of communicating the disease; that previous to that gentleman's attending my patient, a window-sash had fallen on his finger, which produced a sore; that this sore became a venereal one, and infected the ladies before he was aware of its real nature. The child of the first lady was weaned on the 15th, afterwards spoonfed, and continued healthy. The other child was transferred about the

19th to a sound healthy nurse. In a month, a rash appeared on the head of the child, which in a little time spread over the body, and remained anomalous for a month, but afterwards became distinctly syphilitic, and yielded to the influence of mercury."

Such are the facts stated by Dr. Healy; and if credit be attached to them we must conclude that a secondary venereal sore can produce primary ulcers.

I am happy to have it in my power, from personal knowledge and observation, to corroborate Dr. Healy's statement, and thus to remove the doubts of any who might incline to be sceptical respecting it. I was not myself called upon to visit any of the sufferers on this melancholy occasion, nor was I consulted by the accoucheur himself, who was the undesigning author of their sufferings. My knowledge of the circumstances occurred accidentally in the following way :—

Walking at a very early hour one morning in the month of March to my hospital, I met this gentleman in a retired street. He told me that he had been up all night, attending the wife of a friend of mine in that neighbourhood—that he had left her quite safe. He next stated that he had not been in good health for some time past—that he had been afflicted with severe rheumatism, for which his friend, Mr. R., had made him take a great deal of James's powder and other diaphoretics; and that, by his advice, he had also used several tepid baths, but all with very little benefit. However, that within the last few days he had got rid, in a great measure, of the rheumatic pains, by the coming out of a general rash ; at the same time he drew off his glove, and showed me the palm of his right hand. My surprise was very great on beholding a most unequivocally syphilitic eruption. I also noticed an ulcer on each side of the nail of the index finger ; these bore the strongest marks of secondary venereal sores.

I did not at the time express to him my opinion of the nature of the disease, but told him, in a most impressive manner, that I thought he should not lose any time in showing it to his friend, Mr. R., under whose advice he was acting. When we parted, I could not help wondering at the readiness and unconcern with which this gentleman exposed the symptoms of his disease : and I at once concluded that he was totally ignorant of the common appearance of the venereal disease. At the period of this rencontre, and even for two months afterwards, no alarm, no notice of any mischief done by this practitioner in midwifery had occurred among his patients ; and therefore it is plain that all those ladies who were contaminated by him received the infection from the secondary venereal ulcer on his finger.

In a short time after the nature of this gentleman's disease became publicly known, I saw him in consultation with his attending surgeon, and had then an opportunity of examining his symptoms at my leisure ; and I became perfectly satisfied that his finger had been affected with secondary venereal ulcers.

Let us for a moment reflect upon the nature and value of this testimony. A surgeon, deservedly of the highest character in Dublin, deposes on oath before a commission of legally appointed arbitrators that this accoucheur had two years previously been afflicted with primary ulcer on the finger, contracted in the course of his profession ; that he had undergone a complete course of mercury, and used even a larger quantity than usual ; and that a window-sash having fallen on his finger, it produced a sore which proved also to be venereal, as it infected several females before he was aware of its real nature. I can feel no hesitation in saying that the ulcer on each side of the finger-nail was most decidedly a secondary venereal ulcer, and that the eruption was as strongly-marked a syphilitic eruption as any I had ever seen. The readiness with which this gentleman unnecessarily exposed the symptoms of his disease to me tends very much to prove that he himself was not at all aware of its real nature.

When he was affected with the primary ulcer, it would seem that he had not at that time done mischief to any of his patients, probably because he was aware of or strongly suspected the nature of his ailment, and adopted the necessary precaution of covering the part : but at the time that I saw the secondary venereal ulcers on his finger he was certainly ignorant of their real nature, and accordingly it was then that he communicated the infection to a considerable number of his patients. I was totally unconnected with any of the parties in this unfortunate business: chance alone made me acquainted with it in the manner I have mentioned; and I have thought it right to state the particulars which came within my own knowledge, because the corroborating testimony of a witness totally unconnected with any of the parties must have some weight in gaining credibility to a fact which is so contrary to the generally-received opinion, as to require all the support that can be given to it.

In further confirmation of this conclusion, I shall next adduce an article extracted from the 'Edinburgh Medical Essays and Observations,' which, having been written many years before the profession was at all aware of the distinction between primary and secondary symptoms, is on that account fully entitled to credit, as the plain statement of an eye-witness to a curious medical fact.

In vol. iii., Article xxi., page 297 of this work, is an account of a malignant lues venera, communicated by suction, in the city of Cork, in the year 1728. By Edward Barry, M.D., F.R.S. :—

"A woman in this city, who was commonly employed to draw the breasts of lying-in women, had, probably in the course of her business, received the infection in her mouth; which she either did not suspect, or concealed, till she communicated the poison to several persons of distinction.

"I think this affection was still more remarkable for its malignity and the quick progress of the symptoms than for the uncommon manner of its being received. As I had an opportunity of seeing most of the unhappy persons who were thus affected, I thought that a faithful account of its appearance, and the method of cure which I found successful, might not be unworthy of a place in your essays.

"The nipple first became slightly inflamed, which soon produced an excoriation, with a discharge of a thin liquor; from thence red spreading pustules were dispersed round it and gradually spread over the breast, and, where the poison remained uncorrected, produced ulcers. The pudenda soon after became inflamed, with a violent itching, which terminated in chancres that were attended with only a small discharge; and in a short time after, pustules were spread over the whole body. It finished this course, with all these symptoms, in most persons, in the space of three months. This disorder made a quick and dangerous progress in such who first received it, they not being apt to suspect an infection of this nature in their circumstances. The husbands of several had chancres, which quickly communicated the poison, and produced ulcers in the mouth, and red spreading pustules on the body; but such of them escaped who had timely notice of the nature of the disease before the pudenda was affected. Some infants received it from their mothers, and to the greatest part of them it was fatal.

"When I first mentioned my opinion of this disorder to the midwife of a person whom I visited, she said the woman who drew her breasts was a few days before, on such a suspicion, examined at the request of a lady of distinction before she would make use of her, and was declared free from any such distemper; by which means that lady was unhappily deceived, and was one of the last who received the infection. I ordered the woman to be sent to me, and I observed a small ulcer at the root of her tongue and a large recent cicatrix on the inward part of the upper lip. She obstinately denied that she ever had any sore there, but was so much terrified when I told her she would certainly rot away, that she begged I would not suffer her to perish if I suspected so much danger. While she was in a salivation she owned to me and Mr. Osborne, an eminent surgeon in this place, that she had had an ulcer where the cicatrix remained, which she cured by two or three doses of physic and a gargle made of woodbine, and some other ingredients; and said she concealed it because she imputed it only to cold, and was afraid, if known, it might for ever destroy her business and character.

"The woman who communicated this infection to so many had no eruptions on her body; and by what I could find, the infection never made any progress beyond the mouth. Some women, whose breasts were drawn by her, had never any marks of infection; but, by what I could find, the few who escaped were such whom she attended after the large ulcer on her lip was healed; for, while that continued, the nipple was received into a bed of corruption. But the case of a lady was very remarkable, whose breasts were drawn twice a day by her, when she communicated the infection to all others who came in her way. A violent confluent small pox seized this lady immediately afterwards, and she never had any marks of infection.

"The activity of this poison was so great that I immediately directed a mercurial salivation to even such as were but lately and lightly affected, and ordered it to be brought on by repeated unctions in a small quantity, with a few grains of calomel internally, and continued the salivation five or six weeks.

"I have often observed that, when the salivation easily rises on the use of the small quantity of mercury, the cure is uncertain, and the symptoms often return; and that the same inconveniences often attend a large salivation, accompanied with a great inflammation, from which many other dangerous symptoms also flow.

"The venereal disorder returned to some after a regular salivation, but was entirely removed by the following method, which I made use of to all who had this infection in a violent degree."

Here follows an account of his treatment by giving the patients large quantities of decoctum sarsæ, and also subjecting them to be sweated in a closed chair or bath :—

"When they had bathed in this manner five or six times, I ordered a strong decoction of guaiacum to be used instead of the former, and a few grains of calomel to be taken an hour before they entered the bath; and in some cases I have directed two grains of the turbith mineral to be mixed with the calomel, which, though continued for a considerable time, seldom affected the glands of the mouth, being determined with such force to the surface of the body. And, if the expression may be allowed, mercury thus managed produces a salivation through the pores of the skin. When the salivary glands are in the least affected, the use of mercurials is to be omitted till that symptom disappears. I ordered the bath three times in a week; in some cases repeated it five or six times successively every day, and gradually remitted the use of it."

I shall not avail myself of the charitable construction put by Dr. Barry on the manner in which this woman contracted the disease, viz., " that she had probably in the course of her business received the infection in her mouth," as this would incline us to think that the ulcer of the lip was a secondary ulcer; I shall admit that it was a primary ulcer or chancre, without stopping to inquire how it was contracted. It is obvious then that the ulcers of the nipples must also have been primary ulcers or chancres, but from these the whole system became contaminated, for we find that pustules were spread over the whole body; and that these pustules, or that this eruption, had also seized upon the pudenda, we have proved by the statement, "that the pudenda soon after became inflamed, with a violent itching, which terminated in chancres that were attended with only a small discharge."

This general eruption on the surface, as well as those spots of eruption which appeared about the pudenda, it must be admitted were secondary symptoms. This account of the general eruption having seized upon the pudenda will be corroborated by the experience of every surgeon who has been accustomed to treat the syphilitic complaints of females. For in them we almost universally find that the pudenda and

parts in their immediate vicinity are thickly beset with spots
of the eruption, which, however, in general differ from those
on the other parts of the body by being moist, yielding a
discharge having the appearance of an ulcerated surface.
By the way, it is remarkable that, while this takes place in
the venereal eruption of females, we find much fewer spots of
the eruption appear on the genitals of men than on any other
part of their body. From Dr. Barry's account it is plain that
the husbands of several of these ladies were infected by having
sexual intercourse with their wives, and that the disease which
they received was truly syphilitic, for " the husbands of
several had chancres which quickly communicated the poison,
produced ulcers in the mouth, and red spreading pustules on
the body." This surely is a pretty clear account of the
primary and secondary symptoms contracted from the wives;
and I cannot conceive by what ingenuity the above facts can
be explained away, or how it can be asserted that the disease
which was communicated to the husbands was from primary
and not from secondary venereal symptoms.

Let it be observed that at the time when this Essay was
written, medical men were unacquainted with the distinction
between primary and secondary ulcers, and therefore when
Dr. Barry speaks of *chancres* on the pudenda of the females,
we are to understand merely venereal ulcers. It also deserves
to be noticed that at this period no question had been raised
as to the power of any venereal ulcer to contaminate,—this
was then believed to be a property possessed by all and every
venereal ulcer. In my opinion, then, we have in this Essay
as strong traditional testimony as the case can admit of: it is
given by a man who had no particular theory or doctrine to
support, who seems to have written for the mere purpose of
stating facts, and recording an account of the propagation of
this afflicting disease in a manner which is so uncommon.
The only matters like theories in this Essay are his notions of
the causes why mercury fails at times to cure the venereal
disease.

A very striking instance of the power of secondary symptoms
to contaminate and produce the venereal disease is seen in
cases of infants infected with Syphilis, communicating it

first locally and then constitutionally to a sound nurse. I am
well aware that Mr. Hunter is of opinion that these cases are
not venereal. But I will appeal to the candour and observation
of every surgeon conversant with this branch of practice, and
who has witnessed cases of this description, and I ask him
whether the symptoms in the infant are not exactly such as
he has repeatedly observed in the adult. I am certain that
the answer will be in the affirmative: I put it to his candour
to say whether the secondary affections of the nurse differ in
any respect from those secondary symptoms which are the
consequence of genuine chancre? Are not the features of the
disease precisely the same? and are not both similarly affected
by mercury? If, then, the disease of the nurse so closely
resemble that produced by chancre, both being preceded by a
local sore, and if the secondary symptoms so closely resemble
each other that we are not able to discover any difference;
finally, if both be similarly affected by mercurial treatment,
how can we dispute their identity?

I am well aware that Mr. Hunter in his work has laid hold
of some few particulars in which these two diseases seem to
differ; but this difference, I think, may be explained without
involving the supposition that the disease of the infants and
nurse are not venereal. This I shall attempt to do when I
come to speak of the venereal disease in infants.

CHAPTER II.

Preparatory process necessary—Curative action of Mercury is attended
with Ptyalism—Regimen to be observed during a Mercurial course.

It will be readily conceded that the efficacy of any remedy
in the treatment of disease must in a great measure depend
on the accurate knowledge which the practitioner possesses as
to the most judicious mode of administering it; this position,
which is true in respect even to the most simple and innocuous
medicine, will be found to apply with peculiar fitness to the
use of mercury, when we reflect on the astonishing powers of
this substance, its great superiority over every other medicine
in the cure of certain forms of disease, the different results it
is capable of producing, according to the mode in which it is
administered, and, above all, when we consider that in the
hands of the ignorant and injudicious it will not only fail to
remove that complaint for which it had been prescribed,
but may induce other diseases of a still more intractable
nature, and that it even may, as too often has been the case,
cause the sudden extinction of life.

The practitioner about to administer mercury should not
only be well acquainted with the doses suited to the age
of his patient and with the probable effects of certain combinations on particular functions, but he should also bear this
fact in mind, that unless, in general, he prepare the system by
some preliminary attention so as to adapt or fit it to receive
the full benefit of this medicine, he will most probably be
foiled in his attempts to remove the disease, and he may even
entail upon his patient some other and more serious ailment,
and thus perhaps irretrievably injure a naturally good
constitution. Such preparative process is particularly required
in those diseases, and especially in certain forms of Syphilis
in which we consider that a protracted course of mercury may
be found necessary.

The older surgeons were strongly impressed with the necessity of some such preparation, and although I by no means subscribe to all the details of their practice, I yet fully agree in the principle. The preparative measures which were adopted by them were not confined to plethoric and robust individuals labouring under primary symptoms (such persons were soon reduced by bleeding, purging, warm bathing, and low diet), but they also very judiciously extended the principle to those cases where the health appeared broken down, and where the venereal disease was accompanied by some morbid state of the system not necessarily arising from that disease. At the time when I commenced the study of the surgical profession this practice was not entirely extinct; every venereal patient, on his admission into hospital, was then, as a matter of course, ordered to be bled once, to be purged four times, and to have the warm bath twice in the first week, and not until all these directions had been complied with was the use of mercury commenced. In the course of a little time the practice of bleeding was confined to those cases in which active inflammation attended on chancre or bubo, but still that of purging, warm bathing, and low living was in all instances strictly adhered to. In cases of secondary symptoms, in enfeebled and reduced habits, I never at any period saw any sort of preparatory process adopted; at present I may say preparatory treatment is considered unnecessary, for I cannot consider as such that plan which is now generally pursued in such cases. For example, when a consultation of surgeons is held on a patient who is emaciated and hectical, with enlarged lymphatic glands in the neck, &c., and that it is decided that he should pursue a course of sarsaparilla and acids, with country air, &c., I say that such is not a preparatory course in the true sense of the word, because this plan has been recommended in the hope that it may alleviate the venereal symptoms, or even altogether remove them; it has, in fact, been ordered as a succedaneum for mercury, with some flattering expectation that the patient may thereby be restored to health, and that thus mercury may be totally dispensed wi'' ; and, therefore, not with a view of preparing the system more effectually to benefit by its use, or to avoid the dangers which occasionally attend its exhibition.

I do not wish it to be inferred from the foregoing paragraph that it is my opinion that mercury should *never* be prescribed for a venereal patient who is much emaciated and hectical, so long as this latter condition of the system shall remain; on the contrary, I shall hereafter show that many such patients can and do bear a course of mercury, and that too with the greatest advantage, provided it be conducted with proper care and judgment. But all I contend for is, that if in such unfortunate cases we discover any peculiar morbid state super-added to, but not arising out of, the venereal disease, we should then first attempt the removal of that morbid condition before we venture to commence the use of mercury. If, for example, we venture to administer mercury to a patient who has for many days been troubled with a loose state of the bowels, we run the risk of inducing such a violent and obstinate diarrhœa as will be found intractable, and will speedily exhaust the slender remains of strength; or if a patient, subject to habitual bronchitis, should chance to labour under a fresh attack of it, great danger will attend the immediate use of mercury, and, therefore, some preparatory treatment is required. In fine, I am strongly of opinion that the want of a due preparatory process has of late years contributed to bring this valuable remedy into much disrepute, and hence partly it is that execrations against mercury are heard in every part of Europe; hence has it been accused of, and been condemned for, all those serious evils (and they are manifold and severe), which have arisen from its indis-criminate exhibition, all which ought rather to be ascribed to want of judgment on the part of the surgeon, and to incautious and irregular conduct on the part of the patient. In proof of this opinion that mercury does not merit all this odium, I may observe that while we hear it condemned by some in reference to the treatment of the venereal disease, we yet daily see its wonderful powers, and hear of its superior efficacy in the cure of many other diseases, over which its influence has been only of late years discovered. Thus, in acute inflammation of the deeper membranes of the eye, in acute inflammation of the joints, and of the serous membranes of the cavities, it is now lauded as a most invaluable remedy.

It is easy to explain why mercury should be generally so successful in other inflammatory diseases, while it so often fails in primary syphilis, which has generally a much less inflammatory character. The circumstances under which it is given in the two classes of cases are totally different: thus in syphilis its use must be continued for a much longer time than in those acute diseases; as many weeks will be required in the former as days in the latter; again, in these acute inflammatory affections the surgeon has, perhaps unintentionally, used the very best preparatory process, namely, he has freely bled and purged his patient, has enjoined absolute rest, and has put him on the lowest diet. Contrast with this the course ordinarily pursued with venereal patients. A young plethoric man, who has been pursuing a dissipated course of life, and pronounced by a surgeon to have a venereal ulcer and bubo, is directed to commence at once the use of mercury, either externally or internally: no previous measures are adopted, except possibly one dose of some cathartic medicine, no injunctions as to the necessity of rest and quietness and confinement to the house; no alteration in his diet—in fact, no attempt to reduce the inflammatory and plethoric state of his system; but because he is young and healthy, and free from any of those diseases which appear in our nosological arrangements, he is therefore considered as in a fit state for the immediate use of this powerful medicine. Could I now induce such of my readers as have had practical experience to recal to their recollection some of those cases of young men who have been thus hurried into a course of mercury without any preliminary attention to their state of health, or without any particular instructions as to their mode of living; I have no doubt that every practitioner could adduce many instances in which their expectations of a safe and speedy cure have been disappointed; how frequently have they witnessed in such cases that although the mercury acted on the salivary system, the ulcer assumed not only a very unhealthy, but even a very novel and peculiar appearance, one which has sometimes caused them to doubt the correctness of their first opinion as to its venereal character, and has induced them to lay aside the further use of mercury! Again in other cases,

E

how have they been surprised at the inefficacy of this medicine,
though the doses of it may have been increased from time to
time to a considerable amount, without producing any fœtor
of the breath, any swelling or ulceration of the gums, tonsils
or palate, or any obvious effect whatever, except an occasional
diarrhœa which may last for a few hours, and occur at
intervals of five or six days!—and when, at length, the
accumulation of mercury in the system has become very
great, a train of most alarming and dangerous symptoms
suddenly supervened, although not one of its ordinary or
salutary effects have been produced. Every surgeon I have
no doubt must have witnessed the misery that such patients
occasionally undergo even in respect to the local disease: the
inflammation attending their chancre is sometimes so excessive
as to require cold applications, and repeated and often
unavailing leeching; the bubo, too, is often attended with a
peculiarly severe pain of a scalding, burning or shooting kind,
although the swelling and the discolouration of the integuments
may be but to a moderate degree. When the purulent
collection distends the skin the suffering of the patient are
almost beyond endurance, and when the surgeon is forced to
give it vent the pain attending the act of opening it is sometimes so acute as to overcome even the greatest fortitude and
patience; the fluid in such cases is usually thin pus, mingled
with the colouring part of the blood. Although the patient
may have derived some relief from the operation, yet still he
continues to pass sleepless nights and most uncomfortable
days from the continuance of pain; even during this state,
however, he often possesses a good appetite and a fair share of
strength: but no local application succeeds in relieving the
local symptoms until the mercury shall have been laid aside
and the general health attended to; then the suitable topical
means will soon succeed in improving the condition of the
sores, and the system will gradually recover from the several
effects of the precipitate and injudicious mode in which the
use of mercury had been commenced and continued. I feel
certain also that surgeons must have often seen, in cases of
patients affected with secondary symptoms, who have been
worn down and exhausted with an irritable state of the bowels,

induced perhaps, in the first instance, either by a highly stimulating diet, adopted with the view of exciting appetite, or, as in the poorer ranks of life, by ardent spirits and food of an innutritious nature, that this condition of the intestinal canal has been greatly increased, and even urged on to an uncontrollable diarrhœa by superadding the irritation of mercury during this state of the system generally, and of the bowels in particular. Many of such cases have, under these circumstances, sunk into a state of the most alarming and even fatal weakness the moment the mercury has made an impression on their irritable and enfeebled frames.

These then are some, but by no means the whole, of the pernicious effects which may be, and which often are, induced by the administration of mercury when due precaution has not been used to prepare the system for its reception. When speaking of secondary Syphilis, I shall adduce some instances of its mischievous effects on many of the symptoms.

Having now, I trust, shown the necessity of some preparatory course of treatment previous to commencing the use of mercury, I ought next to speak of the mode of administering it; I wish first, however, to offer a few remarks upon the curative action of this medicine.

When mercury is exhibited for the cure of any other disease, as well as for Syphilis, we shall find that its sanatory impression on the disease is contemporaneous with its action on the salivary system, and that when the latter effect has not been produced, neither will the former have occurred. Thus in cases of acute inflammation of a joint, or of the dense membranes of the eye, we find that the progress of the disease is arrested the moment the salivary system becomes affected; and even in cases of other diseases, which cannot be considered as purely inflammatory or acute, the same remark will be found to hold good; thus in cases of orthopnœa depending on disease of the heart, with effusion into some of the thoracic cavities, and in which we commonly prescribe mercury in combination with squill and digitalis, the patient is not at first sensible of any improvement, but almost invariably, as so 1 as the gums become affected, he experiences relief, and perhaps the very next morning after this occurrence he tells

tells us with joy and gratitude that he is considerably better, that he has passed a night of refreshing sleep, and that he has been able to do what he could not have done for weeks previously, namely, rest in the recumbent posture without any of that distressing and alarming sense of suffocation under which he had previously laboured, and which always supervened the moment he sunk into that position.   It is unnecessary to particularize many other diseases in which the same fact occurs : indeed it may be asserted as the general rule.   The contrary also will be found equally true, that is, that mercury will not prove serviceable in any disease for whose cure it has been prescribed when it does not produce its wonted effect on the salivary system.   How often has this been verified during that time when it was the fashionable practice to prescribe a course of mercury in all chronic affections of the liver ?   It then happened over and over again that slight delicate females have been subjected to this treatment ; friction perhaps has first been tried, and this failing to relieve the complaint,—because it had also failed to affect the salivary system,—the internal use of mercury has then been substituted, or probably combined with the former ; thus the medicine has been persevered in and the doses increased even to an extravagant degree, but yet withal no salutary effect has been induced ; on the contrary, the little remnant of strength has suffered so materially that at length the mercury had been lain aside, and the friends, as well as the medical attendant of the patient, have had reason to express not only their disappointment, but also their amazement at the inefficacy of the mercury, of which the patient had taken fully as much as would have sufficed to salivate at least half-a-dozen young and vigorous men.   If, then, it be so very generally found that, whenever mercury exercises a salutary influence over disease, it at the same time always affects the salivary organs ; and if, again, whenever it fail to produce this latter effect, it be also found altogether inoperative in the cure of disease, it is surely a fair and legitimate conclusion to affirm that ptyalism marks the natural and salutary operation of this mineral.

I am well aware that this doctrine has been questioned by

some practical writers on Syphilis, but I trust I shall be able hereafter to show that this disease does not form an exception to the above general position. Let it not, however, be inferred from the foregoing statement that I would wish to measure the efficacy of mercury by the amount of salivation which it excites; on the contrary, the degree of ptyalism that I am always anxious to attain is merely an increased secretion of saliva, accompanied by swelling and superficial ulceration of the gums, and sometimes also of portions of the lining membrane of the cheeks and lips: this I am desirous of attaining as a sort of index which denotes, first, that the mercury is acting in a safe and salutary mode upon the system; and, secondly, that it displays that degree of power or energy of action which will be sufficient to eradicate the disease: so certain do I feel of the correctness of this view, that during a course of mercury for the cure of Syphilis, should this ptyalism be suffered to decline for some days, I should fear that all the additional mercury which may have been subsequently given in this more feeble manner would prove useless, that is, unequal to cure the disease. At the same time I am aware that many surgeons have exerted their ingenuity to devise a mode of administering mercury whereby ptyalism or any obvious effect on the salivary system might be prevented; I have reason to believe, however, that the precepts of such surgeons have not been very generally adopted, and with most practical men have fallen into disuse.

The general regimen which patients under a mercurial course should pursue is a subject of some importance, and one of which I have not spoken as yet. In former times it was the custom to have the patient covered with a thick flannel or woollen-cloth dress, and the head even covered with a sort of hood, so that the face only was exposed; he was also confined to a very warm, close room, from which the external air was most carefully excluded. I believe it was Mr. Hunter who effected a revolution in this practice; and this not so much by any forcible argument as by simply asking a question, which he prefaces by an assertion :—

· The manner of living under a mercurial course need not be altered from the common, because mercury has no action upon the disease which is more favoured by one way of life than another. Let me ask anyone, what effect eating a hearty dinner and drinking a bottle of wine can have over the action of mercury upon a venereal sore, either to make it affect any part sensibly, as falling upon the glands of the mouth, or prevent its effect upon the venereal irritation? In short, I do not see why mercury should not cure the venereal disease under any mode whatever of regimen or diet."

While I dissent from the rigid rules of discipline insisted on by our ancestors during a mercurial course, I do, however, believe that their system was, on the whole, productive of much less mischief than what results from the laxity of modern practitioners in this respect, who, I conceive, err exceedingly in the contrary extreme. Thus, to permit a strong plethoric young man, with primary symptoms, and who has not undergone a preparatory regimen, to indulge in all the pleasures of the table, to eat a full dinner and take freely of wine, is only calculated to increase the tendency to febrile action which, when at all excessive, will counteract the disposition to ptyalism.

With respect to the influence of exposure to the air during a mercurial course, we need only refer to what has been written by various authors on the benefits of free exposure of the system to the fresh air in cases of mercurial Erethism, and also to the assistance which the practitioner can derive from it in the management of those individuals whose idiosyncrasy renders them very susceptible to the effects of even small quantities of mercury.

Should our patient appear debilitated after the moderate and the desirable degree of ptyalism has been induced, then we may safely allow him a full and generous diet, from which, however, I would still exclude all stimulating food or drink: moreover, if about this period the patient's strength fail, or if the local symptoms do not improve as we had expected, we may then combine with great advantage bark and opium with the mercury, while at the same time we should reduce the doses of the latter. The experience of hospital surgeons will, I have no doubt, support my opinions on this subject; for they must have observed on many occasions that if the symptoms of an out-patient who is using mercury should become worse, and that he then be received into the hospital, his complaints have quickly improved, and that, too, while the same doses of mercury have been continued.

I know it may be urged against this strict rule of regimen, in favour of which I have expressed myself, that the daily practice of some surgeons proves it to be unnecessary, inasmuch as they allow their patients free exercise in the open air and a liberal diet, and yet they are able to effect numerous cures. While I admit this, still I think it must be granted that surgeons daily meet with many untoward circumstances and considerable delays in this branch of their practice, that not a few of their patients pass on from one stage of the venereal disease to another; and while some of them are cured in the advanced stages of the complaint, after repeated disappointments and protracted sufferings, many others fall victims to its sequelae, or are carried off by some of those acute diseases to which the deranged state of their system renders them peculiarly liable.

When I call to my recollection the results of my own observations on the treatment of this disease, made thirty years ago, and compare it with that of the present day, I think the comparison speaks decidedly in favour of the plan which was then pursued. For some years after I entered on the study of the profession, a surgeon felt himself rather humbled if he allowed a venereal bubo to suppurate; and if secondary symptoms appeared, he was considered to have mismanaged the case, and not unfrequently lost for ever after the total confidence of his patient.

[Many of the older surgeons held the same view. At this time no distinction had yet been recognized between infecting and non-infecting venereal diseases. Mr. Solly speaks in very much the same terms in his evidence before a Committee appointed to enquire into the Pathology and treatment of venereal disease. "When I was a young man," he says, "I was a pupil of Sir Astley Cooper and of Mr. Travers, and if a man had a case of secondary syphilis then it was considered a disgrace to him, and he was held accountable for it; but at the present day, as every man is in the same boat, it is not considered disgraceful if his tr.· ment of Syphilis is followed by secondary symptoms." The question of how far mercurial treatment prevents secondary accidents is still *sub judice*. Many of the best observers, however, agree that it does not prevent them, although many think that it postpones or modifies them. Amongst these we may mention such distinguished names as Skey, Parker, Hutchinson, Erichsen, Sir William Fergusson, Longworth, &c.]

It is true that at that time mercury was often used in excessive and in dangerous doses, salivations most profuse were excited, and which were attended with all their accompanying evils; but still on the other hand, the patient who escaped these perils was generally freed at once from the disease. The regimen then was as strict as the medical treatment was severe. Lodging-houses were established in Dublin, solely for the reception of young gentlemen who required to go through a course of mercury; and these houses were always fully occupied—so seldom was it that a young man could, while living among his own family, undergo the severe discipline to which the surgeons of that day thought it necessary to subject him.

But from the time that Mr. Hunter's work on the venereal disease came to be generally read and acted upon by the surgeons of this city, the discipline became not only less severe, but actually as lax as Mr. Hunter himself could wish. Young men, pleased at the removal of these restraints, too frequently overstepped even the moderate bounds which were prescribed for them: and surgeons, finding that secondary symptoms frequently appeared in cases so treated, were glad to adduce Mr. Hunter's theoretical opinions of " disposition to diseased action," and so forth, partly to justify themselves, and partly to satisfy their patients; so that at length from this, and perhaps from other causes which I shall not now consider, those cases of primary venereal disease which, under the old practice, required six or seven weeks for their cure, were, under the new plan of treatment, found to require as many months, or even years. By the former the disease was really and quickly cured; but by the latter it is only pursued from one resting-place to another, so that the patient's mind is often kept in a state of suspense and anxiety for very many months,— the symptoms disappearing for a few weeks, and then returning; and so this scene may occur over and over again, until, by some lucky chance in the treatment or in the effects of the medicine, the system shall come to be finally relieved from the disease.

While I freely admit that excessive over-doses of mercury
have inflicted the most severe evils, yet I as confidently affirm
that within the last twenty years the same medicine, when
employed in under-doses, or rather when it has not been
pushed so as to induce its legitimate action, has been
productive of infinitely more mischief; and I feel confident
that this position could be supported both by professional as
well as non-professional persons, were the effects of the latter
practice as obvious and striking as are those of the former;
but they cannot be so fully appreciated by the ignorant, and
they have not been sufficiently attended to by the profession.

The alarm which of late years has been raised against
mercury, as well as the mischief to which it often gives rise
when it happens to disagree with the patient, have rendered
practitioners not only cautious, but absolutely timid and
nervous as to its use,—so much so, that as soon as they observe
the first unfavourable change in the symptoms they at once
suspend the further use of mercury; some even seem not to
be aware that chancre and bubo almost always appear to be
worse on the day or two preceding the occurrence of ptyalism;
about that time the chancre spreads somewhat, and the bubo
becomes more painful and uneasy. Now this is the very
juncture when the disease is, as it were, within the grasp of
the surgeon, provided he only persevere in the medicine with
prudence; but if, through fear or ignorance, he lay it aside at
this moment, he will lose the happy opportunity of expeditiously, and almost certainly, curing his patient; and then
he assures him that the mercury is beginning to disagree, and
that it cannot be continued any longer with safety. A patient
had better not have used any mercury whatever than have
been treated in this way, for the constitution has been
disturbed, and febrile irritation has been excited to no purpose;
and should the patient be of an unhealthy habit the local
disease may assume a very unfavourable aspect, while the
general health may become so much disturbed as to render it
unsafe to resume the use of mercury for a considerable time.

A similar mismanagement of mercury in cases of secondary
Syphilis will be attended with results fully as unfavourable;
and the more frequently the error is repeated, so much the

more serious will be the evils consequent thereon. Under such circumstances the surgeon tells his patient that his case can no longer be served by mercury, and therefore he consigns him to the non-mercurial plan. In this manner, I believe, a very few may be cured, but many more I know have been allowed to sink into an untimely grave by the slow and silent, but certain operation of the venereal disease, the symptoms of which have become so changed or masked that common observers could not recognize the features of the original case. No doubt, observation, experience, and sound judgment are very requisite to enable the practitioner to decide whether those changes which appear at the critical moment when mercury is about to excite ptyalism be really unfavourable and alarming, or whether they may be merely the natural effects of the mercurial action : this, however, is not the only instance which the practice of surgery presents, in which accurate observation, discrimination, and judgment are required to practise the profession with safety and success ; these are qualifications which can only be attained by long and attentive experience : no precise rules, therefore, can be laid down to guide the surgeon at this critical juncture, neither can any words impart to him that practical knowledge without which he cannot conduct the further treatment of the case with satisfaction to himself, or with true advantage to his patient.

CHAPTER III.

ON THE ADMINISTRATION OF MERCURY.

Mode of administering Mercury—Friction—Internal administration—
Ordinary effects of Mercury on the system—Necessity of paying
close attention to the patient, from the time of the first impression
of the Mercury on the system—Cautions to be observed in the use
of Mercury—Eruptive diseases consequent on the use of Mercury—
Erethismus Mercurialis.

I now proceed to consider the mode of administering mercury so as to induce its favourable action on the salivary
system, and at the same time its sanative effects on disease,
particularly on the primary symptoms of Syphilis. In order
to explain clearly my views upon this subject, let us suppose
the simple case of a young man affected with primary venereal
symptoms, but in other respects in perfect health. I select
such a case because it will require a more lengthened course
of mercury than most other diseases do, and therefore more
judicious management is necessary; and also because the
changes in primary venereal symptoms will often assist us in
deciding whether the medicine is acting in a salutary manner
or otherwise, and consequently whether it should be persevered
in or discontinued.

I shall first speak of the mode of introducing mercury into
the system by friction, because in this method of administering
this medicine its various effects are more clearly and unequivocally exhibited. The patient should be apprised of the
necessity of rubbing in each dose of the ointment carefully,
but not violently, and this he should do in the morning rather
than at night, and for these reasons: first, the skin is soft in
the morning and will bear friction better; and secondly, the
sleep will not be postponed or disturbed, as usually happens
when the nightly friction is employed, the patient being
fatigued thereby, and thrown into a state of febrile excitement
which is inimical to sound repose.

As to the mode of rubbing in the ointment, the patient should be directed to divide the whole quantity to be used into four parts, and then rub in each portion perfectly and successively until all are consumed; it is better to apply it to one limb only on each day, as thus that pustular eruption, which is a common consequence of mercurial friction, is less likely to be excited. I prefer making the patient apply the ointment himself, whenever his strength will admit of the exertion, because the friction of his own hand is less uncomfortable, and less likely to produce irritation and eruption, than that of any other person, even though protected by a bladder ever so well prepared.

I do not advise the thighs to be shaved, as many surgeons do, because in a few days, when the hairs grow and become stiff, more irritation is produced, and thus the tendency to pustular eruption is increased.

During the course of mercurial friction, the patient should wear the same drawers both by day and night; he will thus have some portion of the ointment constantly applied to the surface; some of which will probably be absorbed during the intervals between each friction. When the same part of the body has been rubbed two or three times, it is advisable to wash off the remains of the ointment with warm water and soap, and this ablution should be made the night before the next friction.

When we deem it necessary to emply mercurial friction with a patient who is in a very weak state, or more especially with one who is in a constantly feverish condition, we must direct the friction to be performed by some other person, to whom we should give the following instructions:—The servant should be furnished with a pig's bladder, which, after having been well steeped in warm water, should be turned inside out, then well impregnated and softened with sweet oil or fresh lard. After this preparation it is to be tied round the wrist of the assistant, and the ointment rubbed by him according to the directions already given.

The dose for each friction (in this supposed case) should be ℥ss. If we are very anxious to induce ptyalism rather early, we may at the same time direct five grains of Pil. Hydr.

to be taken every night at bed-time; we may thus act upon the absorbents of the internal as well as upon those of the external surface at one and the same time. If we use mercury internally, in cases of Syphilis or other diseases requiring a rather protracted use of this mineral, we may employ either Pil. Hydr. in doses of five grains night and morning, which is equivalent to half a drachm of Ung. Hydr.; or, should we prefer Calomel, it may be given in doses of two grains at bed-time, uncombined in all cases, except in such as are likely to suffer by its purgative effect. For it will be found to lose this effect after the second or third dose. But when purging is to be avoided we should combine with it a small proportion of some opiate.

In ordinary cases we seldom observe any of the usual effects of the medicine until the fourth day, when, by *interrogating* the patient, we learn that he feels a peculiar taste in his mouth like that of copper or brass, or, according to some, like that of iron; that this feeling is most distinct in the morning, or when he has remained for some time silent, or without food or drink: at this period he makes no complaint as to his general health, but on particular enquiry he will admit that his appetite is not quite so good, his sleep not so long; that his bowels have a tendency to costiveness, and that he has a little more thirst than usual. All these changes, however, are to so trifling an extent that they will not be complained of, nor will they ever be mentioned unless the surgeon enquire for them. About this period, too, a very slight mercurial fœtor of the breath may be detected, but there is very little alteration in the state of the mouth.

About the sixth or seventh day ptyalism is fairly established, the gums are swollen, and appear as if inclined to separate from the teeth; they also present a slight degree of ulceration on their edges, especially in the intervals between the teeth; the lining membrane of the cheeks, opposite to the last molares, assumes a leaden colour, and is also swollen, so as to bear the impression of the teeth; the soft palate also is often swollen and more red than natural, as if it were slightly inflamed: the patient states that his mouth is disposed to fill with water, and that during sleep some saliva had flowed so

as to wet the pillow; that he cannot chew any firm substance
without suffering pain in his teeth, and that a sense of aching
and uneasiness in the gums remains for some time after
taking food, even though it should be of a kind easily
masticated.

From the third to the seventh day, or rather from the time
of the first impression of the mercury on the system, until
the full establishment of ptyalism, we should pay very close
attention to the patient, for during this period he is liable to
much suffering and to most danger; and it is at this period
that the attentive and judicious surgeon can be of essential
service by giving a right direction to the medicine, as well as
by counteracting any injurious effect it may produce. Thus,
during this critical period, the patient is liable to attacks of
griping, frequent desire to go to stool, and tenesmus; these
efforts are attended with only slight evacuations, which chiefly
consist of mucus tinged with blood; sickness of stomach and
vomiting also often supervene, the skin is hot, and the pulse
quick; all which phenomena are explained by the fact that
the specific influence of the mercury has taken effect upon the
alimentary canal instead of the salivary system. This
dysenteric affection so generally appears at this period that
the patient should be forewarned and prepared for it. He
should be directed to discontinue the use of the mercury as
soon as he feels this uncomfortable effect, and he should be
provided with draughts, containing each Tinct. Rhei ʒi.
Tinct. Opii gutt. xx. in any appropriate vehicle; one to be
taken after each dysenteric stool. An opiate enema may be
used instead of the draught whenever the stomach rejects the
latter. A gentle diaphoresis also should be encouraged by
the tepid bath, or by bathing the feet in warm water. In the
course of a day or two this febrile excitement will, under this
plan of treatment, have somewhat subsided, the bowels will
remain free from disturbance, and then we shall generally
find that the mouth has become a little more affected; and
should we wish to have it still more so, we may resume the
use of the mercury, and continue it in such doses as the
circumstances of the case shall require.

It is most desirable that ptyalism should be induced without any severe constitutional disturbance, within the first seven or ten days, as it establishes the fact that the mercury agrees with the individual, and therefore it assures us that it can safely be employed, and that there is every probability of its accomplishing in this case as much as mercury can do.

The young practitioner should know that sometimes the influence of mercury on the system manifests itself by a somewhat different state of the salivary system. Thus it will occasionally happen that the patient about the usual time will complain of some fœtor and some soreness of the gums, and yet we do not find any increased flow of saliva after a further lapse of some days. If we now carefully examine the gums they will be found less soft and less swollen than in the regular form of ptyalism, but their edges will be more ulcerated; indeed the ulceration of the edges appear to be the only change which they have undergone. Such cases also are attended with rather too much of fever. Should we persevere in the present doses of mercury we shall fail in inducing ptyalism; the result will be a more severe state of ulceration and a smart degree of fever. I have seen cases of this sort, in which the mercury was still continued, and this state of ulceration of the gums and the fever were kept up for five or six weeks, until, at length, a stop was put to the further use of the medicine, either by the fever rising to an alarming height or by the surgeon entertaining an opinion that the disease was cured, because so much mercury had been used, and during so many weeks. This state of things is what the lower orders in Ireland call a "dry course," and it is an effect of mercury which a little experience will discover to be totally inefficient for the cure of the disease; indeed, in some such cases we shall find that all the symptoms remain unmoved and unaltered. This imperfect attempt at salivation may, however, be converted into the full and legitimate ptyalism by reducing the doses and lengthening the intervals between them, and at the same time using such means as are calculated to reduce the rather high degree of fever which attends this peculiar condition of the mouth.

1 some few patients the first influence of mercury on the system is exhibited in the throat and not on the gums.

Such persons about the fifth or sixth day complain that they
have a sore throat, which they naturally endeavour to account
for by saying that they had caught cold. On inspecting the
fauces, we discover a degree of erysipelatous blush on the
arches of the palate and some inflammatory thickening of the
velum palati; on the tonsil, and generally at its upper
extremity, we see a superficial ash-coloured slough : one side
only may be thus affected, or both may be engaged in it. I
need hardly say that, in such a case, a further persistence in
the full doses of mercury would not only prove ineffectual for
the relief of the venereal symptoms, but would also be
attended with considerable danger to life, by inducing a
sloughing condition of the fauces.

Under such circumstances we must for a short time
altogether withhold the mercury, or lessen the doses of it.
By proper management, I have seen in some of these cases a
regular ptyalism induced, but in a few others I could not
succeed in producing this desirable result, the renewed action
of mercury reproducing the same condition of the fauces;
such cases, however, were conducted safely through the
mercurial course by watching them closely, and taking care
that the inflammatory state should not be allowed to rise high.

I have been thus particular in stating at what period the
surgeon should wish ptyalism to commence when using
mercury for the cure of primary symptoms, because I am
firmly convinced that he cannot count upon a cure if the
salivation occur at a period much different from that above-
mentioned; for should it be suddenly excited, and even
though not very profuse, it will yet leave the disease uncured;
perhaps the primary symptoms may be removed by it,—
oftentimes they are not : but at all events the secondary
symptoms will not fail to make their appearance, sometimes
in full vigour, though often under a more subdued form.
This I have seen repeatedly exemplified in cases of iritis
treated by calomel alone, or by calomel combined with
frictions. I have also repeatedly seen mercurial fumigations
which were used to stop a destructive ulceration of the throat
produce a speedy and profuse salivation, by which the
ulceration has generally been stopped, but in almost every

such instance the other venereal symptoms were only checked for the moment, to return with renewed vigour. And here I must remark that, after such sudden and profuse salivations, I have uniformly found the symptoms very unmanageable, and by no means yielding in the usual way to a subsequent use of mercury.

On the other hand, if the salivation be very late in appearing, not only will the patient have to lament the loss of so much valuable time, but the surgeon will too frequently find that the system of his patient has been so irritated by this protracted, though moderate use of mercury, that the ptyalism excited at this late period is not borne with the same ease by the constitution, nor is it productive of the same improvement in the symptoms that would have resulted from it, had it come on at an early period of the treatment.

If the above statements be correct, we cannot then comply with the wishes of a patient who will tell us, "I will commence the use of mercury at once, although for some days I must be exposed to the weather, dine out, and be out at night; this, however, cannot be any prejudice to me, as there will be but little mercury in my system; but next week I will make arrangements to attend to all your directions." This is a most injurious mode of proceeding, for the great object which is sought for by the administration of mercury is to be attained by due attention during the first six or eight days; it is during this most critical period that the greatest care and even nicety in the management of the case is required; and it is then that trifling circumstances may divert the medicine from its proper course. Whereas if the peculiar action of mercury be once established, then there is but little danger of its salutary course being interrupted; so that a man who is under its legitimate influence may be exposed to cold, or may even commit slight excess or irregularity in living, with much greater impunity than one who has only commenced the use of mercury, and on whose system it has not yet exerted any salutary influence. If in this latter case its action be interrupted by cold or fever, or any other cause, all will go wrong, the constitution will only suffer a certain degree of irritation, and the disease will continue unsubdued;

F

much time, therefore, must be lost in allowing these unfavourable effects to pass away, and greater care and judgment must be exercised when next we resolve on resuming the use of the medicine. Therefore in such cases, when the patient is unable to pay the necessary attention to his health, it will be more prudent to postpone the use of the mercury altogether for eight or ten days rather than commence it under the disadvantages of exposure to cold, or of any irregularities in diet, mode of life, &c.

The remarks which I have ventured to offer on the administration of mercury for the cure of primary symptoms will enable us to adapt the use of this medicine to the treatment of other diseases in which it has been found most valuable. Whenever, then, we employ this medicine for the cure of any other disease which is not actually threatening destruction to the texture and functions of the parts in which it is seated, or which does not endanger life, we should observe the same rules and cautions as have been laid down for its use in the treatment of Syphilis. Thus, in cases of sciatica, of organic affection of the heart, attended with orthopnœa and effusion into the chest, the ptyalism established on the sixth or seventh day may be kept up without injury for ten or fifteen days longer, if the obstinacy of the symptoms should require it ; whereas, in cases of destructive ulceration of the throat or nose, or in very acute iritis, or in cases of inflammation in a large joint where we dread that suppuration will take place, we can only arrest these serious evils by a sudden and violent action of this remedy. Here we run the chance of exciting profuse salivation by the very rapid manner in which we pour in the mercury. Now, in such diseases, we know by experience that a protracted use of this medicine is not called for, and we are well aware that the progress of these dangerous diseases is at once arrested the moment that the system acknowledges the full action of the mercury ; and this check having been once given to the rapid and dangerous progress of the symptoms, we have it in our power by other means to subdue what may have been left uncured by the mercury.

I am thus anxious to state my opinion as to the importance

of ptyalism, inasmuch as it is my firm conviction that the most essential part of a course of mercury is to effect this object with as little injury to the general health as possible. Indeed, I feel satisfied that many of those failures which we hear of in the treatment of Syphilis by mercury, as well as of other diseases by the same remedy, are to be attributed to an inattention to this very important part of a mercurial course. Without a knowledge of the difficulties which he has to encounter while endeavouring to excite ptyalism, the surgeon will not only be unable to steer his patient with safety, through the sudden and unexpected changes which occasionally occur, but he will also have much less prospect of conducting him to a final and a perfect cure. I have no doubt that it is in consequence of the doubts and difficulties which have arisen from this cause, namely, inattention to, or imperfect observation of, the various early effects to which mercury gives rise, and ignorance of the proper remedial means in each case, that there have arisen of late years so much discrepancy and vacillation in the treatment of the venereal disease, and which may justify some in adding its name to the list of those diseases which may be considered as an opprobium to the art and science of surgery. I must add, that for whatever disease, whether medical or surgical, a moderate and continued action of mercury may be employed, it behoves the practitioner to be as well acquainted with, and as watchful of, its early effects, as he should be when he employs it for the cure of the venereal disease.

When once the system has been brought under the influence of mercury, there is but little difficulty in the further management of the case; and in my opinion, there is but little to fear even from its protracted use, provided that mere ordinary attention be paid to the general health, and that care be taken not to allow the action of the medicine to pass beyond the line which has been already pointed out. At the same time, if it be our object to keep up a certain degree of ptyalism, we should be very careful not to allow the action to subside too low or too soon; we are to be guided by the state of the gums, cheeks, and fauces, rather than by the quantity of saliva which is secreted, and we are not to allow the tumefaction of the former to subside very much, but should occasionally renew the use of the mercury or increase the dose as soon as we find that its influence begins to decline. The observation of Mr. Hunter is not to be lost sight of :—

" The constitution and parts are more susceptible of mercury at first than afterwards; if the mouth is made sore and allowed to recover, a much greater quantity may be thrown in a second time before the same soreness is produced; and, indeed, I have seen cases where it could not be reproduced by as much mercury as could possibly be thrown in. Upon a renewal of the course of mercury, therefore, the same precautions are not necessary as at first."

The plan which I have found most successful to reproduce the ptyalism when it has sunk too low is to give pretty smart doses of calomel, viz., three grains, with one of capsicum, two or three times a day, at the same time that the former doses of the ointment are resumed.

I shall next advert to some of the untoward circumstances which occasionally occur during a course of mercury, and which frustrate our plans, as also those more dangerous effects which often arise in the early period of its exhibition, and at the same time point out the means of avoiding or of removing those obstacles and dangers.

Almost every individual possesses some peculiarity of constitution, or, as it is termed, some idiosyncrasy which is often very remarkably exemplified in the effects of certain articles of diet and of medicine, but particularly in those of mercury. As such peculiarity cannot by any degree of foresight be ascertained to exist, it is a judicious rule to observe that, before we commence a course of mercury we should inquire, had the individual ever used mercury on a former occasion, and if so, whether it had produced any peculiar or dangerous effect.

In the first place, we sometimes meet with persons so susceptible of the action of mercury that ptyalism will be induced even by a single dose. As this almost sudden influence of mercury on the system is very seldom serviceable, especially in Syphilis, we should endeavour to obviate it; such a patient should therefore be directed to keep very much out in the open air, to take a nutritious diet, and two or three glasses of wine daily; the dose of the medicine also should be reduced to one-fourth, and an interval of one, two, or even three days left between each, and thus we should endeavour to effect what Mr. Hunter calls "an accumulation of mercury in the system." When we have succeeded so far, we may prescribe larger doses or shorten the periods between them, so as to excite ptyalism, which will then occur without any unfavourable symptoms. Salivation once excited, must then be directed or governed according to the judgment of the surgeon, whose discretion will guide him as to the quantity of mercury to be administered, as well as to the frequency of the doses, and the length of time it is to be continued.

Again, we meet with cases in which a very slight effect only of mercury is perceptible about the fourth day; in such it will be necessary to increase the power of the medicine, consequently we should increase the quantity either externally or internally, or both. There can be no precise rule laid down as to the extent to which the dose is to be increased; it certainly ought not to be less than by one-half, and sometimes it may be doubled at once: the former proportion will probably be the better. If the patient have been employing mercury both by friction and internally, the dose of both should be increased; but if he have been using friction alone, then this should be continued at an increase of one-half, and he should take in addition five grains of blue pill every night. If in such cases we did not *early* increase the doses the system would become familiarised to the mercury, and at a late period we should employ very large doses, and thus run the risk of exciting a profuse and dangerous salivation.

We sometimes find in other constitutions that two or three doses of mercury induce a feverish state of the system without the peculiar fœtor of the breath, or any of those other symptoms that indicate the approach of ptyalism: the skin will be hot, the pulse quick, with great general restlessness; under these circumstances, we infer that the mercury is disagreeing with the system, and should we now either continue its use or increase the dose, we shall only add to the mischief by increasing the fever; and if we still persevere, we may endanger the life of the patient; in general, however,

he is soon reduced to such extreme weakness, and is so
sensible of becoming daily worse, that he will himself oppose
any further use of the medicine.   Whenever this febrile state
of the system, unaccompanied by any sign of its acting on
the salivary organs, is thus early induced, we should at once
desist from the further use of mercury, and endeavour to
allay the fever by exposure to cold air, and by the adoption of
such other means as may be best suited to the circumstances
of each case.   Having accomplished this, we must allow some
time for the recovery of strength before we venture to resume
the mercury; and when we do so, we should begin with
smaller doses, and these at longer intervals; for experience
has shown us that the former quantities were such as
overpowered the system.   We must also endeavour to obviate
any febrile action by occasional purging, by tepid bathing,
and by the frequent use of mild diaphoretics; we are now to
proceed cautiously, increasing the strength and frequency of
the mercurial doses, until we gain the desired object, namely,
a genuine and wholesome salivation.   Here I would observe
that this difficulty is not an uncommon consequence of
plunging a healthy vigorous patient into a course of mercury
without any previous preparation.

Not unfrequently we meet with individuals who appear to
evince an obstinate resistance to the effects of mercury
without having any fever induced.   Such patients, when
questioned on the subject, inform us that they had a slight
mercurial faetor on the second or third day; that it did not
continue longer than for a few hours; that they do not now
find any effect except an increase of appetite, there is no faetor
of the breath, no restlessness, no febrile excitement: in a
word, the medicine seems to have made not the least
impression on the system, excepting, perhaps, a slight
improvement in the appetite.   The surgeon now increases the
dose of ointment, but without any sensible effect, and then
perhaps suspecting that the external absorbents are not in a
state of sufficient activity, endeavours to ensure the full
operation of the mercury by directing some favourite
preparation for internal use, while he still continues the
external application; yet after six or eight days' trial of this

combination, he is surprised to find there is no impression made on the system, except, perhaps, some slight emaciation. Under these circumstances, too, it generally happens that the chancre, or whatever the original disease may have been, pursues the same steady quiet course as we observe when it is left entirely to itself. What is the practitioner to do in such a case as this? There is a choice between two courses only: one is to relinquish the use of mercury, or at least to reduce the dose; the other is to increase the dose, from the idea that those hitherto administered had been too small and too weak for this patient. When the latter course has been adopted, I have known two drachms, and even half an ounce of the ointment, ordered to be rubbed in each night and morning, and at the same time large doses of Pil. Hydrarg. Calomel or Hydr. Calcinat., given internally for some days: the result of this practice has been that, in some cases, it has produced that general weakness and that tendency to faintness which marks the approach of erethismus. The surgeon, alarmed at these symptoms, decides on suspending the use of mercury; to this the patient readily assents, though at the same time he expresses his disappointment at the inefficacy of the practice which has been pursued, as well as his regret at the time which has been lost.

In some of these cases, however, we find that these alarmingly large doses of mercury do excite a ptyalism; but not that gentle, manageable kind which it is our anxious wish to obtain, but rather a sudden, a violent, and an ungovernable action which overwhelms the system and threatens destruction to life. The day preceding the appearance of this violent salivation, the patient announces its approach by informing us that he was feverish and restless the preceding night, that he has great headache or tormina, or dysenteric dejections from the bowels; on the following days his cheeks and lips are enormously swollen, there is a copious and incessant flow of saliva, and the tongue is protruded and swollen, the speech is impaired, and deglutition is so impeded that he cannot even drink without much difficulty. Hæmorrhage from the gums to a pretty large amount in many instances occurs repeatedly; the tongue continuing swollen and protruded; its edges, more

particularly on their lower surface, become indented and
ulcerated from the pressure of the teeth. When awake he
hangs his head over some vessel to receive the saliva, which
flows copiously and incessantly; and when overcome by
fatigue he attempts to sleep, the saliva still flows, and bathes
his pillow with a fœtid moisture; his sleep is broken and
unrefreshing, and is frequently interrupted on a sudden by an
alarming sense of suffocation, induced partly by the swelling
of all the salivary and mucous structures about the mouth
and fauces, and partly by the accumulation of the viscid
saliva which he is unable to swallow. After two or three
weeks passed in this way with but little alteration, the saliva
at length becomes more thick and ropy, the swelling of the
face, tongue, and fauces subsides, and the patient feels a
strong desire for food, but is totally unable to take any in a
solid form, and he suffers exquisitely in attempting to swallow
any, even the blandest fluid; and thus he is harassed on the
one hand by a craving for food and nourishment, and on the
other by the apprehension of acute pain attending every
attempt at mastication or deglutition. At length, however,
the swellings subside, the ulcers of the mouth heal, and a
general improvement in the health occurs; yet even then he
often has to lament the continuance of some of the various
and serious effects which the excess of mercurial action seldom
fails to excite, such as loosening of the teeth, pain, and even
caries and exfoliation of the alveoli, or of the jaw bone; sometimes the tongue contracts adhesions to the cheeks, &c. &c.
It will naturally occur now to inquire, can the patient be
assured, as some recompense for all the sufferings he has
undergone, that the disease for which the mercury has been
administered is cured? In many cases this benefit has been
gained, though doubtless dearly purchased; but in many
other instances, the surgeon as well as the patient observe
with regret, and even with dismay, that although the chancre
has nearly healed, yet during the last few days it has
remained stationary, and also that there is more surrounding
hardness than is consistent with the healthy action in the
part. It would be needless to comment on the judgment
or the feeling of that practitioner who could, under such

circumstances, recommend the patient to resume the use of mercury.

It is fortunate for mankind and for the character of surgery that we possess the means of avoiding many, if not all, these evils, by a due attention to certain principles; in the first place, when we learn on due enquiry that the system of our patient had on former occasions been nearly insensible to the action of mercury, when administered in the ordinary doses, we must conduct the plan of treatment according to such rules as experience has proved can generally, even in such habits, excite ptyalism with ease and safety, and only to such a degree as can be perfectly controlled. In such habits then we should employ mercury by beginning with smaller doses than in ordinary, and at longer intervals, or, if we have already commenced with the ordinary doses and periods of administering, we should not obstinately persevere in the use of this medicine, but should purge the patient repeatedly, enjoin low diet, and the frequent use of the tepid bath. It has sometimes happened that, during the employment of these last mentioned means, a safe and mild ptyalism has supervened a few days after the mercury has been laid aside. This apparently anomalous fact is no weak proof that the cause of our failing to produce the desired effect in the first instance was owing to the mercury having been administered in an injudicious, and, as it were, in too forcible a manner. But if, with this treatment, no signs of ptyalism should appear, we must still persevere in the same constitutional means until we conceive that the system is totally free from all mercurial impregnation; to induce this state may occupy twelve or fifteen days. We then commence a new course of mercury, and, taught by our former experience that full doses of the medicine failed to produce the desired effects, we now direct it in very small doses, but at the same time we persevere in the same constitutional remedies as before: after a few days passed in this way we are sometimes pleased to find the signs of an approaching salivation gradually advancing in the very manner we are so desirous to effect. It appears to me that this very precept which has just been inculcated is the great secret which will serve to guide us in the management of

those cases which have hitherto often excited great uneasiness
in the mind of the surgeon; and it is a remarkable fact that
this obstinate state of the system to admit of the influence of
mercury is not unfrequently met with in those very patients
who, on former occasions, had been salivated rather too
quickly with the ordinary doses of mercury, and that, too, by
the very same surgeon who is now so baffled in his attempts
to produce the same effect. If surgeons of much practical
experience will endeavour to recal to their recollection all the
circumstances of such cases, I believe they will find, that on
the latter occasions in which they had to treat such cases,
they were very anxious that the effects of the mercury should
be quickly established; now, in my opinion, this very anxiety,
by inducing the surgeon to urge the mercury too freely, has
been one source of its failure, and of the disappointment of
his surgeon.

In the course of my practice I have met with eight or ten
individuals who appeared to be wholly exempt from the
influence of mercury as a sialogogue, and yet I must admit
that these persons were cured of their venereal complaints;
and it appears a remarkable fact that their primary symptoms
yielded exactly in the same manner, though at later periods,
as in those patients whose constitutions evinced the influence
of mercury on the salivary system. The greater number of
these irregularities or peculiarities to which I have alluded
were met with in one family of five sons, in each of whom I
experienced the same constitutional character or property of
resisting the salivating influence of mercury; each of these
young men had been under treatment for venereal symptoms
more than once, and had been on each occasion under the
care of a different surgeon, so that the failure of exciting
salivation, or making any impression on the salivary system,
could not be attributed to any peculiar mode of management
of mine, but solely to the idiosyncracy of these individuals:
I may add that each of these young men so far bore mercury
well that they did not suffer any feverish excitement, loss of
appetite, severe dysenteric attack, or nervous debility from its
use. Such constitutions, I presume, are very rare indeed.

I have not any particular remarks to offer as to the

respective merits of the various preparations of mercury. In my own practice I have confined myself to a few of those which I considered the least likely to disturb the stomach and bowels. I have long thought that if as much talent and industry had been applied towards learning the best mode of administering mercury, and of regulating its effects, as have been employed in compounding new preparations of this mineral, we should not at the present day be in such uncertainty, and even ignorance, as to the best mode of conducting the treatment of the venereal disease.

On the manner of using fumigations I shall offer merely a few remarks.

In the first place, I believe it is an erroneous opinion to suppose that ptyalism can be excited by mercurial fumigations, applied to the sound skin only. I have frequently seen it thus applied without producing the slightest effect. * When it has appeared to have excited ptyalism, I fancy that this was caused by some portion of the fumes passing into the mouth and throat. When directed upon an ulcerated surface it will often produce this effect; and I need not add, that when applied for an ulcer in the throat, it will generally cause a pretty smart ptyalism in the course of two or three days.

[Colles's assertion that it is an erroneous opinion to suppose that ptyalism can be excited by mercurial fumigations applied to the sound skin only, completely coincides with my own observations. When a student I frequently ordered, under the direction of the late Dr. Hutton,

---

* Amongst Colles's manuscript notes, apparently intended to be used in the second edition of this work, which, however, he did not live to accomplish, I find the following :—" 2nd January, 1841. Sir Benjamin Brodie's opinion as to the superior efficacy of mercury in the cure of Syphilis. Cusack having commissioned Robert Todd to accompany his patient, Mr. P., to Sir Benjamin Brodie, writes the following as the opinion expressed by Sir Benjamin :—' He thinks that effects of mercury are not to be feared at all in comparison with those of Scrofula, and that the syphilitic poison often excites the scrofulous action unless overcome by mercury. He thinks the safest mode of administering mercury is by fumigation. He subjects the whole body to the process, but finds it always nec ·sary to include the mouth, and consequently believes that it acts mc through inhalation than through the skin. He has no reliance on any other mode of overcoming the effects of Syphilis but Mercury.' "

calomel ointment for the treatment of psoriasis and lepra, not syphilitic. Those who were acquainted with the practice of this eminent surgeon will remember that he frequently prescribed an ointment consisting of one drachm of calomel to an ounce of lard, to be frequently rubbed over the surface of the body in such cases. I was struck by the fact that this mercurial preparation, so active when given internally, never mercurialized when rubbed over the skin in the form of ointment. I have known pounds of it to be used,—rubbed-in in ounces daily over the entire body,—but I never knew a case to be salivated by it. Later, when turning my attention more to the subject, I found that mercurialization did not follow even the very frequent use of the calomel vapour-bath,—provided the patient did not let his head into the vapour and inhale the sublimated calomel. When a patient leaves the calomel vapour-bath, if we scrape the surface and examine in the microscope what has been there deposited, we find the sublimated calomel in crystals—a form ill-adapted for cutaneous absorption. Doubting that it was ever absorbed in this way, I made some experimental observations. Some ounces of finely-levigated calomel were tied in a muslin bag. Every day, after coming out of an ordinary warm bath, the patients were dusted over with the calomel in the bag, from the waist down. I have myself rubbed ounces of calomel over the legs, abdomen, and scrotum, leaving the patient as white as a miller with it, but I have never found it salivate the patient.]

The process of fumigation may be conducted in an easy and comfortable manner by directing the intended dose of cinnabar, or grey oxide of mercury, to be mixed with melted wax, and with a cotton wick be moulded into a small candle. This may be stuck on a common plate, and then burnt under a curved glass funnel, which is to be raised about an inch from the plate. By conducting the process in this way we are certain that all the mercury is consumed, which is but seldom effected in the ordinary mode of throwing it on heated metal; the fumes, too, are thus more gradually brought into contact with the diseased surface; and the patient, if fatigued, may blow out the candle and suspend the process until he feels able to resume it.

As to the general directions which the practitioner should give his patient from the time the mercurial course has terminated, until he is able to resume his usual avocations, I consider the principal caution to impress is to avoid exposure to wet or damp, to cold, or to night air; the bowels should be attended to, and an occasional tepid bath may be enjoined, more particularly as this advice usually accords with the patient's feelings, there being a vulgar, but a very general, impression that it is useful in taking the mercury out of the system.   Patients, after mercury, are often delicate, and more subject to cold; hence a great liability to rheumatic pains, and hence, I believe, to guard against exposure to cold is the best advice can be given to those who are recovering from a mercurial course.   It is a matter of common observation that persons, after a mercurial course, not only recover their flesh very quickly, but even manifest a great tendency to corpulency.

Of late years surgeons have become very well acquainted with a peculiar eruption which is occasionally induced by the use of mercury: this affection is named Erythema mercuriale.

It was, I believe, first described by Drs. Alley and Moriarty, of this city, and more lately and fully by Mr. John Pearson. Of this affection I shall in the first place remark that we may observe some slight indication of an approaching ptyalism at the very time that this rash appears.   It is therefore to be looked for in the early periods of the mercurial course.   Sometimes it scarcely attracts the attention of the patient for the first two days, and not until the uneasy sense of itching, which it excites, shall have deprived him of a night's rest, and then, when he complains first of it, it may be found widely spread over the limbs and body.   During the earlier period of a mercurial course, therefore, we should frequently examine the surface of the body, more particularly the angular fold at either groin, as also those between the scrotum and each thigh; for it is in these situations that this eruption usually makes its first appearance, in whatever form mercury be used. On the slightest symptoms denoting the approach of this affection the surgeon should instantly discontinue the mercury, purge the patient, and, should the disease advance farther, he should expose him to the open air as freely as the state of the weather will permit.

The degree and kind of fever which attends this eruption is very various, and seems to be influenced partly by the previous habit and constitution of the patient, but still more by the severity and duration of the eruption.   I could not find that any particular type of fever necessarily or even generally accompanied it.   Indeed, unless the disease be severe, we take no note of the fever, so very slight is it, and so little different from that which attends the ordinary action of mercury when inducing ptyalism : and here I would remark that instances of a severe case of this disease are so rare that I have not met with more than one continuing beyond eight days during the last three years. In a word, it seems to me that the severity of this disease has been owing to the eruption having escaped notice, and to the use of mercury having been continued for two or three days after the rash had appeared, rather than to any peculiarity of constitution. However, we must expect now and then to meet with a rare instance of a severe case. The treatment of the fever in such cases must be regulated by the existing symptoms.

With respect to local treatment, I believe that dusting the excoriated parts with any of the mild drying powders will be found to afford as much relief as any other application. Sometimes the application of cloths, wetted with black wash, has procured relief, and has appeared to promote the formation of cuticle. In a very severe case it will be necessary to have the sheets, in which the patient is laid, prepared so as to prevent them from sticking to the skin ; I think that this is very effectually done by a mild ointment of rather a stiff consistence. The common one, made of equal parts of suet and bees'-wax, spread as thin as it can be spread by holding the spatula on its edge, will answer extremely well. I do not pretend to say that some other composition may not be discovered which will better promote the healing of the excoriated surface, but whatever it be, I should recommend it to be made of a firm consistence, for the softer ointments, by the heat of the body, are found to run quickly through the sheets, and consequently to leave the surface which is applied to the body nearly dry.

I may next remark that an obvious amendment takes place in the symptoms of the venereal disease on the first appearance of this eruption, and that in a degree more striking than that which attends so slight a degree of ptyalism.

What is most worthy of remark is this, that we never find this eruption to make its appearance while the system is under the influence of ptyalism. So that, after we have ptyalism fully established, we may dismiss all our fears on account of this rash. But let us not be lulled into a false security merely because this symptom may not appear in the early part of a mercurial course; for in some instances it does not appear until the mercury has been used for a considerable time.

I recollect the case of a young woman affected with an induration of one of her breasts, for which I had directed small doses of Pil. Hydr. combined with Extr. Conii. No sensible mercurial effect having been produced at the end of three weeks, I increased the dose of the pills; the result was very speedily a slight degree of ptyalism, and with it very full eruption of mercurial erythema, which proved tedious, obstinate, and alarming.

In this case it is obvious that the eruption attended that slight febrile excitement of the system which mercury so generally occasions when it is just about to act on the constitution. During the first three weeks of its use the mercury had not produced any sensible effect, and therefore had not excited this eruption. We may, in fact, declare that at whatever period of a course of mercury the mercurial fever is *first* suddenly excited, there is danger of the erythema. Hence it should be a rule with those who are conducting a course of mercury to watch carefully the earliest effects of each increase in the doses of the medicine, and to question the patient minutely, that he may get the earliest notice of the presence of this affection. When once the first burst of mercurial fever is over, and ptyalism has been fairly established, then the surgeon may carry on the mercurial process to any length of time necessary (provided he do not allow the action of the mercury to subside), and yet be under no apprehension of an attack of this rash. The following case is well deserving of attention :—

Oct. 22, 1807.—Burrows, a dragoon, was admitted with a chancre on the frænum : on this and the succeeding night he rubbed in half a drachm of mercurial ointment, and on the morning of the 24th he was found affected with mercurial erythema ; this attack quickly subsided without the use of any other medicine than the Mistura Salina Diaphoretica of the hospital.

On Nov. 4th he was ordered Calomel gr. ij sing. noct.; this he continued until Dec. 4th, when he was directed to take the same quantity mane et nocte. On Dec. 5th mercurial erythema again appeared; the mercury was laid aside and resumed on Dec. 19th. Calomel gr. ij omn. nocte, which dose he continued until Dec. 26th, when he was discharged cured.

I need not point out how small the quantity of mercury was which excited the first attack of the disease in this case; but it is worthy of notice that when the dose of calomel was doubled, on Dec. 2nd, it produced another attack of this affection, although his system had been for twenty-eight days accustomed to the simple dose; hence I infer that the attack of this eruption is produced, and is to be looked for whenever the mercurial impetus or fever is expected, whether at the commencement of the mercurial course or at a late period, in a protracted use of mercury whenever the dose is quickly increased.

It may not be amiss to observe that on the declining of the first rash the chancre improved; that it remained stationary, or nearly so, during the entire period of his taking the single dose of calomel, and that it healed very rapidly as the second rash began to subside. This case also proves that the internal, as well as the external, use of mercury will excite this affection.

There are some persons whose skin will become affected with this eruption by an inconceivably small quantity of mercury, such as a single blue pill, or a few grains of mercurial ointment which may have been used perhaps for the mere purposes of cleanliness. It may be asked, how are we to manage the case of such an individual when affected with Syphilis, for we are told that this cutaneous disease is worse than that for the cure of which the mercury is required: it happens, however, fortunately that this peculiarity of constitution is very rare, and it is a still more fortunate circumstance that we have it in our power to bring such a system into a state which will bear whatever quantity of mercury may be required for the cure of even the most obstinate form of venereal disease. The plan to pursue in such cases is this : in addition to purging and warm bathing which have been alluded to before as a judicious preparatory course, we should enjoin the patient to wear lighter clothing than usual, to exercise in the open air during the greater part of the day, to keep the windows of his sitting-room pretty constantly open, to keep on light covering at night, to live abstemiously, and on food of the least stimulating quality. The mercury should be administered at first in extremely small doses, and at long intervals; by degrees the former may be increased and the latter shortened, in proportion as we find the medicine to agree with the system. By this sort of management I have treated some such individuals, both for primary and secondary symptoms, and have at length brought them to bear the fullest doses of mercury. No doubt this plan or process is very tedious, but it is a certain mode of curing the disease, and is free from the risk of inducing this cutaneous affection, which is certainly among the most formidable effects of mercurial action.

The remarks which I have offered on the mercurial erythema will apply only to the generality of individuals who are subject to it, and by no means to those (fortunately) very rare cases in which the susceptibility of this disease is so very great that it is constantly excited by the smallest portion of mercurial medicines. Let us take the following as an example :—

Mr. R. applied to me, 12th August, 1813, for the cure of a chancre. I had treated him in 1810 for a similar disease, and was then made fully acquainted with his extreme tendency to mercurial erythema. Having premised a tepid bath and some active purging medicine, I commenced by directing a pill of Hydr. Acet. gr. ss. and Pulv. Antim. gr. ij. to be taken every night. On the second day a slight rash appeared on the inside of his thighs; of course the mercury was withheld for a day. No improvement in the chancre attended this eruption. The pills were continued sometimes every day, sometimes every second day (according to the appearance of the rash) until the middle of September, when the chancre took an unfavourable turn, and showed a tendency to slough. Then the mercurial medicines were entirely discontinued. The rash having disappeared for four days, the state of the ulcer induced me to recommend the ordinary black wash (Calomel and Aq. Calcis). This had been applied for one day only, when the rash reappeared, and with increased severity. On resuming the mercury, on Oct. 3rd, I directed Hydr. Acet. gr. ss., and Extr. Cicutæ gr. i. No rash appearing, the dose of the pills was doubled: the rash reappeared on 11th October. After this the dose was cautiously increased to three, and ultimately to four pills per diem. The chancre was very slow in healing, but was completely healed on the 1st of November. The mercury was continued some days longer, although a slight return of the rash again obliged us to desist for a day occasionally. During this lengthened course of mercury Mr. R.'s general health continued good. No ptyalism was induced; the gums were not made sore, but the inside of the cheeks assumed a leaden colour, and became a good deal swollen during the last fortnight of the treatment.

Mr. R. gave me the following account of the effects of mercury on his system :—Previously to the year 1806 he had used mercury for cure of a chancre, and did not then experience any but the ordinary effects from the use of this medicine. In 1806 he again had occasion to use it for cure of a chancre, and then he employed it, both internally and externally, in very large doses. On this occasion it did not affect his mouth ; but immediately after he had laid aside the medicine he was attacked with mercurial erythema, which affected him in a very severe and dangerous degree. Since that period (and never before) he has experienced an extreme susceptibility to the action of mercury in producing this rash ; for example, a very small portion of mercurial ointment rubbed on the pubes, for the purposes of cleanliness, has more than once produced it. A grain of calomel combined with purgatives has had the same effect.

In this case I would remark that the venereal symptom did not improve on the appearance of the rash as it does in less susceptible constitutions. The mercury had not the effect of producing ptyalism, though used for such a length of time. The effects of the topical application of black wash affords the strongest proof of the great susceptibility of this individual.

We should carefully distinguish between erythema mercuriale and another, but more partial, eruption arising from the use of mercury. They both come on under similar circumstances ; both seem to be excited by the first impression of mercury on the general system. Our attention is attracted to this latter eruption by our patient informing us that he fears he has got the itch—that he could scarcely get a wink of sleep for one or more nights preceding. He then exhibits on his hands and wrists an eruption beginning with small but very distinct red papulæ, some of which, in a more advanced stage, have vesicles on their apices : they chiefly occupy the anterior surface of each wrist, and of the fore-arm half way up to the elbow ; the backs of the hands and fingers are also thickly beset with them. On first view this eruption closely resembles a form of itch, in which the vesicles are small ; but on more careful examination you discover that the clefts between the fingers are altogether free from the former, while they are known to be the principal seat of the latter. This eruption is accompanied by a slight degree of fever, and generally by marks of commencing ptyalism.

I cannot say what changes or effects on this eruption would be produced by persevering in the use of mercury, because all the patients in whom I witnessed this symptom were also affected with a smart degree of fever, and complained so bitterly of the itching and of the restlessness caused by it that I felt afraid to go on with the mercury until the irritation of this eruption had subsided. A few days' use of the antiphlogistic regimen, and abstinence from mercury at the same time, were sufficient for the desquamation of the pustules, and the removal of this rare effect of the mineral.

I need hardly observe that this eruption differs from the mercurial erythema by the early appearance of the vesicles, by the eruption being much more distinct and less thickly set in the skin, by the parts of the body which it affects, and, we may add, by its not extending to the other parts of the body, and not seizing on the angles of flexion in the limbs, where two skins occasionally lie in contact with each other. I suspect this effect of mercury is observed only in cases where the patient is using mercurial frictions. Here, as well as in mercurial erythema, we observe a decided improvement in the primary symptoms uniformly to occur.

Having for some time noticed this eruption only among the soldiers under my care in the hospital, I at first suspected that it was produced in a great degree by the oatmeal diet, to which they are so much accustomed; but subsequent observation has removed this error, and convinced me that it is attributable solely to the use of mercury. Of course the use of mercury must be resumed as soon as the itching has ceased and the eruption begins to desquamate.

Another effect of mercury, allied perhaps to the foregoing, which the surgeon should watch for, is an excoriation of the skin on the corresponding surfaces of the scrotum and thighs. If this be discovered in its commencement it will be seen as an excoriation in the very angle between the thigh and scrotum; from this it spreads over the entire extent of the opposed integument of these parts, and a profuse discharge of a very fœtid nature takes place.

We cannot well imagine a more distressing state than that of a patient labouring under this affection; at least no persons more pathetically lament their condition. Not only are they deprived of sleep by day and by night, but they cannot attempt the slightest movement in the bed without inducing severe pain and agony, and they most earnestly solicit some relief even of a temporary nature. It should also be carefully remembered that in some instances, in addition to the parts already mentioned, this disease will also affect the skin in the vicinity of the anus, where the integuments of the nates lie in contact with each other. Occasionally, too, we meet with instances where this disease affects only the skin in the vicinity of the anus, while the scrotum and adjacent skin of the thigh remain perfectly free.

The degree of fever and suffering attendant on this excoriation is so very severe that we must at once relinquish the use of mercury. If the patient, through downright stupidity, or from an anxiety not to disturb the course of treatment, should conceal his sufferings from the surgeon for one or two days, the continued use of mercury will have the effect of aggravating very considerably both his fever and his local sufferings, but it does not seem to make the excoriation spread to other parts.

When this excoriation takes place we do not find that the venereal symptoms are improved; they sometimes remain stationary, but more frequently they become in a slight degree worse, and they improve in proportion as the excoriation goes on to heal.

The duration of this affection varies from eight to fifteen or twenty days. For the relief of all these sufferings we find opium unavailing; it fails to procure any continued sleep. The antiphlogistic regimen scarcely moderates the fever. The most immediate and most effectual relief is procured by local means, the most efficient of which is dusting the excoriated parts with equal parts of lapis calaminaris and starch, very finely levigated: this is to be laid on pretty thickly, and then a fold of old linen interposed between the two adjacent skins. Another application, which often procures immediate ease, is the black wash of Aq. Calcis and Calomel; the affected parts are to be kept asunder by lint constantly moistened with this lotion.

The following case affords an example of this effect of mercury :—

Dennis Dempsey, æt. ann. 20, admitted Dec. 13th into Stevens's Hospital, No. 11 Ward. The penis and scrotum are much swollen, red, and excoriated; the prepuce is particularly engaged in this swelling, causing a very great enlargement of the anterior extremity of the penis. On the lower surface of the prepuce are several superficial ulcers ; a number of similar ulcers occupy the skin covering the body of the penis. Excoriations also exist between the scrotum and thighs, obviously caused by want of cleanliness. He is pale, weak, and emaciated : has had ulcers within the prepuce for some weeks past.

From 13th to 30th December he used black wash, tepid bathing, and Pulv. Antimon. The inflammation and excoriation having been then quite removed, he was ordered Ungt. Hydr. Fort. ℥ ss. om. n. and Pil. Hydr. gr. v. om. n. ;

by these his mouth was made sore in six days. On the 7th the mercury was laid aside because his mouth was sufficiently affected, and more especially because he complained of sore-ness about the anus. On examination the surface of the skin of the nates on each side of the anus, and from this along the perinæum, appeared raw, wet, and red; and yet it could not be said to be excoriated, as there was no appearance of the want of cuticle. On 20th January, this affection of the skin having entirely ceased, I again put him on the use of mercury, in the same doses and of the same preparations as before; and on 31st January (in eleven days) his mouth again became very sore, with considerable ptyalism; indeed the action of the mercury proved more severe than I had wished. Yet on this latter occasion there was not the least sign of a return of the excoriation and discharge. He was dismissed cured on 21st February.

I have never seen this effect of mercury in any patient, except when I thought the action of the mercury was too high. I do not pretend to say that too high an operation of mercury is always attended by this effect; I only mean to call the attention of the profession to this as one of the marks by which they may know that mercury has made too severe an impression on the system. It is satisfactory to observe that a repetition of the same forms and doses of mercury may be resorted to without producing the same unpleasant effects.

The occurrence of each of these eruptions points out to us the necessity for suspending the use of mercury; but because the patient has already suffered in this manner we should watch with more anxiety the earliest opportunity for resuming its use. Should we wait until *all* the constitutional tumult raised by the mercury has entirely subsided, we shall not only require a greater length of time and a larger quantity of mercury to reproduce it, but by too long delay we may run the risk of having our efforts to effect a cure again interrupted by a recurrence of the very same symptoms. I have more than once witnessed the ill effects of this mismanagement. I have known cases of mercurial erythema produced by the first three or four doses of mercury; the surgeon then delays

to resume the use of mercury too long after the fever has
ceased; at length he begins it by employing small doses, and,
through fear of reproducing the same state, continues these
small doses too long, until both he and his patient, being tired
of the delay, he finds it necessary to venture on larger, perhaps on double, doses; and the new impetus of mercury
occasioned by its exhibition in these quantities on the system
produces a renewed attack of erythema.

Let us, then, bear in mind these facts—that each and all
of these untoward effects of mercury are owing to the first
impression of the medicine on the system: that on the
subsidence of this state the mercury may be used as freely
and as long as can be required for the treatment of any
venereal symptom (or indeed for the cure of any disease
curable by mercury) without the danger of reproducing the
same condition.   Let us, then, during the early period of
a mercurial course,—say from the second to the twelfth day,—
anxiously watch and guard against these untoward occurrences; but as soon as we have brought the system once
fairly under the influence of this medicine, let us dismiss all
fears and anxiety on this head, and now direct our whole
attention to the changes in the venereal symptoms, to the
degree of salivation, and to the strength and general health
of our patients.   The salivation once fairly established we
may consider ourselves as having escaped the chief dangers of
a mercurial course, and as now being on the high road
to a certain cure.   We now may be confident that the mercury
will not act (as it too often does) as a poison, instead of
its proving one of the most active and beneficial remedies in
the materia medica.

Of all the dangerous effects which may result from the use
of mercury, on the occurrence of ptyalism, the most alarming
is that which has been so well described by Mr. Pearson,
under the name of Mercurial Erethismus.

"This state is characterised by great depression of strength, a
sense of anxiety about the præcordia, irregular action of the heart,
frequent sighing, trembling, partial or universal, a small, quick,
and sometimes an intermitting pulse, occasional vomiting, a pale
contracted countenance, a sense of coldness; but the tongue is

seldom furred, nor are the vital or natural functions much disordered."

Again he adds (page 158) :—

"The gradual approach of this diseased state is commonly
indicated by paleness of the countenance, a state of general
inquietude, and frequent sighing. The respiration becomes more
frequent, sometimes accompanied with a sense of constriction
across the thorax ; the pulse is small, frequent, and often intermitting, and there is a sense of fluttering about the præcordium.
When these, or the greater part of these symptoms are present, a
sudden and violent exertion of the animal powers will sometimes
prove fatal ; for instance, walking hastily across the ward, rising
up suddenly in the bed to take food or drink, or slightly struggling
with some of their fellow-patients, are among the circumstances
that have commonly preceded the sudden death of those afflicted
with the mercurial erethismus."

I imagine that, in general, this dangerous effect of mercury
comes on suddenly, and is actually established before the
surgeon has due notice of its approach. It may, however, be
useful to observe that I have very frequently remarked this
affection to have been accompanied with an intense desire for
some acidulated drink ; and although the patient remarks
that a slight exertion induces palpitation, yet his countenance
remains unaltered, or it may perhaps be somewhat paler than
usual.

Although palpitation of the heart is the prominent symptom
in this disease, we are to consider it only as one of a series of
those effects which mercury produces when it acts as a poison,
and not as owing to any peculiar tendency to injure that vital
organ ; for when mercury acts favourably on the system it is
so far from producing any specific bad effect on the heart
that in diseases of this organ, attended with anasarca,
orthopnœa, and effusion into the chest, it affords considerable
relief ; so much so that the patient himself acknowledges its
utility, as soon as the gums become affected, by joyfully
announcing to us the glad tidings that he is now enabled to
lie down, and even to enjoy sound sleep. Erethismus, then, is
caused by mercury acting in the manner of a poison on the
constitution. I never knew an example of its occurring
after ptyalism was fully established : if a patient once have

a regular sore mouth we may continue the use of mercury to
an indefinite period, and in any doses, without the risk of
producing mercurial erethismus. When I say this I speak of
what accords with my own observation of this affection; for
I am aware that Mr. John Pearson says the subjects were men
who had nearly, and sometimes entirely, completed their
mercurial course. Now, according to my observation,
this affection comes on at a late period of a mercurial course
only in those cases where an increased dose has at length
been employed with a view of inducing ptyalism, or where the
ptyalism was slow in coming on; or where that which had
been first excited has been allowed to subside, and that we are
endeavouring to renew it, or rather to reproduce it. I have
never seen any instance of its affecting a patient who had
entirely completed his mercurial course. Indeed, I feel perfectly confident that it will not affect any person while he is in
a state of moderate salivation. I am aware that now and
then instances have occurred where mercury has appeared to
lie dormant and inactive in the system for two or three weeks
after it has been altogether laid aside, and that after this
internal ptyalism has come on. I imagine that when mercurial
erethismus attacks men who had entirely completed their
mercurial course, it must have been in some one of these very
rare instances.

The treatment to be pursued in a case of this formidable
disease should be as follows:—First, at once discontinue the
use of mercury, and charge the patient to change his dress,
and to lay aside every article of it which can be in any way
impregnated with that mineral; also, caution him to avoid
any, even the slightest exertion, such as getting out of bed
without assistance; for we know by experience that, under
these circumstances, a patient may sometimes expire on
making any slight muscular effort; in the next place we
should exhibit cordials in small, but frequent doses; but above
all we should expose him, in the horizontal posture, to the
free, open air, *both during day and night;* we need entertain
no apprehension of any of those injurious effects which
exposure to cold so commonly induces; it would appear as if
the present febrile state of the system suspended the ordinary

effects of exposure to cold, or at least enabled the system to resist them.

If this treatment have proved successful in rescuing our patient from the imminent danger in which he was involved, we must be careful not to resume the use of mercury for two or three weeks at least, and not even then unless his health and strength are perfectly restored ; and when we have determined to have recourse to it anew, we should observe the following precautions :—We should use the medicine in smaller doses than those which were used when the erethismus appeared ; we should interpose occasional purging, allow the patient occasionally to take mild exercise in the open air, and use every effort suitable in his state to direct the action of the mercury to the salivary system, so as to induce ptyalism.

CHAPTER IV.

ON CHANCRE.

ALTHOUGH every surgeon must admit that Mr. Hunter's description of a chancre is correct and drawn from nature, still I believe few will confine this term, or that of primary venereal sore, to those ulcers only which answer to his description. We sometimes, though rarely, meet with ulcers which possess all the characters of the Hunterian chancre, and yet prove to be not venereal. The only distinction which I can discover between these and the venereal chancre is that the former are of a very diminutive size, and although they heal like the Hunterian ulcer, they do not even approach it in point of size. As the result of long, attentive, and anxious observation, I should say that primary venereal ulcers present an almost endless variety of character. I would define a primary venereal ulcer to be "one which is remarkably slow in yielding to ordinary, mild, local treatment, but which is curable by mercury, and which, if not so cured, is likely to be followed in two or three months by secondary symptoms, which again also are curable by mercury." If then there be, as I affirm there is, an almost endless variety in chancres, how can we decide on the nature of primary ulcers so as to pronounce some to be syphilitic, and others to be mere common sores, or simple excoriations? I reply, that we are to be guided in our decision by observing, first, that many of these suspicious ulcerations cannot be referred to any class of common ulcers, as they strikingly differ from them; and, secondly, by attending to the course which these take, when not interfered with by any stimulant or caustic application, and when treated only with some mild ointment or cold water. If under these circumstances we find that after eight or ten days such ulcers show no disposition to heal, and if at the same time there be a total absence of any cause, such as defect in the general health, to account for this obstinate

condition of the local disease, we may then pronounce them
to be syphilitic.   But I repeat that the local applications
must have been of the mildest kind, for almost any primary
venereal ulcer may be made to heal by the use of stimulating
applications, applied even for so short a time as one or two
days.

It often happens that young men affected with primary
sores will first treat them for one or two days with stimulant
washes, and then apply to a surgeon; he directs the mildest
applications, and finding the ulcers to heal under this
treatment without any hardness remaining, concludes that
they were not venereal; the appearance of secondary
symptoms, however, in a few weeks afterwards, proves the
error of that opinion.   I have known a few such cases, in
which the surgeon has felt so confident in this opinion that he
has allowed the patient to marry in a short time after the
healing of the sores.   I need not attempt to describe the
state of his feelings when called on to treat the wife for
primary venereal sores!!   I am therefore anxious to impress
this as a very important rule, not to pronounce any ulcer on
the genitals as of an innocuous nature which may have
healed under the mildest applications, if at any time, even
during the short space of one or two days, it have been treated
with stimulant applications.

We shall now consider the proper treatment for a case of
true venereal ulcer, the Hunterian chancre:—The local
treatment should, in my opinion, be confined to the most
bland and mild applications, such as cannot in any manner
alter the features of the ulcer, for I am certain that many
useful indications which may serve to guide us in the
administration of mercury are to be derived from observing
the changes which these ulcers undergo through the agency of
that medicine; these changes, and their corresponding
indications, I shall hereafter advert to.   I am further
confirmed in this opinion from having observed that little
or no benefit is to be derived from a country practice; thus,
I have known a chancre completely cut out on the first or
second day after its appearance, yet the occurrence of
secondary symptoms was not prevented; I have also, in

numerous instances, known various caustics and stimulating applications employed, but I have not seen that such cases were rendered thereby more manageable, or that the patient was secured from those untoward changes which too frequently occur in the chancre itself before it is finally healed; nor were such cases less liable to secondary symptoms.

But before we can appreciate the effects of, or draw indications from, the various changes induced in a chancre by mercury, we should first be well acquainted with the wholesome or natural effects of this medicine on such an ulcer. If it be asked how soon may we expect to find the condition of the ulcer in any manner influenced by the mercury? I would say, in ordinary cases, we can observe it from the third to the seventh day. I cannot agree with Mr. Hunter as to the time in which he says a chancre becomes affected by mercury. Page 333, octavo edition:—

" Probably from the before-mentioned circumstances (the variations produced by certain peculiarities of the constitution at the time) it is that a chancre is, in common, longer in healing than most of the local effects from the constitutional disease, or lues venerea; at least longer than those in the first order of parts; and this is found to be the case, notwithstanding that the cure of a chancre may be attempted both constitutionally and locally, while the lues venerea can, in common, only be cured constitutionally. It is commonly some time before a chancre appears to be affected by the medicine. The circulation shall be loaded with mercury for three, four, or more weeks, before a chancre shall begin to separate its discharge from its surface, so as to look red and show the living surface: but when once it does change, its progress towards healing is more rapid. A lues venerea shall, in many cases, be perfectly cured before chancres have made the least change."

Now, I think it may be affirmed that, in ordinary, the constitution is brought under the influence of mercury between the third and the seventh or eighth day; and I feel no hesitation in saying that, as soon as this takes place, we shall observe a striking change in the appearance of the ulcer. The first remarkable change is that the chancre appears a little larger, but, at the same time, less deep. This increase in the size of the ulcer is not very considerable, nor is it attended with an increase of surrounding inflammation or swelling. The first demonstration of the influence of mercury on the ulcer will be eagerly looked for by a surgeon who is acquainted with its effects; and yet it would create uneasiness in the mind of a practitioner who was unacquainted with it. Indeed I strongly suspect that this first change in the ulcer has, in many instances, deterred the surgeon from its further use, and has impressed him strongly with an opinion that, in in such cases, the mercury disagreeing with the ulcer proved that it was not venereal. And thus the patient was debarred, for the present, from the benefit of this medicine, and was so prejudiced against its use that he could not be prevailed on to employ it in any subsequent stage of the disease, until convinced, by long and sad experience, of the inefficacy of other modes of treatment. While an experienced surgeon will not be deterred, by this slight change in the ulcer, from persevering in the mercurial treatment, he will also be well aware that a sudden and considerable enlargement of the ulcer, especially if accompanied with an increase of surrounding inflammation and swelling, would denote a very unfavourable change. We next find that the surrounding hardness declines, that granulations begin to arise, that the discharge becomes purulent, and that the entire surface of the ulcer becomes clean and red. In a few days more the ulcer contracts, a thin cuticle forms on its edges, and this daily increases until the ulcer is finally healed. Some degree of hardness, however, remains for four or five days after the healing of the ulcer, but this also disappears, and leaves the part possessed of its natural softness. Such is the progress of this symptom from the time it comes under the salutary influence of mercury until it is finally cured, in those cases where the mercury agrees well with the patient, and where it has been used with judgment. I shall only add that, in general, it will be prudent to continue the use of mercury not only until all hardness be removed, but even for a few days longer. I think we may lay it down as a general rule that the course of mercury, even when it has been well conducted and has agreed well with the patient, should be continued for not less than one month. I know that some cases have been perfectly cured in three weeks; but I have too frequently seen relapses follow such short courses of mercury when employed for the cure of primary symptoms.

I believe it is scarcely necessary for me to add that I should wish a moderate ptyalism to be kept up from the time the mercury comes to act on the system until it be finally discontinued. The doses in which the mercury should be administered, and the circumstances which require it to be from time to time discontinued and again resumed, must be left to the judgment of the practitioner, and will of course vary in different cases.

From some expressions in Mr. Hunter's work I have known some young men led to suppose that, when a chancre had lost its venereal characters and had got into the state of a granulating ulcer, it was devoid of all venereal virus, and therefore incapable of conveying infection. In the early part of my professional life I have known more instances than one which proved the fallacy of this notion. I had once an opportunity of learning that a chancre, even when recently healed, was still capable of communicating the venereal disease if the cuticle chanced to be rubbed off. A young man, whom I was treating for a chancre, had the imprudence to marry privately in a day or two after his chancre had healed; the cuticle was rubbed off, and I was called upon, in a fortnight after, to treat his wife for a chancre and bubo.

The progress of an Hunterian chancre towards healing often deviates from the description just now given. I shall here notice a few of those interruptions which deserve attention, as they show a necessity for some alteration either in the local or constitutional treatment; and I may here add, that the other forms of primary venereal ulcers are liable to the same interruptions.

First. The ulcer, though its surface may be clean and florid, may yet show no disposition to cicatrize, but will remain stationary for many days, although the mercury seem to agree well with the system. When we see this indisposition to heal attending the full and healthy action of mercury, we have only to apply a single touch of argentum nitratum; this will induce a rapid cicatrization.

Secondly. After the ulcer has spread a little under the

influence of mercury, it will sometimes throw up a whitish
fungus, and continue to enlarge more than we should expect
in a healthy chancre : this condition is most apt to occur in
those cases in which the mercury has been late in beginning
to act on the system, and in which it does not act very kindly.
In such a case the most judicious treatment is either to
suspend the use of mercury for a few days, or at least to allow
longer intervals between the different doses ; and at the same
time to give the patient bark joined with small doses of opium,
and to apply the wash of calomel and lime-water to the part.
This plan will in a few days bring the system more
satisfactorily under the influence of mercury, and put the
ulcer into a healthy granulating state.

Thirdly. The mercury, in some cases, will most unexpectedly take a sudden and severe hold of the system, inducing
profuse ptyalism within the first three or four days, and, as a
consequence, the almost instantaneous healing of the chancre;
such a healing of the chancre, however, is not a cure of the
venereal disease. For we shall find that a bubo will now
make its appearance, perhaps at the very time that the
chancre has thus suddenly healed ; this bubo will proceed
with unusual rapidity to suppuration if we do not at once
desist from mercury. I have not tried by experiment whether
such bubo absolutely requires a farther use of mercury, I can
only say that I have always acted on the supposition of its
being required, and as soon as the salivation had subsided
I have resumed the use of mercury in such a manner as to
affect the mouth again, but more slowly and more gently.

Fourthly. When, during the healing of the first chancre,
one or two new ulcers arise, possessing all the obvious
characters of chancre, I think that we are not called upon to
make any change in our treatment, for I have generally
observed that these new ulcers pass through the different
processes which lead to the healing of a chancre, but with
greater rapidity than the original chancre had done, so that
they are perfectly healed at the time that we should have
desisted from mercury for the cure of the original ulcer, if
these secondary chancres had not appeared ; I do not recollect
having seen one of these accessory ulcers ever undergo an
unfavourable change.

Fifthly. After a chancre has been healed, even without the assistance of active topical treatment, the hardness will sometimes continue too long. When this occurs in those chancres which had been seated on a flat surface, we can generally refer the induration which remains to a late or to an unsatisfactory action of mercury: under such circumstances we must therefore be cautious not to lay it aside too soon, and we should at the same time rub the cicatrix twice a day with mercurial or iodine ointment. But when the chancre has occupied the very edge of the prepuce, the hardness, which will then very generally remain, is merely owing to the peculiar loose structure of the part, and the adhesion produced by the inflammation in the interposed cellular substance now preventing the outer from moving on the inner skin. The hardness, in such a case, must be left to the slow operation of time, and its removal may be expedited by repeated frictions of the part. Here I would observe that the opening of the prepuce sometimes becomes so much contracted by the cicatrix of one or two chancres which had been seated on its edge as so require the operation for phymosis: the division of the skin in this case should be made in a line with and close to the frænum, by which we shall sufficiently enlarge the prepuce, and leave less deformity or inconvenience than if the division had been made along the dorsum or sides of the penis. Mr. Wilmot long since suggested to me a practice in this operation, which I consider a material improvement, namely, to pass two stitches of interrupted suture in each lip of the wound, so as to keep the cut edges of the skin and mucous membrane on either side in apposition with each other, and thus effect the healing of each lip of the incision by the first intention, leaving the angle to heal by granulation.

When chancres become inflamed and paraphymosis takes place, the usual operation for dividing the stricture should be performed.

Sixthly. Although I have repeatedly stated that when ptyalism has been established the further use of mercury becomes comparatively safe, yet it may happen, either from the mercury being continued too long or too largely in a weakly system, or from a state of fever induced by other causes, that a chancre which has made some progress towards healing will take an unfavourable turn, and assume a phagedænic or a sloughing disposition. From whatever cause this change may arise, it is plain that the mercury should be instantly laid aside, and every means adopted to remove the febrile condition, while the local treatment must be varied as the state of the ulcer shall require. In cases where, from peculiar irritability or delicacy of the patient's habit, such unfavourable changes might be apprehended, we may guard against such a change by giving bark and other tonics as soon as the ptyalism is established.

Lastly. The other changes which chancres occasionally undergo are trifling when compared with those which are produced by the supervention of a high degree of inflammation, We occasionally meet with cases in which a chancre has existed for many days, without any remarkable degree of pain or swelling of the parts—yet on a sudden it is attended with a high and dangerous degree of inflammation. This we can generally trace to intemperance, to excessive exercise, or to neglect on the part of the patient. In such cases we find the entire penis becomes much swollen, the integuments are of a bright or purplish red, and there is considerable discharge from the prepuce. These local symptoms will be attended with more or less of fever, the type of which may vary from that of the high inflammatory down to the low asthenic or typhus. If in such a case we press the prepuce all around, we shall detect one spot with an extraordinary thickening and hardness, and here the patient will feel a tenderness beyond even that of all the rest of the prepuce. Should the inflammation continue and increase, we may at length observe a black spot in the prepuce, on its dorsum or one of its sides, the effect of gangrene; this soon opens, the opening from day to day grows larger, until at length the glans appears protruding through it; as soon as this has occurred we shall observe that a considerable remission of pain takes place, the aspect of the ulcerated opening improves, the discharge declines and soon ceases, but the glans continues protruded, and what remains of the prepuce forms such a thickened mass, that to enable the patient to have sexual intercourse all the projecting skin must be removed by the knife. How should such a case be treated before the gangrene of the skin has begun? I believe the safest line of conduct is to endeavour to subdue the attendant fever; and therefore we must employ the antiphlogistic regimen to such an extent as may be required; nauseating doses of tartar emetic appear of the greatest service in all cases which do not require venesection, or where this evacuation has been previously made. Warm or cold applications may be employed, according as either is found to suit best the feelings of the patient; but on no account should we neglect to make him use the black wash very freely: this can be applied most effectually by means of a syringe with a long pipe, introduced between the glans and prepuce, while the orifice of the latter is closely pressed around the pipe, so as not to allow any of the fluid to escape until the prepuce has been fully distended. This partial gangrene of the prepuce must be looked upon as rather a favourable consequence of this very high inflammation; for in many instances it threatens to destroy the whole of the prepuce, and even to involve the penis itself in a similar fate. Under these circumstances I have repeatedly found that the destructive process was at once arrested by washing the diseased surface with the strong white muriate of antimony, and sometimes with the nitric acid; the former, however, I much prefer. This application very generally arrests the further progress of the unhealthy action, relieves the patient from pain so that he can enjoy sleep, and assuages his constitutional fever: however, until this favourable change takes place, we cannot lay aside our fears for the safety of more or less of the penis. In a few cases the process of destruction ceases not until it has destroyed this organ to a level with the pubes; in such melancholy instances, even after the ulcer is healed, the patient is tormented by the manner in which the urine passes out of the distorted and contracted opening of the remnant of the urethra; for it rises so much upwards that the unhappy patient is obliged to direct the stream forward by holding his hand above the pubes.

About twenty years ago, when I had charge of a large

number of soldiers labouring under the venereal disease, who
were received into the attic wards of Stevens's hospital, I
attempted the treatment of such cases by throwing in mercury
largely and suddenly; but whether it was owing to the want
of a judicious plan of using mercury, or to the bad habits of
the men, induced by intemperance and dissipation, I know
not; but I freely admit that, with many, this practice was not
successful. However, the success of the two plans, that by
mercury and that by the antiphlogistic regimen, was so
evenly balanced at the time the military hospital was broken
up, that I was quite undecided which to prefer. About this
time I learned the use of the black wash, which has rendered
such essential service in the early periods of this condition
that I have not since repeated the experiment of administering
mercury through the constitution. Many of the cases treated
with mercury were cured without the slightest destruction of
any part, but this was purchased by the certain and severe
sufferings of a violent salivation. Some escaped with the loss
of part of the glans, and some few had the penis destroyed
down to a level with the pubes. Possibly the mercurial
treatment would have been more frequently successful, had I
more constantly used venesection and other evacuations as a
process preparatory to the use of mercury. There is one
condition of the sloughing penis which I look upon with total
despair of being able to afford any means of arresting its
progress until it has destroyed the entire penis down to the
pubes: I mean that condition in which the sloughing part is
so soft as to resemble melted tallow when beginning to form
into a solid. I have never yet seen the progress of this
arrested, even for a moment, by any local or constitutional
means hitherto employed.

When, in addition to the inflamed and painful condition we
have just now alluded to, we find hæmorrhage to supervene,
we must carefully watch the effects of the evacuation. In
some fortunate cases the patient informs us that the pain is
much mitigated since the bleeding; here we may hope for a
favourable issue, the inflammatory swelling of the prepuce
will begin to decline, as also the pain and fever. Possibly
another hæmorrhage may take place in three or four days

ON CHANCRE.** 101

subsequently, which will be followed by a still more marked decrease of inflammation and fever. But in other less fortunate instances the first hæmorrhage is not followed by any mitigation of inflammation or fever, and a succession of bleedings may take place at intervals of twelve, twenty-four, or thirty-six hours. The type of the fever then becomes changed; it is no longer of the inflammatory, but merges towards the typhoid form, in all its characters strongly representing what some have called an irritative fever. The patient gets no sleep, he looks pale and haggard, is rapidly emaciating, complains of the unceasing and severe pain of the diseased parts, and is in constant dread of a fresh hæmorrhage; the discharge becomes more copious and extremely fœtid, and so acrimonious as to excoriate the neighbouring parts over which it runs. In order to afford effectual relief in this miserable state, we must slit up the prepuce, taking care to carry our incision so as to run near to the chancre, and having thus exposed the sloughy surface, we must apply to the bleeding point a dossil of lint soaked in spirits of turpentine, and then apply moderate compression. The turpentine seems to serve such cases not only by arresting the bleeding, but also by exciting in the ulcerated surface a more healthy action, which soon induces the parts to heal. We should be careful not to have the dossils of lint so wet as to allow the turpentine to trickle along the adjoining parts, as this will be found to scald them very much, and thus add unnecessarily and severely to the inflammation of the penis, and to the sufferings of the patient. I have seldom seen the ligature succeed in preventing a return of the hæmorrhage, as the coats of the arteries appear to slough again almost immediately under the ligature.

In some cases of sloughing of the prepuce the process spontaneously stops, after having completely removed the entire prepuce, leaving the patient as if he had undergone the operation of circumcision. By some practitioners we are recommended to consider this as a perfect and permanent cure of the disease, and not requiring the further use of mercury. I have certainly known some cases which would tend to confirm this opinion, especially when the sloughing

began at the early period of the chancre; but in other
instances secondary symptoms have supervened; and therefore
I should consider it prudent to subject to a course of mercury
all those in whom the sloughing had taken place at a late
period.

Having offered these few remarks on the accidents to which
primary venereal ulcers are subject, I shall make some observations on *chancrous excoriation*.

This excoriation does not at first present any distinctive or
specific characters; after ten or twelve days, however, it
presents new features; sometimes a great part of the excoriated
surface heals, and leaves one or two spots unhealed: these
are usually of a circular form, and rest on a hardened base,
although up to this period no induration had been perceptible.
Another striking feature, often met with in chancrous excoriation, is that while it generally presents a red surface, it looks
as if it were dry, or is like an excoriation which has been for
some time exposed to the air, and this even although it be
dressed with lint soaked in water.

As a proof that such excoriations are primary venereal
ulcers, we shall find that their morbid characters will begin to
improve as soon as the system feels the salutary action of
mercury; and that they will heal, and the hardness be
dispersed, in the same time and manner as the true Hunterian
chancre. We shall also find that if mercury be used in an
injudicious or ineffectual manner these excoriations will, like
ordinary chancres, undergo various unfavourable changes, and
will be followed by the usual secondary venereal symptoms.

A few instances will suffice to show the changes which
excoriations may undergo before they assume the venereal
characters; and to show that this form of primary venereal
symptom requires an attentive and lengthened course of
observation before we venture to decide on its nature.

July 31st. Mr. W——, six days since, immediately after connexion, perceived two excoriations, to which he applied dry
lint: one of them has healed; the other shows no disposition
to heal. This latter is situated on the inside of the prepuce,
at the root of the frænum, and is equal in circumference
to that of the flat surface of a split pea; is very superficial,

with a smooth surface, and edges not at all raised; it appears to yield very little discharge, and is seated on a very hard base, the hardness extending a little beyond the limits of the excoriation. Sumat Pil. Hyd. gr. v. mane noctcque.

Aug. 3rd. The excoriated spot presents more the characters of a secondary than of a primary ulcer; it is enlarged, and now has a white centre, surrounded with an orange-coloured, or reddish circle; the hardness is unaltered. Sumat Pil. Hyd. gr. v. ter. in die.

Aug. 6th. Inner gums of lower incisors are slightly affected: ulcer yields a purulent discharge in sufficient quantity: the white part in the centre is much reduced in extent, as if the orange circle had encroached on it; hardness not at all reduced; an incipient bubo in right groin, which is slightly painful when he sits down or gets up, but not when he is walking. Pergat.

Aug. 9th. Mercurial action more discernible; ulcer has its central part of a less deep colour, and of less intense white-ness; the outer edge of the orange border has healed; orange colour is less deep; the hardness of the base is also less. Pergat.

Aug. 12th. Ulcer is all healed, except a line in the centre, which is quite red and healthy; the groin is easy; mouth as on 9th instant. Pergat.

Aug. 16th. Ulcer perfectly healed, glands of groin quite free from all pain, mouth not more affected. Sumat Pil. Hyd. gr. x. ter in die.

Aug. 20th. Mouth not more affected. R. Calomelanos ʒ ss. Ext. Opii gr. v. Ft. Pil. x. Sumat i. ter in die.

Aug. 22nd. Sumat Pil. unam omni mane. Ungt. Hyd. Fort. ʒ ss. omni nocte.

Sept. 14th. During last three days mouth has been a good deal affected.

Sept. 17th. Mouth rather less affected. Sumat Pil. j bis in die. Rept. Ungt. Hyd. o. n.

Sept. 21st. Mouth a little more affected.

Sept. 24th. Omitt. Med.

Mr. B— applied to me ten days ago on account of a small

circular excoriation on the inner surface of the prepuce; this
was so purely an excoriation, of such a healthy red colour, so
free from hardness and from every venereal or suspicious
appearance, that I did not at that time look on it as venereal,
and accordingly advised him to apply only lint and cold water.
In the course of six days the appearances began to change;
and on the tenth day I observed that the centre of the excoriation had assumed a dirty yellow colour, while a pretty wide
margin retained its healthy red appearance. On the thirteenth
day the surface was as on the tenth, but the entire ulcer now
rested on a base of considerable hardness; judging this to be
venereal I then put Mr. B. on the use of Pil. Hyd.; this did
not agree very well with him, and the ulcer did not proceed
on kindly towards healing: on the contrary, at one period its
central part assumed a blackish tinge, and showed a disposition to slough, whilst the margin remained of a pretty healthy
red colour.

At the expiration of six weeks (the ulcer not being yet
healed) he was affected with venereal eruption and sore throat.
These symptoms yielded, and the ulcer healed, under a rather
protracted course of Pil. Hydr. gr. iii. and Sulph. Quininæ gr.
ii., the doses of which did not at any time exceed three in
the day.

Towards the conclusion of this course, being anxious to
finish the treatment by a more decided action of mercury on
the system, I added to the above Ungt. Hydr. Fort. ℥ ss. omni
nocte, for eight nights. But this proved to be too much for his
system, as was evinced by an attack of mercurial excoriation
about the anus; on the appearance of this of course the
mercury was laid aside, and since that period Mr. B. has not
had any return of venereal symptoms.

Feb. 17th. Mr. P— had been exposed to the danger of
infection fourteen days ago, three days after which he applied
to me with a pretty large excoriation of the prepuce behind
the glands; this was attended with scarcely any hardness,
and it healed in the course of eleven days under the application of spermaceti ointment. At present the surface,
though healed, looks unhealthy, and this spot resembles a
venereal tubercle; there is considerable hardness, fully as great

as that which attends the true Hunterian chancre: Pil. Hyd. gr. v. mane et nocte.

Feb. 20th. Surface is again ulcerated, or rather excoriated, with a purulent discharge, and considerable hardness of base. Sumat Pil. Hydr. gr. x. omni nocte, et gr. v. omni mane.

Feb. 26th. Sumat Pil. Hydr. gr. x. nocte maneque.

March 3rd. Yesterday had an attack of mercurial dysentery; mouth is now sufficiently sore; for the two or three last days has taken only three pills a day: ulcer is much smaller, and almost free from hardness. He himself remarks that it differs from every other chancre he has had, or has seen, in its never having gone deep.

March 9th. Mouth still sore; ulcer is perfectly healed, not leaving the slightest degree of hardness. N.B.—He left town the following day, with directions to keep his mouth slightly sore for a fortnight longer. He has since enjoyed good health, and has had no appearance of secondary venereal symptoms.

[It is to be regretted that in two of the foregoing cases the dates are not given up to which the patients continued in the enjoyment of good health, and free from venereal symptoms.]

A chancre on the skin of the penis, or external surface of prepuce, is quickly covered by a scab. Perhaps the best mode of treatment (at least the most comfortable to the patient) is to let the scab remain, and whenever it is accidentally rubbed off to expose the part to the air, and cause a fresh scab to form. The drying in of the scab, its gradual contraction, and its less intimate adhesion to the surface of the ulcer, will point out the gradual improvement in the concealed ulcer, and enable us to judge of the favourable action of the mercury; while the contrary changes in the scab will indicate that the mercury is not acting in a salutary manner.

I deem it right here to notice an appearance in chancres healed by local means only, which we occasionally meet with in cases where mercury has been injudiciously employed—an appearance which I have never, except in one instance, seen. The appearance I allude to is a remarkably hard lump, as large as a filbert, situated on the prepuce in the site of the original chancre; the skin which covers it is of a peculiar purple colour. In one instance, that of a young gentleman, whose chancre had been healed for three or four weeks, this appearance was attended with most alarming symptoms, viz., extreme emaciation, thirst, loss of appetite, and profuse night sweats. The suddenness with which he had been reduced to this state from a tolerably full habit of body, and from apparently excellent health, excited in my mind great apprehensions for his safety, especially as some of his family had died of hæmoptysis succeeded by phthisis. I never saw any case in which the venereal virus acted in a manner so like to what we might fancy should be the effect of a slow poison on the system. Yet by means of a mercurial course gradually raised from very small doses, and continued for a longer period than usual, the hardness and discoloration were removed, and he was soon restored to good health, which he still enjoys. This case occurred many years since, at the time when the non-mercurial treatment was first introduced into Dublin. In all the other instances of this large purple tumour, where it has formed after the healing of a chancre by local means only, I have seen secondary symptoms follow in the same time, and pursue much the same course, as when these have succeeded to chancres which have been only imperfectly cured by the injudicious administration of mercury.

I shall next consider some of the peculiarities of chancres, depending on the peculiar structure of the part in which they are situated. It is scarcely necessary to say that the quantity of hardness is much less in chancres situated on the glans than in those on the prepuce: but still a degree of hardness sufficient to guide us in the treatment attends even the former.

There is a form of venereal chancre on the edge of the prepuce, of which I do not recollect ever to have read any description. The account we receive from the patient is that he thinks he tore himself in the act of connection; we see that the lacerated part shows no disposition to heal, but is not painful or swollen, and appears in the form of a fissure. When the prepuce is partly retracted this fissure appears as

if it had been divided by a long incision; ths edges are hard
and deep, not swollen or inflamed, and yielding scarcely any
discharge; it is not at all painful, unless when an attempt be
made to retract the prepuce. This is certainly a primary
venereal ulcer: its peculiar characters may probably depend
on the infection being ingrafted on the torn surface; it requires
the use of mercury to the same extent as does the circular
Hunterian chancre; nor does it need any peculiar local treatment. I cannot recollect any instance in which this form of
ulcer took an unfavourable turn, although I have known some
patients who did not apply for assistance for three weeks after
connection, being lulled into a feeling of security by observing
that this ulcer was unattended with swelling, inflammation,
pain, or hardness.

[No doubt we have here a period of about three weeks' incubation, as
it would now be designated, during which time, if any sore at all
existed, it was so far "unattended with swelling, inflammation, pain, or
hardness," that it escaped the patient's notice.]

We sometimes see the edge of the prepuce beset with five or
six circular ulcers; these, if left to themselves, will first
granulate, then become fungous, and finally will heal spontaneously. The form, as well as the slow and indolent
character of these ulcers, might dispose us to conceive that
they were syphilitic; the appearance, however, of several of
these at once along the margin of the prepuce, and their
being destitute of any hardness, will serve as a criterion
whereby we may make the distinction. These ulcers may
always be cured by suitable local treatment alone: I have
never seen them take an unfavourable turn. Sometimes there
may also be present, in conjunction with these, others of a
true venereal character, and for which mercury must be
administered; we shall then have an opportunity of remarking
that this medicine exerts no influence on the ulcers at the
extremity of the prepuce. Sometimes this circle of ulcers is
accompanied by a single one of the same character far within
the prepuce.

It appears clear from the observations made in the foregoing para-
g⋅ ph that Colles had caught hold of the fact, that besides the Hunterian
chancre there exists another variety of venereal sore, occurring on the

genital organs. When he speaks of five or six circular ulcers, granulating, then becoming fungous, destitute of hardness, and always capable of cure by local treatment alone, he in fact gives most of the symptoms regarded at the present day as characteristic of the sore now described under the following names by different authors, viz., soft chancre, simple chancre, non-infecting or suppurating chancre, chancroid, chancrelle, false syphilis or local syphilis. When he says that we can have an opportunity, in cases where the true infecting chancre co-exists with those local sores, of observing that mercury exerts no influence on the latter, he forestalls the assertion of Lancereaux, who mentions this as one of the distinguishing features between true and pseudo-syphilis, viz., that the former is "susceptible of being influenced by certain agents, as proto-iodide of mercury and iodide of potassium," the latter (soft chancre) "being entirely unaffected by alteratives, if not aggravated by the administration of them."]

When a chancre is seated close to either side of the frænum we find it invariably happens that the ulceration passes through this process of integument, at first making an opening through it, and in a few instances the ulcer has healed, leaving this hole; but in the majority of cases the ulceration goes on until it has divided the frænum entirely, then each side of the ulcer heals separately, the part on the glans showing a tendency to advance slowly, making a superficial groove in the anterior part of the glans. The only direction to attend to in the management of these chancres is to cut through the frænum as soon as the ulcer has once perforated it; for we shall generally find when this has not been submitted to that the sore will not assume the granulating state until the ulceration has completely divided the frænum. The posterior part of the ulcer having generally the larger part of the frænum will require repeated touches of caustic to prevent it from healing with a large knob, while the anterior part will be benefited by the black wash or some other gentle stimulant. Sometimes the frænum (having been divided by the surgeon or by ulceration) presents a furrow, showing that the part is composed of two layers of cuticle; this may be a little more tedious in healing.

A chancre sometimes, though rarely, is seated at the very orifice of the urethra; in this situation it is always slow in going through the different stages preparatory to healing, and it too often assumes an eating disposition, passing all round the orifice; there is one very serious ill consequence which always results from this, namely, that in some time, perhaps in one or two years, the patient will suffer severely from symptoms of stricture. In treating such a case we should use every effort to prevent the extension of the ulceration to the entire circle of the orifice, for unless it entirely encircle the orifice contraction will not follow; this can with certainty be accomplished by touching the ulcer, as soon as it begins to extend, with the colourless muriate of antimony, or with nitric acid: these applications are no doubt severe, but the evil they avert is one of great magnitude; for I will venture to assert that of all forms of stricture this is one most apt to recur; indeed, it does not in any instance admit of a cure by the ordinary treatment of strictures.

I am happy to say that I have lately discovered a mode of treating this stricture which has proved eminently successful in the few cases in which I adopted it. This plan of treatment consists in this simple operation :—Having detached the skin from the end of the urethra, to which it is generally intimately adherent, I divide the urethra below to the length of more than half an inch. I raise the mucous membrane from each lip of the incision, then cut away a portion of the bared corpus spongiosum to such an extent as will allow the raised mucous membrane to cover the cut edge. I stitch down this membrane upon the corpus spongiosum, and thus, having covered each lip of the wound with mucous membrane, I have effectually guarded against the possibility of reunion of the lips of the wound, or subsequent contraction of the opening. The opening of the urethra thus produced is of course of a size larger than natural.

Of late years, notwithstanding what Mr. Hunter asserts to the contrary, I am confident that I have seen cases of chancre seated altogether within the urethra; such cases have been frequently mistaken for mild gonorrhœa, and have for weeks together been treated as such: in some of these cases the surgeon has not been apprised of his mistake until the ulceration has actually laid open the fore-part of the urethra, or has extended forwards to its orifice, so as to become visible. A very little attention will enable us to discover the real nature of this case, even in its earliest stages, for we can externally feel a hardness in that spot of the urethra where the chancre is seated, and the symptoms of gonorrhœa are so slight that this alone should excite our suspicion, and induce us to search for a chancre in the urethra.

[It was probably a case of urethral chancre from which John Hunter took the virus for his experimental inoculation, the result of which induced him to believe the gonorrhœal and syphilitic poison to be one and the same, and which led to such unfortunate results in consequence of his strongly maintaining this view. It is not impossible that the experimental inoculation made by Hunter (and stated by H. Lee to have been performed upon himself) may have been made with matter from a urethral chancre mistaken by Hunter for gonorrhœa. Lee, in his turn, adopts a different hypothesis; he believes that in a syphilitic subject "the irritated and inflamed mucous membrane of the urethra may furnish at the same time a secretion resembling that of a simple gonorrhœa, and also the poison which may give rise to an infecting chancre." Such he supposes to have been the "venereal matter from a gonorrhœa " in which John Hunter dipped the lancet for his experiment. The absolute truth must remain for ever in obscurity. All that is certain is that the experiment led Hunter to conclude that the virus of gonorrhœa and syphilis was identical, and that it required many years of observation and experiment to prove that he had been in error.]

I shall not attempt any further description of the various primary venereal ulcers which are daily to be met with ; much less shall I undertake any classification of their endless varieties, or of the course which each peculiar form has been supposed to run. They will be found to differ so constantly that very rarely shall we find any two of them to correspond accurately with each other ; not merely at their commencement do they present such dissimilar characters, but in their different stages towards healing they will be also found to deviate most strangely from each other, more particularly so when they have been treated by mercury. One remark, however, as to their treatment deserves notice, viz., that patients affected with these irregular ulcers have been found in general very difficult to salivate. This cannot be ascribed to the medicine having been used too sparingly, for in the reported cases we find the doses were rather too full than the reverse. Some have said that it was owing to a degree of febrile excitement in the system, which the irritable state of the ulcer had induced and maintained. But this remark can truly apply only to a few of these cases, for the majority of them will be found to possess rather an indolent character. The explanation which I would offer on this subject is, that the general health of such patients has not been in a good condition at the time these ulcers have made their appearance, and on that account mercury was more likely to disagree and to exert its poisonous than its salutary operation on their system, especially if used in full doses. Inattention then to the existing derangements of the system, ignorance of the means to remedy these, and probably, in most instances, the exhibition of mercury in too high doses, have all conspired to bring into discredit the plan of treating such ulcers by mercury.

There are some, I know, who assert that mercury is an unfit remedy, " because it disagrees with the ulcers;" but such persons altogether overlook its influence on the system, and imagine that mercury disagrees with such ulcers only because the latter have been derived from a peculiar kind of virus, to which they suppose mercury is inimical. That it is not with the ulcers, however, but with the constitution that mercury disagrees, we may learn from this fact; that when treating a patient affected with any other irregular and obstinate ulcers, and who may chance to be regularly salivated by the mercury, the ulcer is not found to be injured by this medicine; moreover, such an opinion is against all experience of the effects of mercury upon almost every other description of ulcer. For my own part I cannot recollect an instance where mercury, when it produced a healthy ptyalism for the cure of other diseases, had disagreed with a fresh wound, or with an ulcer of any ordinary class; on the contrary, I am sure that every surgeon has at times been agreeably surprised on finding that a salivation, excited for the cure of some acute or chronic disease, had at the same time, most unexpectedly, caused an obstinate ulcer to heal. This fact has been so often forced on my observation that I have in some instances of very obstinate ulcers subjected the patients to a course of mercury solely for the purpose of healing these ulcers. I shall adduce some instances of the success of this practice, when treating of diseases not venereal, which are curable by mercury.

CHAPTER V.

ON BUBO.

To Mr. Hunter is the profession indebted for correct notions respecting the nature of bubo, which he has properly classed among the primary symptoms of the venereal disease. It is not a little surprising that the most extensive opportunities have not as yet furnished practitioners with a single instance in which this symptom can be traced to a secondary venereal ulcer; and yet buboes cannot always be attributed to an inflammatory condition of the chancre, as they more generally arise when the inflammation in the latter has subsided, or when it does not exceed that which usually attends the secondary venereal ulcer: hence it would appear, in the latter case, as if the venereal virus had undergone some change which tends to assimilate it to the human solids and fluids.

The period at which bubo occurs after the appearance of chancre is not confined to any definite time, nor is it a necessary attendant on all chancres, many chancres being cured which had never at any time been followed by bubo; nay, it sometimes happens that chancres are cured without the use of mercury, and secondary symptoms supervene, and yet the passage of the venereal virus into the system had not been delayed or announced by the formation of bubo, or by any tendency to it. The venereal virus, therefore, can pass unaltered into the system, although in this course it must have passed through one or more lymphatic glands.

When chancre and incipient bubo co-exist, it is admitted that the treatment by mercury should be precisely the same as if the chancre alone were present: and accordingly we find that as the latter improves so the swelling and tenderness in the former subside: even should the bubo, in consequence of neglect, have acquired great size, and have advanced so far toward suppuration that the integuments shall have become

red, still the treatment should be the same as for the chancre
alone. When the ptyalism is commencing the inflammatory
condition of the bubo will for a day appear rather aggravated;
but when the ptyalism is established, then the inflammation
obviously begins to subside. Some advise the use of mercury
to be postponed until the gland has suppurated, and the
inflammation has been reduced by letting out the matter; I
do not think this delay is necessary, though I admit we should
first adopt antiphlogistic means to subdue or mitigate the
febrile symptoms which have been excited by the inflamed
condition of the bubo; this being accomplished there is no
reason why we should longer defer the use of mercury. But
should we meet a case in which, while the chancre is improving
(ptyalism being established), the bubo becomes larger, and is
attended with an increase of inflammatory redness, and a
painful sensation like that of scalding, we are then to infer
that the mercury is disagreeing; we should, under such
circumstances, discontinue it for a few days, direct brisk
purgatives, and the most strict quietness, also the horizontal
posture. The amendment in the bubo will assure us that
such a line of practice was most judicious, and we may then
resume the mercury as soon as we observe its effects on the
system beginning to decline, using it perhaps in less powerful
doses.

It happens in some few individuals that mercury, when
rubbed upon the thigh for the cure of chancre, will excite
a slight swelling of the inguinal glands. In such cases it is
prudent to desist from the use of mercury on that side for a
few days; during this time the swelling will subside, and
on resuming the mercury there will be no return of this
inconvenience.

When we come to speak of secondary symptoms we shall
find that the cervical glands sometimes swell on the approach
of ptyalism; but this swelling will also subside on discon-
tinuing the mercury, and on allowing the salivary system
a few days to unload itself. Thus it appears that mercury
may cause a swelling even of those lymphatic glands which
are emote from the route the mercury takes to reach the
circulating system; or it may be that the lymphatic cervical

I

glands are only sympathetically affected in consequence of the excited state of the salivary organs.

If mercury, administered for the cure of a chancre, should chance suddenly to produce profuse ptyalism, we shall find the chancre will heal almost instantaneously, while at the same time a bubo will begin to show itself. The state of the mouth in such instances prevents us from falling into the error of giving more mercury. Purging, and such other evacuations as may be employed, with the view of reducing the ptyalism, will also tend to produce a resolution of the inflamed gland.

Should we meet with a case in which no traces of a bubo are discoverable at the commencement of the mercurial course, but in which it may arise after the mercury has fairly affected the system, in such a case also we should for a time discontinue the use of mercury.

In all the above cases we should, in addition to purgatives, endeavour to restore the system to a healthy state by giving bark or opium, or such other medicines as we may judge necessary. Sometimes carriage exercise will be found highly advantageous.

If, notwithstanding our endeavours to the contrary, the bubo shall have suppurated, are we to allow it to open spontaneously, or when, and how, ought we to open it? The opening of the bubo should be delayed until the tumour has pointed and the skin has become very thin: an earlier opening will induce an increase of pain and fresh inflammation which will not subside in less than three or four days; by delay, also, there is a chance that the matter may be absorbed, and the patient be thus saved from all the annoyance attendant on an ulcer in the groin.

In most cases I advise the opening to be made in the thinnest part of the integuments, and only large enough to admit of the easy escape of the matter.

There is one particular form of bubo in which a very different rule for opening it must be adopted; I allude to that which has acquired a larger size, in which the tumour does not point or become conical, but assumes a flat surface, the whole of the integument becoming very thin and discoloured:

in such a case we should cut through the entire length of this thin covering of the abscess. Were we only to make a small opening, we should find that the discharge would continue very long, that it would be thin and of bad quality, and that at the end of five or six weeks there would be no prospect of healing; for the integuments will have lost so much of their vitality that they will either be removed by ulcerative absorption or will have curled and turned in, and then the surgeon will be compelled to apply the caustic or the knife for their removal. Formerly, surgeons were in the custom of using a large curved scissors to cut away all this loose skin, and thus leave a very large ulcerated surface: such an operation would be considered at the present day not only useless, but injurious and cruel. When a quantity of loose, thickened, and corrugated skin forms the sides of a deep ulcer, we should rub these flaps, or their edges, with the kali purum: thus a portion of this superfluous skin will be removed, and a vigorous healing action will be excited in the remainder, and the healing will be much promoted by pressure applied over the entire ulcer.

Should a bubo become phagedænic, and threaten to open the femoral artery, or should it give rise to constitutional fever and irritation, its further progress can often be almost instantaneously arrested by brushing over the edges with the strong muriate of antimony; and however large the surface, it will begin to heal, even if the edges alone have been touched. The pain from this application must be severe, but it does not last in general more than from half an hour to an hour; and the patient then often finds great relief, and falls into a sound refreshing sleep of some hours' duration; although, perhaps, for many nights previously, he had been unable to procure rest, even with the assistance of large doses of opium.

After a bubo has opened spontaneously, we sometimes find that one of the lymphatic glands rises up from it, and protrudes beyond the surface of the surrounding skin, the loose edges of which overhang its base. This, though formidabl in appearance, is not very painful: it gradually subsides, and becomes covered by skin in the process of cicatrization, like the rest of the ulcer, and yields as certainly, though not as quickly, as the other primary symptoms when once ptyalism is fairly established. Some have advised the destruction of the protruded gland by caustics and other means; but I have never seen a case in which any such severe measures were necessary.

We sometimes meet with the following condition of a bubo, —a large, indurated, indolent mass, the integuments but little or not at all discoloured, with sometimes a slight tendency to inflammation and suppuration, and yet the patient has used mercury so as to affect his system : in such a case I think blisters will be found most efficacious : they should be applied repeatedly, but not retained longer on the part than five or six hours. Some advise us, if we feel a softened spot in the centre or in any part of this tumour, to apply the kali purum, with the hope not only of letting out the matter, but also of exciting a healthy inflammation in the part : this practice, however, entails the additional inconvenience of an open ulcer without accelerating the dispersion of the bubo as efficaciously as blisters do. Many cases of this sort are ascribable to an injudicious use of mercury, and therefore we must at the same time apply ourselves to restore the system to a more healthy state.

A bubo not unfrequently leads to another tedious local affection, that is, a sinus commencing at the pubic corner of the ulcer, and descending in the angle between the scrotum and the thigh ; we observe the skin in this situation to be red and raised ; by pressing on this we force out a little matter into the ulcer; from day to day we can see this sinus increasing in length. This disease teases the patient for a long time, as it continues even after the accompanying chancre has healed under the influence of mercury. Pressure has no influence in arresting the progress, or in effecting the cure of this sinus. The treatment to be pursued in this case is either to leave it to the restorative powers of the system, or, if these prove inefficient, we may divide it in its entire length, or make a small opening into it inferiorly, and daily inject some stimulant fluid.

Another, and a still more troublesome consequence of bubo, is a superficial spreading of the ulceration along the inside of

the thigh, in some cases even to the anus; in others it extends
upwards on the abdomen, and in some it occupies both
situations; one edge of this ulcer is deeply and slowly
increasing or eating away, while the opposite is thin and may
be healing: this is what has been described as the horseshoe
ulcer; it is often very sensitive. Mercury in general does not
serve this symptom, it sometimes even seems to excite its
rapid extension : lotions of different kinds, especially the black
wash, appear to me to agree with it better than any other
applications, yet in some cases very minute doses of mercury
will be found most useful in disposing the ulcer to heal. Let
the following serve as a striking, but not an unparalleled,
instance of the extent to which this kind of ulcer sometimes
spreads, of the great length of time during which it may
continue, and of the difficulty with which it is at length
brought to heal.

October 13th, 1826. Mr. H—, aged 30 years, contracted a
chancre on the frænum three years since. Argentum nitratum
was rubbed on the ulcer, and he used mercurial ointment
under the care of a physician who made him repeatedly
suspend the use of mercury the moment his mouth became
ever so slightly affected. Before he had ceased the course of
frictions, a bubo appeared in his groin—this was punctured
with a lancet. In five months after the appearance of the
chancre he consulted an eminent surgeon in Cork, who
prescribed some form of mercurial pills which slightly
salivated him: he used only twelve of these. This surgeon
then told him he should not use any more mercury, and
advised sarsaparilla, lotions, and ointments of various kinds.
Twelve months from the commencement of the disease he
again consulted the same surgeon in consultation with
Mr. Evans, the author of the 'Treatise on Ulcerations of the
Genital Organs.' Repeated venesections and low diet (the
results of their deliberations), he states, induced great
weakness, and nearly brought him to death's door.

When recovered from the debilitating effects of this plan, he
repaired to London, where he remained for one month under
the care of the late Mr. Abernethy. Blue pill every second
night, and a sort of blue wash, constituted the plan of treatment.

He next applied to another very eminent surgeon in London, who tried blue pill for a short time, but laid it aside because he thought it caused erysipelatous inflammation, and he cautioned him *never again* to use mercury. Under this gentleman's care he remained for two months, trying a great variety of external applications, never employing the same beyond four or five days. He also used a variety of internal remedies, among which were acids, bark, and arsenic.

He next applied to the late Mr. John Pearson, and during six weeks that he remained under his care he took sarsaparilla-syrup, and used a variety of local applications, of which he can only name carrot poultice.

In the summer of that year he drank the Harrogate waters for one month, having been told that the sores were prevented from healing by a scorbutic state of his blood. He derived no benefit from these waters.

During the last year he had gone through long courses of sarsaparilla, and repeatedly.

At present he has an ulcer, the size of a shilling, on the left outer ancle: this does not possess the character of a secondary venereal ulcer. In the left groin is a cicatrix, which, beginning about the anterior spine of the ilium, is continued down the groin and passes to the back of the thigh, where it joins a prodigiously extensive ulcer; this ulcer reaches from the anus down along one-third of the back of the thigh; below the fold of the buttock it covers the entire breadth of the posterior part of the thigh; above the fold of the buttock it is less wide, being in one place as narrow as one inch and a half. It is surrounded by a very deep edging of skin, which is cut very irregularly into knobs. The surface of this extensive ulcer is everywhere devoid of granulations, and presents three spots of deeper ulceration in parts remote from each other. The entire of this ulcer is exquisitely sensitive and sore. Another branch of the cicatrix extends along the front of the thigh, and, having traversed the upper third of the limb, it there ends in an irregular deep ulcer, the size of half-a-crown. The skin of the forehead near the roots of the hair is a deep red or copper colour, and the scalp is in a very scruffy condition. The general health is apparently very good. For three years he has been prevented by this disease from attending to any business.

October 13th, 1826. Ordered Ung. Hydr. Fort. gr. x. omni nocte femoribus infricanda. ℞ Aq. Calcis, ʒviij. Hydr. Oxym. gr. ij. M., ft. lotio ulceribus applicanda.

23rd. Repet. Ung. et Lotio.—Ung. Hydr. Nitr. ʒi. Ung. Zinci, ʒi. ft. ung. quo curentur ulcera omni nocte.

28th. No sensible effect on the mouth; bowels costive. Ung. Hydr. Fort. Ɔi. o. n. Lotio Acet. cupri ulceribus; Pill. Purg. pro re nata.

November 5th. Mouth slightly affected, which induced him to omit the mercurial ointment the last three nights. Anterior ulcer of the thigh had been covered with Hydr. Præc. Rub., which has acted too severely, having carried off everything like granulation.

7th. Ɔi. Ung. Hyd. Fort. alt noct.—℞ Ung. c. Lapid. Calam. ʒi. Ung. Ærug. Æris ʒij. fiant ung.—quo curentur ulcera omni nocte.

13th. Ulcers are all healing, but without throwing up granulations. In the ulcer in front, as also in that behind, some new ground has been broken up. Ung. Hyd. Fort. Ɔi. alt. nocte.

16th. Mouth not affected; ulcers improving. Ung. Hyd. Ɔi. omni nocte.

22nd. Mouth slightly affected; ulcers are all improved; that on the outer ancle is not larger than the surface of a split pea; that on the front of the thigh is not as large as a sixpenny piece, but not granulating. The ulcer on the back of the thigh has the greater part of its surface covered with very small granulations, and altogether shows a strong disposition to contract and heal. In one part it is only half an inch across, and in a narrow neck only one-quarter broad. There is still too great a thickness of the scales in the scalp. Repetatur Ung. Ɔi. o. n. Sumat. Sulph. Quinæ. gr. ij. ter in die.

26th. Mouth not more affected: ulcers are all slowly but steadily improving; the large one is obviously healing, more by contracting its surface than by granulations. Repetatur Ung. Ɔi. o. n. et Sulph. Quinæ.

December 1st. Mouth more affected: the small ulcer on
the front of the thigh is nearly healed; since 26th ult. it has
healed much more rapidly than I had expected; the large
ulcer is also contracting rapidly, its edges are much less
raised. Pergat.

7th. He has now rubbed for twenty-four successive nights
Ung. Hyd. Fort. Ɔi., by which his mouth is affected, but this
affection does not increase according to the increased number
of rubbings. The two small ulcers have scabbed; the large
ulcer is very much reduced in extent, is much less irritable, and
its edges are very healthy. Omitt. Ung.—repet. Sulph. Quinæ.

13th. Mouth is more affected since he left off the rubbings.
The large ulcer is diminishing rapidly, its edges are much
thinner, its surface is more healthy, being red and covered
with very minute granulations.

28th. General health is very good; the scab has dropped
off the ulcer on the ancle: that on fore part of the thigh is
very small. The large ulcer is healing fast by contraction,
and is not skinning from the edges, although the edges are
thin and the entire surface is covered with small firm
granulations; no bad spot appears on its surface; the
greatest length, which is along the internal edge, is four
inches and a quarter, and the greatest breadth from angle to
angle is three inches. Ung. Hyd. Fort. Ɔss. o. n.—omitt.
Sulph. Quinæ.

January 14th, 1827. Greatest breadth of the ulcer is two
inches and three-quarters; greatest length, three inches and a
half. The surface of the ulcer is not quite so healthy; it shows
more of a yellowish surface. The gums of the lower incisors
are slightly affected by the mercury. General health is good.
Repet. Ung. Ɔss. o. n.—℞. Ung. Ceræ Flavæ ʒi. Hyd. Præc.
Rub. ʒi. Fiat. Ung. omitt. lotio.

10th. The ulcer is not contracting; it has become more
irritable; in different spots the granulations have been swept
away; no sloughing. Ung. Hyd. Fort. Ɔi. o. n.

19th. Mouth very slightly affected; ulcer is rather more
unhealthy than on 10th instant. Und. c. Lapid. Calam. ʒi.
Ung. Ærug. Æris ʒij. fiat ung. Omitt. Ung. Hyd. Sumat.
Sulph. Quinæ gr. ij. ter die.

23rd. Ulcer much contracted; surface much more healthy, with only one or two spots of a yellow colour; the edges are remarkably diminished in height.

February 1st. The ulcer is three inches in length and two inches in breadth; its edges are perfectly flat and level with the granulations. The ointment does not smart it; he dresses it twice a day. Repet. Sulph. Quinæ.

21st. The ulcer is one inch and a quarter in its greatest length; the edges and the surface are perfectly healthy, except in the centre; here it is of a dirty dark green colour, bleeding on the slighest friction. He has laid aside the quinine.

March 2nd. Ulcer has now assumed a circular form, and is less than half an inch in diameter; the edges are not at all raised; the granulations are very healthy. It will be healed perfectly in six or eight days. He returns to the country to-morrow.

Although a period of four months and a half might, at first sight, appear very long for the healing of an ulcer, yet when all the circumstances of the case are considered a different conclusion will probably be drawn. When we take into account the very great extent of ulcerated surface, and the increased difficulty of healing any sort of very large ulcer; when we recollect the very obstinate nature of this, which had proved so intractable in the hands of so many eminent surgeons, some of whom had devoted themselves very much to the treatment of the venereal disease and its sequelæ; and also when we consider the great variety of local and constitutional remedies which had been so ineffectually employed, it must be admitted that this was an ulcer difficult to heal. It is pretty plain also that sarsaparilla had not the power of inducing it to do so, for in addition to the use of it under the late Mr. John Pearson's care the patient himself had employed it repeatedly during one year. Without presuming to say that this ulcer could not have been healed without the use of mercury, this much is pretty clear, that it had resisted a vast variety of other modes of treatment. When I undertook to employ mercury in this case I proposed to myself to ‘ eat it as I would one of those cases of secondary venereal ulcers which require a very sparing use of mercury, and of

which I shall speak when treating of secondary venereal
symptoms.

In Mr. H's case the system resisted the action of mercury
very strongly, as his mouth was not in any degree affected
until the 5th November, and did not at any time exhibit
symptoms of ptyalism. Yet as soon as the system acknowledged the action of mercury, then have we the first account
of improvement in the ulcers. On December 7th the mercury
was omitted because I feared to persevere longer in its use,
seeing that the soreness of the mouth did not increase with
the increased rubbings, and fearing that this mineral would
soon begin to act as a poison. The mercury was resumed on
28th December, but it now appeared to disagree, for on
January 10th, in different spots, the granulations had been
carried off. On 19th January, the ulcer receiving a still more
unfavourable aspect, the mercury was relinquished, and in a
few days the ulcer put on a favourable appearance, and
proceeded slowly, yet without interruption, to a final healing.

I am fully convinced that had the mercury been employed
in the ordinary manner of beginning with $\mathfrak{z}$ss. or $\mathfrak{z}$i., and
afterwards increased, it would have had a deleterious effect on
the ulcer, and have caused it to spread and to slough ; so
mainly is the benefit of mercury dependent on our apportioning
the doses to the state of the system, and to the peculiar
condition of the existing symptoms.

Now, although this ulcer had not the characters of secondary
venereal ulcer, and although it was not attended with any
other venereal symptom, if we except the suspicious state of
the forehead, yet I am disposed to think that it had been kept
for two years from healing by some infusion of the venereal
poison ; this, however slight, the constitution was not able to
subdue or expel, even though assisted by various reputed
antivenereal medicines, until mercury was tried and persevered
in, in the manner I have detailed.

I beg it may not be inferred that I would recommend the
use of mercury in all cases of this horseshoe ulcer, the
consequence of ulcerated bubo ; or that I conceive that such
ulcers cannot be cured without the use of mercury. On the
contrary, I have often seen them yield, though slowly, to

other treatment, and I have even known some to have been made worse by the use of mercury. But surely when other means have been tried fully, and without good effect, we should not refuse our patient the chance of being benefited by mercury, especially as we can protect him against any of its mischievous effects by using it in very small doses, and by carefully watching its influence on the system and on the ulcer. In the case I have just stated, I preferred the ointment to any internal preparation of mercury, as I was certain that this would agree with his system fully as well as any other form of the medicine, and especially as the last internal preparation which he used seemed to have disagreed with him so much as to have made his surgeon caution him against ever using mercury again.

The history of bubo would be incomplete were we not to mention that this symptom occasionally, but rarely, precedes by a few days the appearance of chancre : such cases require great caution on the part of the surgeon ; he should be slow to pronounce them venereal, and he should not commence a mercurial course until the chancre has appeared : this delay cannot be productive of any bad consequence, but on the contrary will encourage both patient and surgeon to enter on the mercurial course with greater confidence as to its propriety.

[The form of bubo above alluded to is that recognised by French writers as "Bubon d'emblée." Hunter admitted its existence ; but in more recent times, Lagneau, Vidal de Cassis, and Castelnau have more particularly discussed *the bubo at the first onset.* There can be little reasonable doubt that true venereal buboes may exist either without preliminary chancre, or where the chancre has existed in so slight and transient a form as to escape the notice of the patient, and have healed before advice is sought ; it is, however, also probable that *buboes at the first onset* may be instances of simple adenitis connected with excessive coition.

Between the virulent adenitis of soft chancre and the multiple, indolent, indurated, non-suppurating glandular engorgements of true syphilitic bubo, Colles makes no distinction. The chronic form of bubo next mentioned and particularly dwelt upon in Chapter XII. of this work belongs to a later period of the disease.]

There is a chronic form of bubo which sometimes appears

in connexion with secondary venereal symptoms, the consideration of which we shall reserve until we come to speak of the secondary symptoms of syphilis.

The lymphatic vessels of the penis are sometimes affected not only with inflammation along their course, but in a few cases this inflammation will in some one part of the vessel go on to circumscribed suppuration, forming what has been termed lymphatic bubo. In a few instances we may find two of these buboes in the same individual.

In an hospital patient, in September, 1833, one of these buboes took place a little anterior to the pubes, and another in the central line of the pubes, a little above the penis. These arose from rather a large chancre on the external surface of the prepuce, near to its orifice, and on its dorsal aspect.

In a private patient I lately saw one small bubo a little anterior to the pubes, and another between the pubes and left groin. So that there is no determined point at which these must occur. It is curious that, although one of these buboes may go on to suppuration, especially when mercury disagrees with the system, yet we rarely, if ever, see the second run into this state—it slowly retires.

CHAPTER VI.

ON A DISEASE OF THE LYMPHATIC GLANDS OF THE GROIN, ATTENDED WITH PECULIAR SYMPTOMS.

The following account was published in the Dublin Hospital Reports in the year 1818: I have thought it advisable to introduce it here, in reference to the subject of the preceding chapter.

One of the lymphatic glands of the lower or femoral range is generally the seat of this disease. Sometimes, however, it is seated in one of those of the upper or inguinal row. I have never had an opportunity of seeing the disease in its incipient state, the patients not having applied to me until the swelling had attained the size of a walnut. At this period the integuments are not in the slightest degree discoloured, nor is the surface shining. The inconvenience of which the patient complains is a slight pain which he experiences in walking or in making any considerable exertion with the lower limbs. The progress of the tumour to suppuration is uniform, though slow; the integuments become red, but not pointed; and the matter is spontaneously discharged at a period varying from the fifth to the eighth week. The cavity of the abscess is small in proportion to the extent and hardness of the tumour. The matter is in general of a tolerably good consistence: not unfrequently a second, and sometimes even a third, collection of matter forms in the neighbourhood of the first, the tumours exhibiting the same indolent character. The openings by which the matter escapes are narrow, and spread not to a large size, preserving rather the appearance of fistulous orifices than degenerating into broad ulcers. In general they heal spontaneously in the course of two or three months from the period of ulceration; but I have met with some few cases in which they became complete fistulous ulcers, and remained open for the space of even twelve months, the patients having refused to submit to the surgical treatment which fistulous ulcers require. A very striking feature of this disease is the trifling degree of pain which attends it : the patient suffers so very little as to be capable of walking about without any perceptible lameness. I have known some of them, in the situation of merchants' clerks, continue to lead a very active life under this complaint, doing the out-business of the house, as they term it, during the entire progress of the disease. In fact, the patients generally complain more of the bulk than of the pain of the swelling. One case only occurred to me where the pain was such as to require the abscess to be opened with the point of the lancet ; and in this instance, for three or four days after the opening had been made, the patient experienced the most unaccountable soreness and pain from this very trifling operation. In some few cases, while the tumour is approaching to suppuration in one groin, the glands of the other begin to swell, and in a rare instance, now and then, the tumour having arrived at suppuration, remains for a time stationary, the matter is then gradually absorbed, and the swelling at length slowly removed.

This disease usually occurs in men between the ages of twenty and forty, but in general nearer to the former than to the latter period of life. I have met with only one instance of it in a female, who was about thirty years of age.

From the very earliest period at which I have had an opportunity of observing this disease, the constitution is found to be engaged. The patient is affected with headache, which is more severe in the morning, and which is increased by stooping : he also admits, when questioned, that he feels more fatigue than usual from long-continued or violent exertions; his pulse is quick, being in no case, when he is out of bed, under 100, and generally beating 120 in the minute. This quickness of pulse appears the more extraordinary, as it is obviously not produced by a high degree of pain, nor is it accompanied by a discoverable derangement of any other of the functions; on the contrary, the countenance is natural, respiration easy, skin of temperate heat and not very dry, tongue clean, appetite as good as usual, and scarcely ever nocturnal sweats : the patient, however, feels himself more comfortable in the open air than when confined to the house.

I have had an opportunity of examining one patient only while lying in bed in the morning. His pulse was then only 72, but on his rising and dressing himself it rose to 110. The tumour at this time was as large as half a hen's egg, and the integuments were not discoloured.

The patients generally conceived that their health was improved by this disease ; for, before the final healing of the ulcerated opening, they have informed me that they felt themselves in better health than they had enjoyed for some months previous to the attack.

In the treatment I have confined myself to those means which I have conceived to be calculated to mitigate the severity of the symptoms, and to promote suppuration, which in general indeed seemed an unavoidable and always a salutary termination of the disease. The headache appeared to be alleviated by no class of medicines but by purgatives. These were repeated every day, or every other day, until the symptoms were completely removed. Very large doses were often required to produce the desired effect. The removal of the headache was not attended with a diminished frequency of the pulse. Poultices, warm fomentations, and gum plaisters were the only topical applications to which I had recourse. Leeches had been applied in two instances before I saw the patients, but apparently without any salutary effect. Cold, and as they are termed repellent applications, when used for a few days in the earlier stages of the disease, did not appear to produce either benefit or injury.

A knowledge of this disease may probably assist us at some future day in developing the pathology of the lymphatic system, which remains still involved in considerable obscurity, and will at all events be of some use, I trust, in practice.

I cannot say from observation what consequences would result from the exhibition of mercury in this complaint. But I apprehend that we have everything to fear from administering that medicine to patients affected with such an extraordinary quickness of pulse. Indeed, the very apprehension of the evils that might thence result has deterred me from ever putting the matter to the test of an experiment.

We should be careful to distinguish cases of this disease

from examples, no doubt very rare, of truly venereal enlargement of those glands wherein the swelling of the gland
precedes, for some days, the appearance of the chancre. By
attending to the constitutional symptoms, so characteristic of
the disease here described, we shall with certainty avoid this
error.

I may here observe that, in some instances, when an
enlargement of the inguinal gland arises from the drying up a
venereal chancre without the use of mercury, the patient will,
as in this complaint, be affected with severe headache and
quickness of pulse. Such cases, however, will scarcely be
confounded with the disease here described; for the previous
ulceration of the genitals in the one case, and the absence of
that symptom in the other, are sufficiently characteristic, and
though the quickness of the pulse and the severity of the
headache be common to both, yet these constitute almost the
only constitutional symptoms in the one case, while in the
other they form but part of a series of alarming derangements
of the system, such as remarkable prostration of strength,
loss of appetite, and profuse night sweats. In short, the one
case presents us with a striking picture of general disease and
debility; while the other exhibits every character of general
health, except the affection of the head and the extraordinary
quickness of the pulse.

Since the preceding observations first appeared I have had
many opportunities of treating this disease, and I have met
with four or five cases in which the patient was reduced to a
state of extreme weakness, emaciation, and hectic, by
repeated suppurations and profuse discharges; and I am
happy to say that I found a certain and rather speedy remedy
for all this alarming extent of disease in sea air and tepid salt-
water baths. Of course tonic remedies formed a part of the
treatment, but in my opinion the credit of the cures must
chiefly be ascribed to the sea air.

CHAPTER VII.

WHEN a primary symptom has been long neglected, or when treated injudiciously or carelessly, it may happen that the first series of secondary symptoms shall make their appearance before it has been removed, and thus the patient may labour under both primary and secondary disease at the same time. In some cases the condition of the primary symptoms does not appear to undergo any change in consequence of this accession of disease; but in other instances we find it to be materially affected. It is not very uncommon for the surgeon, in treating a chancre of an old date, or an ulcerated bubo, when he has brought them into a healthy state to find his hopes disappointed for the present, and to see this ulcer most unaccountably change its apppearance and take an unfavourable turn, either by sloughing or assuming the phagedænic character; and this unfavourable change will, in spite of the most judicious treatment, continue for eight or ten days, until the secondary symptoms are fully established.

Nor yet is it very common for a primary symptom which has been obstinate, and has for many days undergone a very unfavourable change, suddenly and unaccountably to assume a healthy aspect, and for a few days make rapid advances to healing; this progressive improvement, however, is checked as soon as the secondary symptoms are fully developed. I freely confess that I am unable to explain why such different and opposite effects should result from the same cause, viz., the development of the secondary symptoms. This is a point which deserves close and attentive observation; the discovery of it may lead to some useful hints to guide our treatment of obstinate primary symptoms.

When this combination of primary and secondary symptoms is treated by mercury we shall observe that both forms of the

K

disease will come under the influence of its action at the same
time. No doubt some secondary symptoms, viz., eruption
and sore throat, are apparently much more early affected by
it; thus an eruption will have sensibly declined before the
chancre or ulcerated bubo will have assumed the aspect of a
granulating ulcer. But if we closely watch the primary ulcer
we shall see as early, though not as signal, an impression
made upon it. The assertion that the secondary are cured
sooner than the co-existent primary symptoms appears to me
to be derived from a hasty and superficial view of the subject,
from mistaking the great amendment in the former for a
cure, and that only in two symptoms—sore throat and
eruption. For if we chance to meet with the rarer case in
which primary and secondary ulcers in other situations co-exist,
we shall surely see the primary healed some time before the
secondary ulcer, and that too in cases where both are nearly
of equal size.

In treating of the history of the venereal disease I expressed
my regret that Mr. Hunter had not followed out the course of
the disease beyond the appearance of secondary symptoms,
omitting all those constitutional changes and diseases which
intervene between this period and the fatal termination.
Again we have to regret that he has described the secondary
symptoms in such a manner as to give only a single description
for each class. Thus, in speaking of venereal sore throat, he
has described only one form of ulcer; no doubt he has hinted
at one or two varieties under which the venereal eruption
appears, but he has glanced at these in such a manner as can
convey very little information on this subject to an inexperienced person; while on the subject of venereal sore
throat he has not even once alluded to the vast variety of
features which this class of secondary symptoms presents,
according to the modifications which the disease may have
undergone from peculiarity of constitution, or from the
influence which previous courses of mercurial or other treatment may have induced.

It is remarkable that Mr. Hunter has given only one form
of each symptom, whether primary or secondary,—e.g., one
form of chancre, one of venereal sore throat, the same of

venereal eruption; and on the subject of nodes he is equally limited. Perhaps he intended only to sketch the great outlines of the symptoms, reserving for a subsequent opportunity his observations on the varieties which each order of symptoms occasionally presents.

Every person, however, who has paid even moderate attention to the secondary forms of the venereal disease, will admit the accuracy with which Mr. Hunter has described the order of parts successively attacked. No doubt, as he has remarked, we occasionally meet with exceptions to this rule, and some of the parts second in order will be affected before those which are generally the first. He has not attempted to account for this phenomenon: from a good deal of observation I am disposed to ascribe it, in some cases, to the maladministration, and in others to the excessive use, of mercury, by which both the healthy and morbid actions in the system have been disturbed and deranged.

When we inquire into the time at which the secondary disease has succeeded to the first inoculation, we shall find it to differ in different individuals, and we shall also find that it is influenced by the previous treatment of the primary symptoms. Where these have been treated on the non-mercurial plan we generally observe that the secondary symptoms are later in appearing, and that they are also preceded by less previous disturbance of the system; but when mercury has been used for a short time only, or has been discontinued as soon as the chancre had healed, we shall find in such cases that the secondary symptoms will appear more early. It is generally admitted that if febrile action be excited by ordinary causes, this very action will eliminate the virus and make secondary symptoms display themselves at an earlier period. In some persons, however, secondary symptoms do not occur until after a very long interval of time, perhaps six months.

[The question of how far the secondary disease is influenced as to the time at which it succeeds to the first inoculation, first by the individual differences of the patients, secondly by the previous treatment of the primary symptoms, is one still open to careful observation. Colles had reason to believe that secondary symptoms are later in appearing in cases treated on the non-mercurial plan, and that when mercury has been used for a short time only that secondary symptoms will appear more early. Some excellent observers, as Langston Parker, Hardie, Comrie, Gascoyne, Longmore, &c., take a different view: they believe that those symptoms are postponed and modified by mercurial treatment. Among many of the most reliable practitioners of the present day the opinion prevails that secondary symptoms cannot be prevented by mercurial treatment of the primary disease. Among these may be mentioned Marston, Longmore, Perry (Royal Artillery), Comrie, Nelson, Sir W. Fergusson, Langston Parker, De Meric, and others. Mr. Jonathan Hutchinson says after mercurial treatment "the secondary symptoms, I think, are rather lighter, but I feel certain that it does not prevent their occurrence: they come out just about the same time, I think." Mr. Skey says:—"I think it is clear that the primary sore treated by mercury is as likely to be followed by secondary disease as the primary sore not treated by mercury."* Colles's opinion, as stated at pages 55 and 56, is exactly the reverse of this. "Hunter," says Ricord (and precisely the same may be said of Colles), "had observed that the chancre was on certain subjects a purely local affection. He affirmed that *infection might be prevented* by early and well-directed therapeutical means. But that which he attributed to the influence of medicine I attribute to the special form of the primitive ulceration. Without doubt there are chancres which infect the economy, whilst there are others which do not infect it, and that, let it be understood, independently of every therapeutic influence. Many practitioners adopt in good faith the mercurial treatment for the simple non-infecting chancre. How many honest and upright individuals believe they have cured Syphilis which in reality never existed."]

For some days before the secondary symptoms appear the patient feels himself out of order, is languid in the evening, unwilling to use either mental or bodily exercise, and often complains of pains in the shoulders or legs; these do not remain fixed, but at different times they affect different parts; his sleep is unrefreshing. These fugitive pains all cease as soon as the eruption or sore throat appears.

The latter affection is at first attended with slight uneasiness in swallowing; in the course of a few days this amounts to pain; a dryness is also felt in the throat when the patient awakes at night, and his efforts to swallow saliva are productive

* See evidence and "Report of the Committee appointed to enquire into the Pathology and Treatment of the Venereal Disease with the view to diminish its Injurious Effects on the Men of the Army and Navy," 1867.

of more pain than his attempts to swallow food or drink; a
slight external tenderness and swelling are now discoverable
in the site of the tonsil. On inspecting the throat we observe
one tonsil or both swollen, with increased redness, and we
readily discover an ulcer which Mr. Hunter describes as
"a fair loss of substance, part being dug out, as it were, from
the body of the tonsil, with undermined edges: this is
commonly very foul, having white thick matter adhering to it
like a slough, which cannot be washed away." The patient
complains of pain running upwards in front of the ear and
occasionally down the side of the neck, even to the point of
the shoulder.

This form of venereal ulceration of the throat may be
looked upon as the type of the genuine venereal sore throat.
However strongly the characters of the ulcer may be marked,
still we must not rely solely on the present appearances; we
should trace back the history of the previous disease, look to
the interval which has elapsed, inquire into the premonitory
symptoms, as also into the treatment which had been employed
for the cure of the primary disease.

Having satisfied ourselves of the nature of the disease, we
next enter upon the treatment, which must be conducted
according to the rules already laid down for the treatment of
secondary symptoms. I need not repeat the admonition
of taking care to excite the mercurial action in the manner
most likely to insure its sanatory operation, to produce the
least possible injury to the constitution, and yet to excite
and keep up such a mercurial action as we know is most
likely to effect a permanent cure of the venereal disease.
This action, I need scarcely say, will require, in general, to be
kept up for eight or ten weeks. In this, as in primary
chancre, I would abstain from local applications to the ulcer,
as the changes which the latter undergoes will, in various
instances, assist us in discovering when mercury is beginning
to disagree with the system; besides, we must be guided in our
opinion as to the proper duration of the mercurial course by
a reference to the period of the healing of the ulcer in
the throat. Should we be so unfortunate as to have merely
caused the ulcer to heal without at the same time thoroughly

eradicating the disease, we shall find that the ulcer will return, in the same or the other tonsil, and with precisely the same characters as the first ulcer in the throat had done.

The perfect cure of the disease is not at all times enough to satisfy the patient; for we are often called upon by a patient whom we have cured of a venereal sore throat, but who is in great alarm, and states that he is not cured; that he feels his throat as bad as ever: that he has the very same pain in swallowing, and the very same uneasy sensations which accompanied the former ulceration; yet, on inspection, every part of the throat appears in a healthy state. On the closest examination we cannot discover anything morbid, except occasionally some very slight appearances of inflammation. Now all this alarm is groundless; it arises from the facility with which morbid sensations are renewed. We sometimes trace this alarm to a local excitement caused by a fresh cold, sometimes to a derangement of the stomach or bowels, and in some few cases it is the offspring of slight hypochondriasis. I think the practice of washing and rubbing the throat externally will prove the best means of preventing or removing this disquietude. Country air and sea-bathing are also of the greatest service.

The venereal ulcer of the throat, though generally seated in the tonsil, will yet in some instances fix itself in a part where it is very much concealed from view; and therefore we should be prepared to search for it when the sensations of the patient lead us to apprehend the existence of such an ulcer. In one or other of the following spots we shall discover these concealed ulcers. Sometimes we meet with a patient who complains of very severe pain whenever he attempts to swallow, and which occasionally is so constant as to deprive him of sleep; yet on inspecting the fauces we are surprised to find that all within our view appears in a healthy state; no ulcer can be seen, nor even any appearance of inflammation in any part. This first view, however, should not satisfy us; we should consider in what points an ulcer may exist and lie concealed. The most common place for such to occur is on the back part of the pharynx, and there it may be concealed by the interposition of the velum palati. We should therefore desire the patient to inspire as fully as he can; in attempting this he raises the velum, and if we then look into the pharynx we shall generally discover the lower part of an ulcer: this, of course, leads us to make a more full examination; to effect this we must depress the tongue, and with a curved probe raise up the velum.

The ulcer, which now becomes more fully exposed to view, is of a circular form, is sunk deep in the substance of the pharynx; the surface is rather foul, but not at all sloughy; the surrounding inflammation extends a very short distance beyond the margin of the ulcer. The present sufferings of the patient loudly call for speedy relief, as he is not only deprived of the power of taking nutriment, but he is sometimes kept in a high state of irritative fever. Here we must use topical applications, although by so doing we lose the opportunity of watching the progress of the ulcer towards healing. We should rub the entire surface of the ulcer with muriate of antimony; the pain which this occasions is no doubt severe, but it is of very short duration, and the patient will soon enjoy sleep, and swallow without any suffering.

The best mode of applying the muriate of antimony is by means of a little lint rolled pretty firmly on the eye-end of an aneurism-needle, and dipped in the liquid. We should take great care that the application be confined to the ulcer, for should any of this fluid trickle down towards the larynx it will cause a most distressing sense of suffocation, one which sometimes alarms the surgeon as well as the patient and his friends.

All the misery attendant on ulceration of the throat will sometimes be found to proceed from an ulcer of the pharynx, situated below the level of the base of the tongue, and therefore, when we cannot discover any ulceration of the tonsils, or back of the pharynx, high up, we should carefully look for it in the lower part of the latter; indeed we should be doubly solicitous to discover it, as an ulcer situated so low down may creep still lower, and spread to the sides and top of the larynx, producing there such mischief as must lead to a certain, though often a slow, death.

The ulcer low down in the back of the pharynx generally

presents a foul, and sometimes a sloughy surface, seldom assuming the venereal characters described by Hunter, and has this remarkable feature,—that its lower edge is very sharp, while the upper part of the ulcer is superficial. A pretty constant symptom complained of by the patient is that when he attempts to take any food the morsel stops at a certain point, and can only be got down by his taking after it a sup of liquid: in proportion as the ulcer improves the power of swallowing solids improves also.

The ulcer once discovered should be very well rubbed with muriate of antimony; taking very great care, however, to prevent it trickling down to the larynx.

Another position in which we sometimes find a venereal ulcer, which causes not only pain in swallowing, but also pain shooting up one side of the head and face, is close to the insertion of the anterior palatine arch into the tongue. If the latter symptom be absent the ulcer will be found about the central part of the base of the tongue. In making an examination of the throat in such a case we are frequently warned of the existence of the ulcer by the severe pain which we cause to the patient, when, in attempting to depress the tongue by the spatula, in order to search the throat as low down as possible, we chance to lay the instrument on the ulcer.

The ulcer will be found deep and foul, but not at all sloughy; and whether it be seated on the dorsum or near the edge of this fold, its exquisite sensibility should be immediately destroyed by touching the surface with strong solution of nitrate of silver, or with muriate of antimony.

In a few instances we observe that the voice of the patient is rendered very nasal, and this sometimes even on the first appearance of the venereal sore throat and eruption; and yet neither the situation nor the condition of ulcer visible in the fauces will enable us to account for this symptom. In this form of the disease we find that the patient not only suffers severe pain in any attempt to swallow, but he is also teased by frequent desire to draw down the mucus from the back of the nares; and this secretion, when coughed out, is often found slightly tinged with blood. A smart degree of fever also generally attends.

The ulcer in this case is seated behind the velum, high up in the angle between the upper and back part of the pharynx, or at the juncture of its occipital and vertebral portions.

In this case we should endeavour to allay these sufferings as soon as possible; no benefit whatever can result from our waiting in the expectation of its being healed during the mercurial action; and as the ulcer is out of sight we cannot derive from it any information to guide us in the administration of the mercury.

The patient will be materially benefited by our rubbing to the ulcer some liquid caustic, such as a strong solution of Argentum nitratum, ℈j. ad ℥j., by means of lint wrapped round the end of a silver aneurism-needle. Here let me caution those who adopt this practice against the danger of this lint slipping off, and remaining behind the uvula when the aneurism-needle is withdrawn. Should this occur the patient will be perfectly miserable as long as it remains there; and yet such is his impatience that he will not, or cannot, submit to the ordinary means in use for dislodging it. To avoid such an accident we have only to pass one end of the lint through the eye of the needle, and then wrap the remainder about it.

When the ulcer is seated on the posterior surface of the velum—rather an uncommon case—we are not only directed to its exact seat by an appearance of thickening, together with a blush of redness on the anterior surface of the velum, which appearance corresponds to the seat of the ulcer; but we can gain a positive knowledge of its situation by carrying behind the velum an aneurism-needle armed with lint, and rubbing it to the suspected spot; if there be an ulcer the lint will be covered with the discharge from it. Here also the immediate application of liquid caustics to the ulcer is indispensable.

Although I have spoken of these as pure original venereal ulcers of the throat, yet I am doubtful whether this arrangement be strictly correct. For although I can recollect instances in which the original ulcer first fixed itself in some of those concealed situations, yet I cannot speak with certainty of them all; and I freely admit that in the relapsed sore throat the ulcer will frequently be discovered in some of the above positions.

There is an affection of the throat of which those who suffer from it complain repeatedly. It is by no means confined to those who labour, or have previously laboured, under venereal disease, and is occasionally met with in females of most unblemished characters ; still, when it occurs in a patient who has recently been afflicted with Syphilis, it is apt to create much mental uneasiness. I allude to that condition of the mucous membrane of the back of the pharynx in which it is seen to be covered with an uniform thin crust of whitish or yellow hardened mucus : this, of course, is more troublesome in the morning, when it is found to be perfectly dry and hardened than it is in the after part of the day, when it has been broken into detached pieces, some of which have been removed in the act of deglutition.

This affection, as far as my observation goes, is very obstinate and very tedious. I have known it continue not only for months, but for years. However, though very obstinate, it does not prove dangerous ; for at the end of some years we find the parts in precisely the same state as that in which we saw them when the case first came under our notice. The nature of this disease is very readily ascertained; it is only required to rub off the mucous crust gently with the probe wrapped round with lint, and then we shall see the membrane perfectly free from ulceration, and in no way morbid, except that it may be a little more muscular, rough, and dry on the surface than it is found to be in a healthy state. The diagnosis, of course, is then easy, but the means of cure are by no means either easy or certain. I have known some cases much relieved by gargling the throat with sea-water, and others to be, for the time at least, apparently cured by sea-bathing ; but I have known both fail in many instances. I have, however, never known any case in which the disease degenerated into a worse condition, although I have watched some for ten or twelve years.

I have already said that ulcerated venereal throat is often materially changed in its characters by an unsuitable line of treatment : this assertion will be found to be strongly corroborated by the appearance which the parts present even when healed. How frequently may surgeons observe, on looking into a throat which has suffered severely or repeatedly by venereal ulcers, that these now exhibit that smooth, shining, silvery surface which we always see in those spontaneous ulcers occurring in young people, and which are by many called scrophulous ulcers of the throat ; whereas those venereal sore throats which heal kindly, and which have not undergone any very bad change in their course, will be seen to present a healthy surface, similar to that of the rest of the mouth, with only occasionally a slight depression at the part.

For practical purposes we shall find it of much advantage to attend to these appearances. Suppose, for example, that a patient comes to us complaining of a relapse of his sore throat, and that we discover an ulcer in a fresh part, while the silvery cicatrix is seen in the site of some former ulcer, we are at once led to suspect that this patient's system must have been a good deal broken down by the previous treatment ; and we are warned to avoid the dangers of precipitating him again into a similar state by employing mercury in a manner suited only to an unbroken system. Before, however, we come to any positive determination in such a case we should make particular inquiries as to what had been the local treatment of the ulcers, for the application of a very strong caustic to an ulcer of the throat will be followed by a silvery cicatrix of such ulcer.

The patient who is labouring under venereal sore throat has not only to complain of the pain which he suffers, but is also sometimes afflicted with a flow of saliva, which pours almost incessantly from his mouth by day, and which, falling back on the larynx when he lies down, threatens instant suffocation, and altogether deprives him of sleep. I know not why it is that this incessant flow of saliva attends some cases and is absent in others. I think this symptom does not depend on any particular situation, or any particular form of the ulcer. I need scarcely say that the diminution of this discharge is one of the first signs of amendment.

Another distressing symptom, which occasionally attends on venereal sore throat, is the regurgitation of the patient's drink through the nose. This alarms the patient excessively; we can, however, relieve his anxiety by assuring him that this will cease when the ulcers have healed. Indeed it is surprising to see how very well many of those unfortunate persons can ultimately swallow, even though they have lost a considerable portion oi the soft parts about the palate and fauces. The powers of deglutition in these persons often become wonderfully restored, partly from the contraction which occurs, during the healing process, in all the surrounding parts, and partly from the power which the muscles possess, either of accomodating themselves to their altered condition or of acquiring some new or peculiar mode of action.

Perhaps in the wide range of surgical diseases we cannot find any one so liable to be materially changed as secondary Syphilis ; I shall not say in its nature, but certainly in its appearance, and in the course of its symptoms, as also in the line of treatment suited to its cure. These extraordinary changes some authors have endeavoured to account for by referring them to a scrophulous, or to some other cachectic state of the system of the individual. In my opinion too much stress has been laid upon this explanation ; for if we confine ourselves to those cases in which, from ignorance on the part of the patient, the disease (suppose a venereal ulcer of the arm) has been undisturbed by any active measure, and has been treated only by purgatives, poultices, and mild ointments,—in such instances we do not observe that the genuine characters of the venereal sore have been modified or altered by the state of the general system ; that is, it still remains a venereal ulcer.

I would merely ask a surgeon, who entertains this opinion of the influence of the habit in modifying secondary venereal ulcers, whether he would undertake, by examining a secondary venereal ulcer in any patient, to pronounce from the appearances there exhibited that this individual is of a scrophulous habit.

In my judgment the great varieties in the appearances and nature of venereal sore throats are occasioned by the manner in which mercury has been used, either for the cure of primary or for the treatment of the secondary symptoms. No doubt when mercury is used injudiciously, and in a manner unsuited to the general health and condition of the patient, we shall find that it causes more mischief, and produces more strange changes in the venereal ulcers of the throat, if the patient be of a highly scrophulous or of a very delicate habit, than if he be of a vigorous and healthy constitution.

Among the secondary symptoms of Syphilis we cannot find any which are more strikingly influenced by injudicious treatment (conjoined with peculiar states of the general system) than ulcer of the throat. To describe all the varieties of these, which are thus produced, would be a useless and almost an endless task; but I think it may not be without benefit to say a few words in particular on one which is the most commonly met with, differing, however, in its site and in its extent, in different instances; as also briefly to notice one or two other varieties.

In a case where a genuine syphilitic ulcer of the throat has been treated by an irregular or excessive course of mercury, and has healed under this treatment, we shall often find the patient suffer what he terms a relapse of sore throat. The new ulcer, whether it appears in the site of the former one, or whether it occupy some new position, will be found to differ most strikingly from the original ulcer. This appearance, which may be very frequently observed, is that of a superficial ulceration of rather a whitish colour, with a good deal of surrounding redness and some slight degree of swelling; in fact, it presents many of those characters which might lead some to call it an aphthous ulcer. If we watch the course of this ulcer we shall frequently see that it creeps along from place to place; that unless it be seized by phagadæna, or sloughing, it appears to be rather indolent in its nature and mild in its character; thus it admits of being healed, or almost healed, by various topical means, assisted, perhaps, by some tonic or alterative internal medicines.

A vast variety of ulcerated throats will be seen between this well-marked type of the Hunterian sore throat and this aphthous-like ulcer. To describe each of these would be both a pain and a useless task, for the judgment of the surgeon alone must direct the treatment, and suit it to each particular

case according to the existing circumstances in each. Thus
an aphthous ulcer of this kind, in the person of a robust man
not labouring under irritative fever, will require and will bear
such a course of mercury as would exasperate a similar ulcer
in a patient whose system was affected with a high degree of
this fever; nay, the misapplication of this treatment might
even endanger the life of the patient. I need not add, that
the state of the ulcer itself should also influence our line of
treatment; for, if phagedæna or sloughing should seize upon
it, we must prescribe for, or according to, these altered
conditions of the ulcer, instead of prescribing according to
the name of the disease, a "relapsed venereal sore throat."

One of the most alarming aspects which venereal sore
throat exhibits is the following :—On inspecting the fauces we
find the entire of the velum and both tonsils in a state of
sloughy ulceration, and the back of the pharynx appears
converted, as it were, into a soft pultaceous mass; indeed,
sometimes we see all parts of the throat covered with this
soft slough : by rubbing these parts with lint wrapped round
the end of a probe we ascertain that this covering adheres
tenaciously to the surface. The patient is scarcely able to
swallow even a sup of the most bland fluid; by day a constant
flow of ropy saliva issues from his mouth, and he complains
that by night he is not only deprived of sleep by the pain of
his throat, but that he is prevented from lying down, for in
that position he is in danger of being suffocated by the viscid
saliva passing down his throat. His general health is
completely broken up, he is emaciated to an extreme degree,
his strength is quite prostrate, pulse extremely quick, and
skin hot; in fact he is in a very high state of fever which is
of the hectic type. In a word, i do not think that a more
pitiable and alarming combination of symptoms is scarcely
ever presented to the notice of the surgeon.

When we inquire into the history of such a case, we learn
that this unfortunate man had undergone repeated slight
courses of mercury, or one or two very severe ones, for the
cure of the venereal disease; that these, for a time, relieved
the symptoms, but that on desisting from the use of mercury
the disease of the throat had relapsed, and had assumed the

present appearances. I believe that the usual mode of treating such cases is to give sarsaparilla in decoction or in broth,* and at the same time to endeavour to arrest the ulceration of the throat by various topical applications : under such treatment many recover, but with no inconsiderable mischief done to the throat, and after tedious and severe sufferings.

Now this is precisely one of those cases in which *very small* doses of mercury will be found of the most signal service ; although I make no doubt but many surgeons will be startled by the mere proposal of using mercury under such discouraging circumstances.

I shall adduce the following as one instance out of many, to prove the safety and the benefit of this line of treatment :—

James Johnson, æt. 40, 12th ward, admitted January 10th, 1836. A number of ulcers with dark brown scabs, but not of a conical or rupia form ; these scabs are of an oval figure, are surrounded by a red defined zone, between which and the edge of the scab runs a yellow ring containing pus. One of these ulcers is on the left side of the nose, one on each eyebrow, two on the forehead, one on the chin, one on the left shoulder, one on the sternum, and one on the left leg. The arches of the palate, uvula, and tonsils are deeply ulcerated ; the entire surface of the pharynx is converted into, or covered over by, a soft slough, like half-boiled flummery. Has great pain and difficulty in swallowing ; there is a constant copious flow of thin saliva ; he complains of pains of shoulders ; sleeps very badly. P. 144. He is very much emaciated and very weak. He states that thirty weeks ago he first perceived a bubo in his left groin ; at this time he had neither ulcer on penis nor discharge from urethra. He took pills, by which his mouth was made sore ; the bubo having spontaneously opened in three weeks. After this he remained well for about seventeen weeks, when he felt his throat sore. For this he

---

* To three ounces of sarsaparilla sliced, add three pints of water ; let them simmer on a slow fire until reduced to two pints ; take out the wood, and bruise it in a stone mortar ; return it into the liquor, with half a chicken, or half a pound of raw beef without fat. Boil slowly for an hour, and pour off the liquor.

was treated in the infirmary at Liverpool, using pills, by which his mouth was made slightly sore, and his throat was cured. In ten weeks after leaving the hospital, his throat again became sore, and again healed under a more severe salivation. Throat remained well for three weeks only. The ulcers began to appear six weeks ago.

January 10th, Throat to be brushed every second day with Solut. Arg. Nit. Decoct. Sarsæ ℔ i. Acid Nit. Dil. ʒi. indie.

January 16th. Throat rather better; is scarcely able to swallow the decoction. Scabs stationary. Pergat.

January 26th. Two ulcers, one on each eyebrow, show the pointed form of rupia. Two others of the ulcers, which have been dressed with ointment, exhibit a flabby surface very different from ordinary secondary venereal ulcer, and are devoid of the yellow ulcerated margin of such ulcers; the flow of saliva continues unaltered. The throat as on admission. Repet. Decoct. Sarsæ et Garg. Chl. Sodæ. Ung. Hyd. Fort. gr. x. o. n.

January 30th. Mouth is sore from the mercury. Complains of weakness; can swallow fluids only. Ulceration of throat less sloughy; scabs on the face appear as if of less extent; those which had assumed the rupia shape have dropped off. Repet. Decoct. Sarsæ, Garg., et Ung. Hyd. Porter ℔ i. indies.

February 6th. Throat now presents one clean ulcerated surface; left tonsil seems to have been entirely removed; scabs drying and dropping off, those that remain are of much less extent; has some purging, yet his countenance is good. Since 2nd inst. has used the ointment only every second night. Mixt. c. Decoct. Hæmatoxyli Creta Ppt., Tr. Opii, et Tr. Catechu urgente diarrhœa.

February 9th. Purging continues; says that he was up twenty times last night. Countenance good; scabs are smaller. Haust. Anod. c. Tr. Opii gut. xxx. Vini Rub. ʒiv. indies. Mist. Cretæ c. Tr. Opii urg. diarrhœa.

February 16th. Throat quite healed; exposed ulcers healing rapidly; scabs have almost all fallen off; bowels regular; ptyalism continues.

February 20th. He is evidently stronger and now out of

danger. Although his throat is healed he complains of soreness lower down, and he points to a part on a level with the top of the larynx as the seat of soreness. Says that solid food is apt to stop there, and it requires to be assisted by taking a sup of fluid. When drinking, some of the fluid passes up through his nose : ulcers are much more contracted and with surfaces perfectly healthy; general appearance vastly improved.

From this time he was ordered Pulv. Ipecac. Comp. at bedtime, a repetition of his porter, and Inf. Cinchonæ c. Tr. Cinch. Comp. et Tr. Opii, until April 5th.

On March 1st, swallowing improved, but some of the drink comes through his nostrils; all the ulcers are nearly healed.

March 28th. Swallows quite well; none of the fluid regurgitates through the nostrils when he drinks. For ten days past has complained of pains of his upper limbs, and of his knees. Profuse night-sweats; sleeps badly; appetite declining.

It is unnecessary to prosecute the details of this case farther. I shall only add, that a subsequent use of mercury, carried to salivation, did not cause the ulcer of the throat to break out again. I shall have occasion hereafter to explain by what principles I was led to use mercury in this case.

It not unfrequently happens that these relapsed ulcers of the throat spread along from part to part of both the fauces and the palate, and at times occupy a very considerable portion of both. One rare and unpleasant consequence of this extensive ulceration of these parts is, that in their closing together and healing they form so complete a partition between the mouth and nose that not even the air can pass through, and thus the patient is for ever after prevented from blowing his nose. Those who have not reflected on the strange approaches that ulcerated parts make towards each other in the act of healing, may doubt the possibility of this occurrence; I have seen at least four or five instances of it, and one of them happened to a patient while under my own care.

William Clarke, æt. 18, a delicate-looking boy, admitted into 11th ward, March, 1836. Twelve months ago he was

L

treated for ulcerated tonsils, an ulcer on inner surface of prepuce, and a small node on forehead. Mercury was used pretty freely, but it seemed not to act favourably; in three months after this supposed cure the throat again became sore. His present symptoms are an ulcer on the back of the pharynx, the upper end of which is pretty much on a level with the dorsum of the tongue; it extends downwards for an inch; it is full half an inch in breadth, being nearly as broad as the back of the pharynx; all its edges are steep; surface covered with a thick, tenacious, purulent discharge; this ulcer is void of surrounding swelling or inflammation. He complains rather of difficulty than of pain in swallowing, for he feels that the morsel is stopped at a certain point, and requires a sup of the fluid to make it go down; node on tibia is remarkably small, and gives very little pain.

April 12th. Appearance of the ulcer is very much improved, and he says that his power of swallowing is so much better that now the food scarcely stops at all; general health very good.

April 19th. Mouth is now quite sore enough. Says he feels his throat quite well, and has no pain on the site of the node, the swelling of which still remains. He now for the first time complains of a stuffing of his nose, and says that he had felt it for ten or twelve days past.

April 28th. Mouth nearly well. He now complains that he cannot blow his nose; on examining with a curved probe I find that the communication between the mouth and nose is quite closed up. This has taken place within the last fortnight, since the time the ulcer low down in the pharynx began to cicatrize.

This result is the more remarkable, because since his re-admission into the hospital I could not observe any ulceration at the upper part of the pharynx, and because the highest end of the ulcer in the back of the pharynx did not reach within an inch of the opening leading to the posterior nares. He was discharged April 30th, no attempt to open the communication having been made.

That means may be discovered for preventing this perfect closure of the parts, however impatient they may be of the presence of any foreign body, I can readily believe; but when once the mischief is completed, I fear that we must despair of being able to render any service, and for this reason, that the natural texture and properties of the mucous membrane towards the mouth are quite destroyed and altered, so that instead of the soft velvet-like surface, we see a shining white surface; and thus whatever assistance we might gain from having to deal with the mucous lining to the part is here altogether absent. Monsieur Cullerier, with whom I lately had a conversation on this subject, informed me that he made some attempts by operation to restore the communication, but without success; the parts closed again in spite of all his efforts.

I shall not dwell at length on what every surgeon knows, viz., that venereal ulcers of the throat have in some instances induced ulceration of the carotid arteries, terminating in sudden death by hæmorrhage. But I cannot refrain from stating the following case, as illustrating in a rare manner the very mischievous effects resulting from repeated and injudicious courses of mercury employed in the treatment of venereal ulcers of the throat.

Mr. A., an apothecary, æt. 34, was affected with a venereal sore throat, for which he applied to a surgeon of his acquaintance, under whose care he underwent not fewer than three salivations, and all of them very severe; some ulceration of the throat returning in a short time after each salivation. He then consulted another surgeon, who either directed or permitted him, at his own urgent entreaty, to subject himself to a fourth salivation. The soreness of his throat still remaining in some degree, and his surgeon having positively refused to renew the use of mercury, he became dissatisfied with him, and went through a fifth salivation without the advice of any surgeon. The throat still remaining sore, he sent for me, evidently with the hope that I would order him to resume the mercury in accordance with his own wishes. On inquiring into his symptoms, I learned from him that for some weeks past he was not only constantly confined to bed, b⋅ almost to one posture, for he could not turn himself in the bed by any exertion unless he applied both his hands to steady and support his head; that he was totally unable to raise himself in bed for the purpose of taking a drink, in consequence of the impossibility of supporting his head; and lastly, that he was suffering under severe pains of the arms. On examining the throat, which could only be done as he lay on his side, I observed a very extensive foul ulcer at the back of the pharynx.

This gentleman seemed very much displeased and not a little disappointed when I told him he should not use any more mercury. It is unnecessary to add that I directed for him a tonic plan of diet, but without the benefit of country air, for his removal was absolutely impossible. In about three weeks from the time of my first visit to him, on entering his room one morning he said, "Sir, I have here something to show you: it is a piece of bone that came from my throat, and caused such a fit of coughing as nearly choked me." On examination, this proved to be a portion of the ring of the first vertebra, with on one side the one half, and on the other one third of the articulating processes. My alarm for his safety may now be readily conceived. I apprehended of course that in some unguarded movement of the head, either by his own voluntary efforts, or perhaps by the awkwardness of an officious nurse, the remaining ties of the head, with the first and second vertebra, might be snapped across, and with these the thread of his life.

But contrary to my expectations, he felt himself daily growing stronger: he became conscious of some acquired power of moving the head; this gradually improved, until at length he was enabled to hold his head nearly erect, but with so very little power of rotatory motion that he was obliged to turn the trunk of the body when he wished to view any object on either side. However, his flesh and strength returned, he became corpulent, he married, and survived the loss of half his atlas by five or six years.

Before we dismiss the subject of what may be termed the pure or original forms of venereal sore throat, I shall only observe, that when any one of these affections is making an alarming progress, and by sloughing or by phagedæna threatens to destroy some important portions of the throat, or suddenly to endanger life by laying open some great blood-vessel, or more slowly, but not less certainly, by the ulceration extending to the top of the larynx, I say that in such emergencies the surgeon has but two resources left, but two means of arresting the dangerous career of the ulcer, viz., the application of muriate of antimony, or some other equally powerful liquid caustic, or the use of mercurial fumigation. His judgment must decide which is best suited to the circumstances of each case. I shall only remark, that if we employ fumigations we cannot expect them in general to produce their sensible effects in a shorter period than two or three days; then they are apt to excite very profuse and uncontrollable salivation, which although it is almost certain to arrest the further ravages of the ulceration, yet does not accomplish a permanent cure of the disease; but on the contrary it leaves the system in a state badly able to bear the subsequent use of mercury.

All those cases which have been profusely salivated by fumigations, and in which a relapse has occurred, requiring a fresh use of mercury, will be found very difficult of cure.

One caution relative to the application of caustics I must not omit. In cases where the ulceration threatens to destroy the bony partition between the mouth and nose, we should confine the application of the caustic to the border of the ulcer; if we apply it to the centre we shall be in danger of having it reach to the bone, and thus produce the very evil we are so anxious to avoid.

There are a few affections of the throat and mouth which sometimes follow shortly after a course of mercury employed for the cure of the venereal disease. It is a very remarkable fact that these affections appear only under such circumstances. For if mercury have been employed for the removal of any other disease, ex. gr. inflammation of the synovial membrane of the joints, disease of the liver, &c., we do not witness any such troublesome consequences. Of these affections the following is rather a rare one :—

The patient complains of a tenderness of throat rather than o sore throat, says that if he eat some particular articles of food, ex. gr. currants or gooseberries, his throat feels more uneasy, not at the time of eating, but on the following day. Either one or both tonsils present a very irregular surface, of a peculiar yellow colour, sometimes without any ulceration, and sometimes with a few small very superficial ulcerations; at times a similar ulceration may be seen at the junction of the anterior palatine arch with the tongue. This peculiar colour of the tonsils is sometimes accompanied with one or more white spots on the edge of the tongue, across the diameter of which a slight fissure may run. This affection of the throat may, after a few weeks, disappear and remain away for some days, then return nearly as before. With these local affections the system does not appear to sympathise, for the general health continues undisturbed. The patient, if particularly questioned, admits that there is no soreness of either the tongue or throat. This affection will continue for many months gradually retiring, by leaving the patient free, at first for a few days, and then for a week, and at length for a month or two; finally it disappears altogether. I have not seen any other suspicious symptom attend this affection.

For the cure of this affection I have known most severe courses of mercury employed, so as to excite very profuse salivation, and these even repeated more than once, but without effecting any other than a momentary improvement in the state of the symptoms, and this improvement has ceased as soon as the ptyalism began to subside. Most assuredly this state of the throat does not require, nor will it be benefited by, the use of mercury. The treatment which I have found most efficacious is to rub the parts very freely with Sulphas Cupri, or with Argentum Nitratum daily, or every second day, generally alternating these applications. The more active caustics do not succeed better, indeed not so well as those I have mentioned. I have known different preparations of iodine to have been employed; but I regret to say without affording any decided benefit. This affection is a striking example in justification of the phrase of a disease " wearing itself out;" for in my judgment time alone has effected the cure.

Mr. R—, in the month of January, underwent a full course of mercury under close confinement, for the cure of a venereal sore throat. Early in March he was in a profuse salivation, which his constitution bore with comparatively slight disturbance. He desisted altogether from the use of mercury not until 28th of April. On 3rd June he complained of his throat and tongue. I find on the right side along the edge of the tongue, opposite to the molares teeth, two small fissures surrounded with white circular spots. At the point of junction of the anterior palatine arch with the root of the tongue is a small superficial ulcer of a brown or buff colour—right tonsil is enlarged. He says that at the end of six weeks, after each course of mercury, his throat has become sore ; he brushes over the affected parts with a strong solution of Argentum Nitratum every or every other day.

10th August, 1834. Now three months since he relinquished the use of mercury; complains that his throat, during the last fortnight, has become more tender, so that after eating fruit, especially gooseberries and currants, he finds his throat more sore on the next day. Surface of the tonsils is of a buff colour, with an uneven surface, which gives it the appearance of ulceration. I have observed that, for some days together, the tonsil changes to the natural red colour. He rubs the affected parts with solid Argentum Nitratum occasionally, when the sensations become troublesome.

30th August. During the last fortnight he has not used the Argentum Nitratum, either in a solid form or in solution, as heretofore; he has merely gargled with sea-water. On the right edge of the tongue is one small white spot with a fissure running across it ; left anterior palatine arch has a very superficial white ulceration or excoriation along its edge ; right tonsil, with a very irregular surface, has a yellowish superficial ulcer on its upper end, and another superficial sore is seen where the anterior arch comes in contact with the tonsil above ; says that he has not any soreness of the throat or tongue, except occasionally; general health very good.

12th October. Superficial yellow ulceration of left tonsil still continues, although he has lately used the Argentum Nitratum as often as the uncomfortable sensations of his throat prompted him to do so ; his general health is excellent;

he has gained weight during the last five months. This
morbid state of throat has now continued for five months
from the time he laid aside the use of mercury.

24th November. Ulceration of both tonsils superficial, and
of buff colour; he has been for days and weeks together
without feeling any pain or uneasiness, and yet with the
ulceration visible; at other times the ulceration becomes
painful, and continues so for some days; he occasionally
touches the ulcers with Argentum Nitratum, perhaps once in
eight or twelve days. It is now seven months since he gave
up the use of mercury.

15th January. His throat looks much better to-day than I
have ever seen it; says that it remains well for seven or eight
days, then begins to grow sore, and when this soreness has
increased a good deal, he rubs it very hard with a pencil of
Argentum Nitratum; the sloughs do not separate for two or
three days; and so this course is run over and over again.

29th April. In January last, Mr. R— consulted two of the
most eminent surgeons in London, who assured him that his
disease was not venereal, and could not be cured by mercury;
he used iodine, but without the smallest benefit; still his
complaint has not been removed; for now about the centre of
the right tonsil there is a circular ulcer about the size of a
split pea, edges very slightly raised, surface rather flocculent
and of a yellowish colour, no surrounding swelling or redness;
this is not attended with pain, even in the act of deglutition:
he merely feels at times a dryness of his throat; general
health is very good; he has not touched the throat with
caustic during the last fortnight; observe he had relinquished
the use of mercury on the 28th of last April, exactly twelve
months ago. N.B.—In three months more his throat got
perfectly well.

Another and more common affection of the throat following
a mercurial course, employed for the cure of some venereal
affection, is the following:—Over the anterior surface of the
velum palati, or on one of the tonsils, or perhaps on some
part of the inside of the cheeks, there appears an irregular
wrinkled surface of a very white colour, looking not very
unlike the mark or track of a snail, or as if the part had been

brushed over with a brush dipped in milk. Sometimes the whitened surface retains its natural smoothness, and sometimes it is in parts wrinkled, resembling, in some degree, the skin of the hands when long steeped in hot water. The discoloured parts are sometimes free from ulceration, while in others we may observe a very superficial ulceration, or rather excoriation, and sometimes a fissure.

However, in some cases we find a distinctly ulcerated spot in the discoloured part, while in others again we find an ulceration of some part of the mouth remote from the white spot; and this appears with a raised or swollen white border, the enclosed ulcer being of a red colour, and more resembling an excoriation than a real ulcer. This affection of the mucous membrane is found in a few instances so very widely spread, that the entire surface of the mouth, throat, and gums partake of this morbid state.

A very remarkable, and yet a very constant, attendant of this state of mouth and throat is a very superficial ulceration or excoriation around the anus, so that almost as often as we question the patient as to the existence of such an affection, we may be pretty sure to be answered in the affirmative; and scarcely does any patient pass through this disease without suffering at one time or other from this excoriation around the anus. This affection is rather troublesome than painful, and it seems to disturb the general health very little, if at all; for we find the patient in the enjoyment of good health and good looks, and without the slightest disturbance of any one of the functions.

If unattended to, this disease may continue for a great length of time, certainly for many months, subject, however, to occasional variations; it may remain unchanged in one part, while getting well in another, or even fixing itself upon some new spot. What the ultimate result would be, if left entirely to nature, I cannot say; but I imagine that it would finally be overcome by the powers of the system, or, as some express it, "would ultimately wear itself out."

However great may be the extent and duration of this affection of the parts within the mouth, if free from actual ulceration, nay, even though it be accompanied with a corresponding affection of the skin at the anus, still we should not be tempted to treat it by mercury administered so as to induce salivation, for such practice will end in disappointment. I have seen the most full and long continued, nay, even repeated salivations fail in effecting a cure, and this whether the mercury had been employed by inunction or had been given internally : neither does the muriate or any of the other preparations of mercury possess any peculiar power over this disease. No doubt, during the time that the salivary system is under the full action of mercury, the mucous membrane assumes a much more natural appearance, the whitish colour of the ulceration is removed, an unnatural degree of redness alone points out the seat of the disease ; but as soon as the high action of the mercury begins to subside, and even before the mercurial soreness of the mouth has ceased, we observe with chagrin and vexation a return of the original affection in its pristine vigour.

I have known this affection treated by protracted courses of sarsaparilla alone, and combined with mineral acids, without the patient deriving the smallest benefit from their use. The application of solutions of Argentum Nitratum, or of Sulphas Cupri, has appeared to me the most useful line of treatment in this, as it is in that yellowish ulceration of the tonsils already described as a sequela to a mercurial course, when employed for the cure of the venereal disease.

I have described this affection as being sometimes free from any ulceration, while in other cases ulceration may be seen either in the discoloured patches or in some other remote part totally free from this discolouration. Now, while we are assured that this affection, if unaccompanied by ulceration, is most certainly not a venereal symptom, and neither requiring nor benefiting by the use of mercury, yet we must admit that, when it is attended by ulceration in the discoloured parts, we cannot give a very positive opinion on this important point. After long and anxious endeavours to discover some feature in the ulcer or accompanying whiteness, which might enable me to decide this question, I freely admit that I have not been able to succeed, and that I am obliged to be guided more by the absence or co-existence of other symptoms which are unequivocally venereal. I need not say how very particular and minute we should be in our inquiries for other symptoms, for, should we form a wrong opinion, we may subject our patient to a course of mercury which is absolutely unnecessary, and which will prove unavailing; or, on the other hand, were we to declare that the use of mercury is unnecessary, we might thus give the patient unfounded confidence in his exemption from the venereal disease, throw him off his guard, and thus allow syphilitic symptoms to produce serious evils before he could think it necessary to apply for surgical assistance. Thus syphilitic iritis might, if unattended to, go on to the destruction of vision by inducing opacity of the crystalline lens, and closure of the pupil.

In the first of the following cases we have an instance of this affection existing along with undoubted venereal symptoms, and in the next we find this affection appearing without any such concomitant.

Mr. M— applied to me on 26th September, 1833. He informed me that six weeks ago he perceived a chancre on the inside of the prepuce, close to the frœnum; for this he used twenty pills, which produced mercurial fœtor of the breath, but no soreness of the mouth. Caustic was applied to the ulcer. For the last three weeks he had not used any medicine. At present the site of the chancre is marked by a red spot, the size of a split pea: this is rather an excoriation than an ulcer; it is of a very red colour, moist, and very hard. On the middle of the flat surface of the tongue is a small superficial ulcer, with a fissure running across it. On the inner surface of the lower lip is a very superficial ulceration, of a circular form, the size of a shilling: a great part of this diseased surface is of a milky-white colour. There is a superficial ulceration on each tonsil, of the same white description as that of the lip. Sumat Pil. Hyd. gr. v. bis in die.

Oct. 6th. Mouth slightly affected; ulcer of the penis has less hardness, and is of a more healthy red; tongue is not much altered; throat appears to be well.

Oct. 18th. Mouth still slightly affected; ulcer of prepuce · uch better, though still moist. The ulcer of the tongue appears rather more extended; this I now rubbed with

Argentum Nitratum. Ulcer on the inside of the lip is also more extended, and the white border which surrounds it shows a greater degree of softness, being more raised. Appetite good; he sleeps well. P. 112. Rept. pil. una bis in die.

Oct. 24th. He is now very fully affected by the mercury. Mouth very sore; fissure in tongue is healed; surrounding ulceration is rather more extended, and is now of a dirty-yellow colour; ulcer of lip is also of a dirty-yellow colour. He complains of weakness, but has no palpitation. Omittatur Pil. Hyd. ℞. Sulph. Quinæ gr. x. Aq. vj. Acidi Sulph. Dil: ℈i. Tinct. Card. Comp. ʒ ss. M. ʒ i. ter in die sumenda.

Oct. 29th. Ulcer of prepuce not quite healed, although very much reduced in size, and some hardness remains; tongue unaltered: ulcer of lip has its border raised and hard, the centre skinned over and of a red colour. Fissure of tongue rubbed with Argentum Nitratum; ulcer of lip with saturated solution of Sulphas Cupri. Mouth is less sore.

Nov. 7th. He resumed mercury on the 3rd inst. Pil. Hyd. gr. v. bis in die. Ungt. Hyd. Fort. ʒ ss. om. nocte. Now he has an iron taste on his mouth; mercurial fœtor of breath is pretty strong.

Nov. 13th. Since 8th inst. the ptyalism has been very copious, with considerable swelling of left cheek, and ulceration of left side of tongue. P. 126. Ulcer on surface of tongue is healed, that on the lip is more clean, with cuticle over part of it. Ulcer on prepuce clean, bleeding on the slightest friction, but the hardness continues.

Nov. 16th. He spits four pints daily. Ptyalism prevents sleep; thirst urgent. P. 90 only. Ulcer of lip as on 13th inst. Ulcer of prepuce is healed, very slight hardness remains.

Nov. 24th. Ptyalism less; saliva viscid; tongue and left cheek still ulcerated; ulcer of lip as on 13th; ulcer of prepuce is healed, and perfectly free from hardness or thickening.

Dec. 4th. For last eight days has used Ungt. Hyd. ʒ ss. omn. nocte, et Pil. Hyd. gr. v. mane et nocte, by which his mouth is again smartly affected. Gums are swollen and ulcerated, but there is not much ptyalism; thirst urgent;

skin not very hot. P. 98. Ulcer of lower lip not healed; a smaller but similar ulcerated spot now appears on the middle of the lip, near the frœnum. Ungt. Hyd. ʒ ss. o. n. Pil. Hyd. gr. v. bis in die.

Dec. 7th. Original ulcer of lip quite healed and of nearly natural colour; the small ulcer of lip remains unaltered. Omitt. Pil. et Ungt. Hyd. Sumat Mixt. c. Sulph. Quinæ.

Dec. 27th. Mouth free from mercurial influence. P. 96. App. good; place of the larger ulcer of lip can be distinguished by its higher colour; the new small ulcer remains. For three days past he has felt some uneasiness in swallowing; on left tonsil is a broad very superficial ulcer of a dirty-yellow colour.

Jan. 1, 1834. Appetite good: general health good, except that his pulse after walking is too quick. Ulcer of lip is enlarged; left tonsil as on 27th ult. ℞. Hyd. Cor. Sub. gr. ij. Resinæ Guiaici ʒ ss. ft. pil. octo. Sumat. unam. sing. noct. Pulv. Sarsæ. c. Carbon. Sodæ mane et meridie.

Jan. 10th. Both ulcers of the lip are healed, but the surface is raised and slightly discoloured: ulcer of tonsil a little more deep, in other respects unchanged.

Jan. 19th. On right anterior palatine arch a superficial ulceration exactly the same as that on the left tonsil—the ulcer of this unchanged.

Jan. 31st. Ulcer of tonsil has now a white surface, with soft white raised edges; ulcer of right arch unchanged; general health good. He is now in good flesh and colour.

Feb. 9th. No sensible effect from the medicine. P. 82. Complexion florid; general health excellent; ulcer of lip rather more deep, or edges more raised and white. Appearances of throat unaltered, but he feels less soreness in it. ℞. Pil. Hyd. gr. vj. Ext. Cicutæ gr. ij. ft. pil. duæ h. s. sumend.

Feb. 23rd. Copperish taste in mouth; gums slightly sore; ulcer of lip more extended, although in parts disposed to heal; ulcers of throat unaltered; all the ulcers touched with Acid. Acet. Fort. P. 120 after walking.

March 9th. Mouth sore from the mercury; ulcers of the throat and lips much improved; appetite bad; headache; bowels confined; sleeps well.

March 16th.   Mouth nearly well; one small ulcerated spot on each anterior arch; ulcer of lip has been subdivided into three small spots by the healing of interposed portions. Sumat pil. duas mane nocteque.

March 23rd.   Weaker, and more easily thrown into perspiration; feels his throat more sore; ulceration of it and of lip as bad as ever; gums sore.   Rept. Pilulæ.

March 30th.   Mouth still sore; throat and lip same as on 23rd.

April 20th.   Ulcerations of lip and throat as bad as ever; they have lately been repeatedly touched with Solut. Arg. Nitr. Sumat Acid. Nitr. Dil. gutt. xvi. ter in die.

May 25th.   Symptoms unaltered.   Sumat. Acid. Nitr. Dil. gutt. xxx. ter in die.

June 1st.   Throat a little better; lip much better; appetite and strength good.   Fumigentur ulcera quotidie c. Rubri. Hyd. Cinerei ℈ ss.

June 26th.   Since 22nd has been fumigating daily with Hyd. Ciner. ℈ j. made up in a candle.   Gums a good deal swollen, with their apices between the teeth in a sloughy state: mercurial fœtor of breath; feels himself weak; ulcer of lip healed, but has left a discolouration which marks its former site and extent: ulceration of throat healed, but natural colour is not restored; spot on dorsum of tongue still remains naked and destitute of papillæ.

July 13th.   Mouth smartly sore.   P. 78.   Thirst urgent; appetite indifferent; sleeps well; bowels costive.   He is easily thrown into perspiration by exercise; lip and throat as on 26th June.   Sumat. Sulph. Quinæ gr. ij. ter in die.   Omit. Hyd. Ciner.

Aug. 17th.   Tonsil with a superficial venereal ulceration. Within the last few days a blister appeared on the under lip; this spot now presents the character of the former disease in colour, but is not ulcerated.   He has now also a superficial ulceration at the verge of the anus obviously venereal.   He does not suffer pain from this when going to stool, but whenever he is warm either in bed or by exercise.   P. 78. Lotio Nigra ano applicanda.

Aug. 24th.   Symptoms as on 17th.   Has used the black

wash in a very careless manner. Spot on lip rubbed with Sulph. Cupri, and he is directed to wash it three or four times a day with Solut. Sulph. Cupri.

Sept. 21st. Spot on lip scarcely discoloured; ulceration of tonsil less deep; the integuments around the anus are healed wherever the lotion has been carefully applied, but a superficial ulceration still exists in the folds of the skin close to the orifice of the anus.

Oct. 19th. Left tonsil is still slightly ulcerated, with a yellow surface; on the inside of right cheek there is now a white spot. P. 90. General health good. Repet. Solut. Sulph. Cupri.

Nov. 16th. Superficial white ulceration of both the anterior arches; also about the right dens sapientiæ adjoining part of the left cheek. Repet. Solutio Sulph. Cupri.

Nov. 23rd. Symptoms as on 16th inst. Sumat. omni nocte Sulphureti Hydrargyri Rubri gr. v.

Dec. 14th. Throat much better; only one or two small white spots on the right anterior arch and around the right dens sapientiæ. Sulphureti Hydrargyri Rubri gr. x. omne nocte.

Jan. 4th, 1835. No sensible effect from the pills. Throat nearly as before in appearance and in feeling. Hyd. Calcinati gr. i. Opii. gr. ss. omn. nocte.

Jan. 11th. Throat better in appearance; ulceration of gums not better; no sensible effect from the pills. Repet. Pilulæ.

Jan. 20th. Mouth sufficiently affected; throat not better in appearance; ulceration of gums as before. Omitt. Pilulæ.

From this time all mercurial treatment was discontinued, and he went to the country. He returned to me on the 22nd of April, 1835, and informed me that an appearance of the ulceration of his throat has since repeatedly returned, but that he always removed it pretty soon by the application of a lotion of Sulphas Cupri. His general health is now excellent.

In the Report of 27th December, 1833, it is stated that the r'cer on the left tonsil is of a dirty-yellow colour. This might ad us to believe that the white condition of the mucous membrane produced by mercury, employed for the cure of

venereal symptoms, is nearly allied to the yellow ulceration
which arises under similar circumstances. After the use of
mercury continued with but little interruption for twelve
months, we find on August 17th, 1834, "there is now a
superficial ulceration at the verge of the anus, obviously of a
syphilitic character." This I advert to, as it establishes the
very intimate connection between certain syphilitic affections
of the mouth and throat, and the skin around the anus.
After a perusal of the details of this case, I think we are
justified in concluding that this condition of the mucous
membrane of the mouth and throat (though a sequela of
syphilis when treated by mercury) is itself not to be cured by
mercury: and yet mercury will, at the same time, cure any
other truly venereal symptom which may exist along with it.

Alexander Hamilton, æt. ann. 40, of a healthy appearance,
was admitted into 12 ward on March 10th, 1835. He complained only of the state of his mouth and throat. The inner
surface of each lip was occupied by large patches of a white
colour, which on a superficial view appeared to be ulcerated,
but on closer inspection these patches were found te be raised,
to be of a milk-white colour, and traversed by fissures in
various directions. The right side of the tongue, the inside
of the corresponding cheek, and the anterior surface of
the soft palate and tonsils, were similarly affected. He
had a small and slightly indurated cicatrix on the glans
penis ; but no other venereal symptoms. His general health
very good.

Eight months previous to his admission he contracted a
gonorrhœa, and soon after the sore appeared on his penis.
For cure of this ulcer he immediately took some pills, which
produced salivation, but the sore did not get well until about
five months ago, when it was healed while he was a patient in
Mercer's Hospital. In one month after he had been discharged
from the hospital, the ulcer broke out again ; he then applied
to an apothecary, who ordered him pills, four of which
produced profuse salivation, and caused the sore to heal ; but
since that time his mouth has assumed the appearance above
described. He was ordered a tepid bath and purging mixture,
and then to take every night two pills consisting of Pil. Hyd.

gr. v. c. pulv. Jacobi, gr. iii. and these pills he continued to take until 14th April. A very slight mercurial action was ultimately produced by these pills, but without any permanent improvement in the state of the mouth or throat. No doubt, at times, one part would seem to improve for a few days, but, at the same time, some other part appeared to be as much disimproved.

He was next ordered burnt alum, washes of chloride of soda, and a variety of topical applications ; but none of them, except such as were possessed of some escharotic quality, appeared to make any impression. The Sulphas Cupri, rubbed on strongly, would often cause an evident improvement, but only for two or three days, after which these parts fell back into their former state.

At one time all the diseased parts of the mouth simultaneously put on an improved appearance, while all the throat seemed to grow worse, for the notes on April 11th say there is a very evident amendment in the symptoms ; the tongue has now its edges thin and natural in appearance, having now only two or three spots indented by the teeth. He himself says that he feels the tongue quite well; the lips are less swollen, and the white swollen mucous membrane is less raised; but he feels his throat more uneasy than usual; it has a good deal of the white coating, or superficial ulceration on the anterior arches. Mouth is scarcely affected mercurially.

May 20th. Although his general health has continued in excellent order, yet his symptoms are nearly unchanged. Altogether, we have not made any decided advances towards curing this man. I now was determined to try if it would yield to minute doses of mercury, and accordingly ordered R. Hyd. Ciner. gr. i. fiat pilula ter in die sumend.

This plan having failed after a full and fair trial, I wished to try the effects of change of air, and therefore I dismissed him from the hospital on 29th May, and desired him to attend as an out-patient, in order that different kinds of escharotics might be carefully employed. This plan was regularly and steadily employed; the result was a very gradual and slow improvement, the steps of which do not require or admit of verbal description. The Argent. Nitr. in solid, and saturated

M

solutions of Sulph. Cupri, were the applications generally used.

July 19th, 1835.   The throat was yesterday well rubbed with Sulph. Cupri, and this day it appears a good deal better.

July 26th.   Throat is now all but well ; at present there is only one small circular spot, not deeply white, not at all ulcerated, lying in the centre of a red patch which covers the front of the right side of the velum pendulum palati.

Aug. 4th.   The ulcer of throat seems disposed to heal, but he now complains of tenderness of tarsal bones of the left foot, and also of pains of his shoulders and limbs.   Sumat. Olei. Tereb. Gutt. xxx. ter in die.

Aug. 8th.   Throat is perfectly well ; pains of shoulders and of limbs are also better, except that of the tarsus.

Aug. 17th.   Pain of foot has been much relieved by a blister, but still some slight tenderness remains.   Pain of shoulders entirely removed.   Says he feels some uneasiness in his throat, but I cannot on the most close examination discover any cause for it.

Sept. 15th.   An ulcer in form of a fissure has appeared on the side of the tongue ; no other appearance of disease in the mouth.   Pain of instep had returned, but again removed by another blister.

By further perseverance in the use of mild caustics, this disease was finally removed.

This affection of the mouth and throat some may imagine is merely an effect of inflammatory condition of these parts left on the subsidence of that action.   Were this the case, we should find it make its appearance immediately on the subsidence of the high mercurial action.   But we will often meet with cases in which this symptom has not shown itself until long after the ptyalism had ceased.   Of this the following is an instance :—

Captain B— applied to me 7th July, 1835.   Says he had a chancre in August, 1834, which was treated by Pil. Rhei. internally, and by black wash, the only external application. Under this treatment it was healed in the latter end of September.   The chancre in a week or two opened again ; he was then treated by Mr. Lawrence, of Brighton, who made him use mercurial friction twice a day. By this process his mouth was not made sore, nor was he much reduced in strength and flesh. He continued free from any symptoms of disease until the latter end of May, when he felt a soreness of his throat. He did not feel any uneasiness at the anus until three weeks ago. The arches of the palate are pretty generally affected with the white rugose state of the mucous membrane, in which a few slight fissures appear. At the anus is a small condyloma, with a very superficial ulceration or excoriation. He is much surprised to learn that these two symptoms are in any way connected, and he remarks that this disease ought to have first made its appearance at the anus, because, as a cavalry officer, he has so much riding exercise; but perhaps it seized first upon the throat because he was so much addicted to smoking.

CHAPTER VIII.

OF VENEREAL DISEASES OF THE MOUTH, ETC.

Under this head we propose to consider all the venereal ulcers of the mouth and tongue, as well as of other parts liable to be secondarily affected.

The tongue is subject to two or three different forms of venereal affections, some without and some with ulceration. I have already mentioned the ulceration on the root of the tongue at the point of connexion with the anterior palatine arch, and also that which is found about the central point of the base, on a range with the termination of this arch. In addition to these we sometimes find the point, and sometimes the edge of the tongue, the seat of an ulcer, which is strictly a secondary venereal ulcer, and it is remarkable we seldom have more than one other secondary symptom accompanying it—perhaps only five or six spots of papular eruption, or a solitary venereal ulcer on one of the limbs or on some part of the body. This, like other forms of ulcers of the tongue, is sometimes attended with profuse ptyalism, and sometimes with scarcely any extraordinary flow of saliva.

The characters of this ulceration are not constant; in some cases we see the point of the tongue (when this is the seat of the disease) broad and truncated, the surface covered with a thin and rather a soft slough, the extremity appearing to the eye much swollen, and presenting to the touch a very considerable degree of hardness. In other instances of this species of ulcer, in which a similar degree of swelling and of hardness are present, yet the surface may exhibit merely an ulcerated and a somewhat foul appearance.

When the ulcer is seated on the side of this organ, we find it present appearances similar to those of the ulcer on the point, but seldom attended with an equal degree of swelling, although the hardness is not less. Occasionally, an enlarged lymphatic gland under the jaw attends the venereal ulcer of the tongue.

We find it a matter of much difficulty to distinguish the venereal from the cancerous ulcer of the tongue : the surfaces of both are subject to vary materially at different times, and also either in the entire or in parts. Both are attended with considerable hardness ; but I think that the hardness in the cancerous form gives much more strongly the idea of a stony hardness. There is one symptom which, when present, strongly indicates the cancerous nature of the ulcer ; and that is a slightly elevated narrow ring, of considerable hardness : if this include an ulcer, with a surface so clean as at first view to resemble an ulcer which is about to assume a granulating state, then we may unequivocally declare it to be cancerous. However, it must be admitted that the surgeon is unable to decide from the appearances, especially if he overlook some other concealed venereal symptom. In all such cases the surgeon should give the patient the benefit, or the chance of a slight ptyalism : if this be induced quickly, it cannot be prejudicial to the cancerous ulcer, and it will so speedily induce such a favourable change in that which is venereal that every doubt must be removed by such a test. I have no doubt that every surgeon must have occasionally met with cases on the nature of which he could not pronounce until he witnessed the improvement induced by the mercurial action. When a fissure is suspected to be caused by the sharp point of a tooth, we must have this cause removed ; and then, if the ulcer heal, we are certain of its real nature.

In a few cases we may find a venereal ulcer on some portion of the dorsum of the tongue, anterior to its base ; the place of this ulcer having probably been determined by some accidental irritation of that part. Ulcers in this situation are generally of a circular form, large as a fourpenny piece, and present pretty strongly the characters of a secondary venereal ulcer of the skin.

In some cases we see a long narrow stripe of ulcer on the under surface of the tongue ; this is of a white colour, and, were it sunk below the level of the surrounding part, might be supposed to be formed by an exudation of lymph : this lies so far away from the edge of the tongue that we see at once it cannot be mistaken for an ulceration caused by the pressure of the teeth.

The chronic, or, as it is generally denominated, scrophulous ulceration of the throat, is found in some few cases to spread to the tongue; and, when once established there, is very apt to spread forward, by a very slow but rather destructive process of ulceration. In some of these cases the ulceration of the tongue is not a continuation of that of the throat, but exists independently of it. In general, however, that of the throat has nearly ceased at the time that this of the tongue is established.

These chronic ulcers of the tongue are to be distinguished from those that are venereal; which may be done by observing that the former are much less foul, but, with still more certainty by looking at the throat, and finding there those silvery cicatrices which invariably follow this peculiar form of chronic ulceration.

Before we dismiss the subject of venereal ulcers of the tongue, I would observe that we occasionally meet with instances of superficial ulcerations of this organ, coming on after a course of mercury for the cure of either primary or secondary symptoms, and which we should take care not to mistake for venereal ulcers. I know not by what characters of the ulcers themselves we could be enabled to decide this point; but I think we shall avoid falling into any mistake if we refer to the history of the case; for those ulcers which are not venereal will be found to make their appearance in twelve or eighteen days from the time the mercury has been laid aside—indeed, before we can be assured that the action of this medicine on the mouth has entirely ceased.

Let it not, however, be forgotten that, in some instances where salivation had been excited for the cure of either primary or secondary symptoms, the ulceration of the tongue, caused by the mercury, has not healed; and, in the space of a few weeks after the ptyalism has subsided, such ulceration has assumed all the characters of second venereal ulcers. I need not add that such require another and more judicious course of mercury.

I have already alluded to that appearance on the dorsum of the tongue, in which patches of different sizes (but usually of a circular form) appear, as if deprived of papillæ; the surface perfectly smooth; these spots may be termed bald. Such are sometimes found along with other symptoms, purely and strongly syphilitic; and sometimes, as I have already mentioned, they attend that white condition of the mouth and throat which occasionally succeeds to a course of mercury employed for the cure of syphilis; but never, I believe, when employed for the cure of any other disease. So that, in forming our judgment as to the necessity of using mercury in such cases, we must be decided by the nature of the accompanying symptoms, rather than by any particular condition of these spots.

I have, in a few cases, known an ulcer on the side of the tongue arise merely from deranged digestion and a vitiated state of the stomach ; and yet so strongly resemble a venereal ulcer of this organ, that it might, by a hasty observer, be pronounced syphilitic. Such appearances I have seen, in a few individuals, repeatedly produced in a few hours by the use of vegetables, or of acids. In all such instances the absence of other syphilitic symptoms will prevent us from resorting to mercury; and, in a short time, the use of medicines calculated to improve digestion and strengthen the stomach will cause the ulcers to heal, and put an end to all doubt about the true nature of these ulcers.

Venereal fissures are occasionally met with on the edges of the tongue, and may run across the edge either in a straight line or in an angular direction: the edges of these fissures are remarkably hard, and we almost constantly find the fissure surrounded by a superficial white blister, somewhat like an ulcer; these ulcerated spots seldom exceed in extent the surface of a split pea. As the edges of the tongue are subject to fissures from other causes—*e.r. gr.* from the irritation of the sharp point of a broken tooth—we shall sometimes be at a loss to decide upon them. If, however, we recollect that the venereal fissure is generally surrounded by the small superficial white ulcer, we shall acquire a confidence whenever we see the accompanying white ulcer.

Not unfrequently we find small circular excoriated spots on the hard palate ; these, sometimes, are accompanied by other unequivocally venereal symptoms ; and in other cases they

appear in conjunction with the yellow or the white superficial
ulceration of the mouth, consequent on a mercurial course.

The nature of the accompanying symptoms must decide
whether or not we shall employ mercury in the treatment of
each individual case.

Venereal Ulceration of the Gums.

The gums are sometimes, though not often, affected with
venereal ulceration; this, if extensive, and allowed to remain
long uncontrolled, may induce disease, and exfoliation of some
of the alveoli, and the loss of one or more teeth. This
ulceration in some patients assumes all the characters of a
secondary venereal ulcer; while in others it is more of a pale
white surface, with considerable redness, and some swelling of
the adjoining gums. In this latter case we shall find the
accompanying ulcers of the throat, or of other parts, marked
by similar languid characters. It is a curious fact, that the
ulcer of the gums is generally confined to the external, and
that the inner gums remain unaffected until the disease shall
have injured the interposed bone. We may see the alveoli
laid bare, to some extent, by this disease; and yet the teeth
remain firm and apparently unaffected.

When ptyalism has been induced by a course of mercury,
which has proved insufficient for the cure of any venereal
symptoms, we may find after the ptyalism has ceased, and
after the rest of the mouth has become completely well, that
the patient complains of soreness of the outer gums belonging
to the last molares. On examination, we discover ulceration
of these gums; and, at the same time, ulceration of the cheeks
opposite to these teeth; and these ulcers are strongly marked
as secondary venereal ulcers in other parts. When we observe
these appearances, after all the other ulcers of the mouth
(occasioned by the mercury) had healed, and that these ulcers
exhibit venereal characters, we do not hesitate to pronounce
that they are venereal. Of course they are then to be treated
on the same plan as other secondary venereal ulcers are treated.

We must take care not to mistake for a venereal ulceration
of the gums, that ulcerated state of them which we sometimes
find attendant on scrophulous thickening and ulceration of the

upper lip, and much more frequently a concomitant of lupus of the nose or lip. A very little attention will enable us to discriminate; for here we observe that the ulcerated part assumes the appearance of a mass of fungous yet healthy granulations, terminated very suddenly by the natural membrane of the gums; this raised surface is not covered with pus, but looks rather as if the granulated surface did not secrete any fluid; nor is there any line of ulceration between the limits of this raised portion and the adjoining sound membrane. In these cases, no doubt, we shall find the teeth more or less loosened.

Again, we sometimes find a considerable length of the gums affected with a swollen, soft, spongy condition, to the depth perhaps of one-eighth of an inch from their edges. This soft, swollen part, has also a different colour from the rest of the gums; having a more highly red colour, with an orange tint pretty strongly marked. This state of the gums is not to be mistaken for venereal ulceration; it not unfrequently seems to have been produced by previous courses of mercury.

Ulcers of the Nose.

Ulcers of the alæ nasi sometimes commence in the angle between the nose and cheek, from a cluster of papular eruptions degenerating into an ulcer. Whenever this ulcer shows a tendency to phagedæna, or sloughing, it should be treated with some very active caustic. For although this has been at first withheld, lest it should produce a loss of substance, yet I am certain that a much greater destruction of parts will be caused as certainly, although rather more slowly, by the progress of the disease.

I am not in possession of any distinctive characters which would enable us to pronounce an ulcer in the cavity of the nostril to be a venereal ulcer. The symptoms of what is called scrophulous ozæna correspond very much with those of the venereal ozæna. No doubt we often find the scrophulous ozæna has been preceded by scrophulous sore throat, which has ended in the formation of those tense and silvery cicatrices already alluded to, and this alone will sometimes serve to distinguish them.

But let us not mistake for a venereal affection appearances which are often met with in some who have never been exposed to venereal infection; I allude to those instances of ulceration of the nose, in which we discover in the septum nasi, about a quarter of an inch from its anterior extremity, an opening through the septum; this is uniformly, I believe, of a circular form, is as large as the surface of a split pea, and has a slight degree of ulceration on its edges. This state of the nose is not unfrequently established before the patient is aware of its existence, so very trifling is the uneasiness which it causes either in its formation or when fully formed. And what is equally remarkable is, that it will be found for years to remain in precisely the same state; at least I have not met with the slightest alteration in any of those cases which I have watched, and I have had opportunities of observing some few for eight or ten years.

We must then hesitate to pronounce an ozæna to be venereal, unless we find it existing along with some other decidedly venereal symptoms: and that we can find its history inter-woven with that of other secondary symptoms. It requires a good deal of experience and close observation to enable us to distinguish between that which is termed scrophulous and the venereal, although it is not, I believe, possible to describe by words the distinguishing marks.

Venereal Iritis.

On the subject of Venereal Iritis I have very little to offer. I shall only say that this is a symptom which cannot be charged to the account of mercury; for it occurs in cases which were treated by the non-mercurial plan. I think that in the treatment of it we should not use mercury in such a manner as may suddenly plunge the patient into a profuse salivation; for in general the disease is not so acute but that it will admit of some delay, and will allow a sufficient time to produce this effect of mercury according to the ordinary mode of exhibiting it. The great disadvantage of a sudden and profuse salivation is, that although it arrests and cures the iritis, it renders all the other symptoms more slow and much more difficult of cure.

When iritis takes place at a time when the system is beginning to throw off a smart salivation, perhaps while the mouth is still smartly sore, although the flow of saliva has begun to decline, or when it takes place in a case where the mercury, though used in large doses, is not acting in a kindly manner on the system, I say in either of these instances we cannot attempt to cure the affection of the eye by mercury. And again, when we have used mercury for the cure of the iritis, and have induced ptyalism thereby, but without effecting much improvement in the state of the eye, we must then have recourse to other means of relief. In these cases we sometimes succeed almost beyond our expectation, by giving bark largely. Sulphate of quinine is the remedy which I have been most frequently in the habit of employing, and generally with the happiest effect. I have also seen oleum terebinthinæ taken internally, as also colchicum, apparently with much benefit.

We not unfrequently see in some of those who are afflicted with secondary syphilis, a venereal ulcer on the edge of the eyelid, and it is remarkable that this appears generally in cases where many other symptoms of this disease exist along with it. This ulcer of the eyelid, on a superficial view, appears to occupy only a very small portion of the edge of the eyelid; but, on turning out the lid, we see it passing down on the conjunctiva palpebrarum to half the depth of the eyelid, presenting rather a raised surface towards the globe of the eye. The deformity and the injury to the globe of the eye, which such ulceration may produce if allowed to extend itself, loudly call for our active interference with local treatment; and accordingly we should employ caustics of such strength as the condition of the ulcer may seem to require, in order that we may as speedily as possible bring it into the condition of a granulating ulcer. I have seen one case in which this kind of ulcer took a very unfavourable turn; it spread along the conjunctiva palpebrarum, as far as the globe of the eye, seized on it, and completely destroyed this organ.

The lacrymal sac and ducts not unfrequently suffer when venereal ulceration has seized upon the nostril, and that, in some cases, to such an extent that fistula lacrymalis is the consequence. It is unnecessary to say that we cannot undertake any thing for the relief of the lacrymal passages until the disease of the nose be cured; and I may add that when this has been accomplished, we generally find that there is no occasion for our interference, for the disease of these passages will subside in proportion as the ulceration of the nose advances to healing, and the former will be cured even before the latter is perfectly healed.

Secondary Venereal Ulcers.

Secondary venereal ulcers, which supervene on the eruption, generally assume a circular form. When the scab is first removed, the ulcerated surface is uneven, foul, and yellow; in its progress towards healing we remark that it begins to cleanse first in the centre; then granulations arise in that situation; these extend towards the circumference where the skin forms a deep edge, between which and the granulations is a ring of the same yellow ulceration as appeared when first the surface of the ulcer was uncovered. In proportion as the ulcer proceeds the granulations encroach on this yellow ring, until at length they reach the edge of the ulcer. The central granulations, to a large extent, will have actually cicatrized before the entire surface of the ulcer is cleaned and healthy.

These ulcers are also remarkable for the strong tendency which the central granulations have to assume a fungoid character: so that, unless particular care be taken to prevent it, they will leave, when healed, a very high and prominent cicatrix. The ulcers also often heal from one side only, so as to resemble a kidney bean or a horse-shoe.

If a patient be afflicted with a number of these ulcers, we may observe that many of them may be induced to heal by applications of rather a mild kind, while two or three will remain unhealed. This fact might lead us to suppose that these ulcers were but little under the influence of the system. Yet, again, under the most mild applications all will sometimes simultaneously and quickly heal, when the system has been brought under the sanatory action of mercury, and this would lead us to suppose that all these ulcers were very much under the influence of the system.

When a venereal ulcer has been healed before the disease

has been eradicated from the system, it will be observed to break out again either by a number of pustules forming in a ring on the borders of the cicatrix; or by a pustule covered with a scab appearing on the edge of the cicatrix, this scab being rubbed or falling off, the fresh ulceration is established.

However numerous or extensive venereal ulcers may be, and however long they may have existed, whatever local or constitutional treatment they may have undergone, still I think we always find in one or more of them some appearance indicative of their nature and origin; this appearance is not such as would singly induce a surgeon to pronounce them venereal, but certainly is sufficiently strong to make him inquire into the history of the case, and try whether he could trace them up to a venereal origin. In forming an opinion on this difficult and very important point, he ought to be influenced more by the present appearances than by the consideration that mercury had been employed, and perhaps repeatedly, without effecting their cure. The difficulty of getting mercury to act favourably on some individuals, the judgment required to employ this medicine at a fit time, and to suit the doses and action to the existing state of the patient's health, and to the peculiarities of his constitution—all these considerations will weigh with him, and incline him to suspect that mercury had not had a fair trial, if the venereal aspect of any one of the ulcers favour the opinion that there is (as the phrase goes) something venereal in them.

Although I have said that the cicatrix of a secondary venereal ulcer often opens again either by a ring of pustules forming around its border, or by a pustule and scab rising in the centre, and each of these degenerating into ulcers, still I do not mean to say that this is the only manner in which a second or third crop of these ulcers appears. For not unfrequently a tubercle of rather a large size forms in the skin and proceeds to ulceration, many such ulcers exhibiting in their cavity a slough which seems to extend to some short distance under the skin. These ulcers continue forming successive crops, even for years, if the disease be but partially cured, either by imperfect courses of mercury, by sarsaparilla, or by other reputed anti-venereal medicines. In illustration of

these remarks, I shall adduce the following out of many
similar cases :—

Mr. P. contracted the venereal disease three years ago.
The secondary symptoms were, in the first place, a rash ; next,
a sore throat ; then tubercles of rather a large size, some of
which went on to ulcerate, and became foul ulcers ; and lastly,
swelling and induration of the right testicle.   For these
symptoms he was treated in June and July, 1833, by calomel,
which he used without caution, exposing himself to the
weather, and living sometimes very freely, and always rather
full.   By this medicine a very smart mercurial dysentery was
produced, with slight affection of the mouth.   At this time all
the ulcers healed, and for one month remained well.   At the
end of the month the ulcers began to break out afresh.   At
present there are, near the upper end of the right fibula, two
small ulcers, of rather a healthy appearance, one on the dorsum
of the foot, with considerable extent of surrounding inflammation with a foul surface, a shred, apparently of ligamentous
texture in a state of sloughing, being extended across it.
Another ulcer in a foul state, but not larger than a sixpenny
piece, is seen on the calf of the left leg ; swelling and hardness
of the testis remains.   There are four or five small ulcers on
arms and trunk.   He is not at all emaciated, nor has he a
sickly appearance.   Such is his state now on the commencement of a mercurial course to be used under confinement.—
Nov. 2nd, 1833.   R. Ung. Hyd. Fort. ʒss. divide in chartas
sex.   Utatur una omni mane.   Sumat mane nocteque Pil.
Hyd. gr. v.

Nov. 6th.   Two of the slighter ulcers are disposed to heal ;
the more severe ones are improved ; mouth is not sore, but he
says that he feels it soon will be so ; has used only two papers
of the ointment.   Pergat.

Nov. 9th.   Ulcer on dorsum pedis is more painful, edge of
the skin overhanging the ulcerated surface ; this now presents
a good deal of florid granulation, the surrounding inflammation
not so much extended, but the colour of it more intensely
high and brighter : ulcer on the calf of the leg unaltered in
its edges, surface more glazed and smooth, with but little
of surrounding hardness ; mouth is becoming more sore.

Repet. Pilulæ ut 2nd inst. Utantur Ung. Hyd. Fort. Ɔ iv. in die.

Nov. 15. Has had mercurial tenesmus for last three days, and mouth is now pretty smartly affected; ulcers are all cleaner and have less of surrounding hardness, but still their improvement appears to be rather slow. Omitt. Ungt. et Pilulæ.

Nov. 18th. Ulcers improving, though slowly; I do not think his mouth sufficiently sore. Ung. Hyd. Fort. ℈ss. bis in die. Pil. Hyd. gr. v. ter in die.

Nov. 22nd. He has rubbed (by mistake) a drachm of ointment twice a-day until yesterday, when he used only one drachm and one pill; mouth is now smartly sore, with a strong tendency to ptyalism. P. barely 90; app. good; sleeps well. Ulcer of dorsum pedis considerably better; that on calf of leg not so much improved; all the other ulcers vastly better. Utatur Ung. Hyd. Fort. ℈ss. in die et Pil. Hyd. gr. v. omni nocte.

Nov. 28th. Mouth very sore; pretty deep ulcers on each cheek opposite the dentes sapientiæ; appetite failed him yesterday and this day; gets five hours of uninterrupted sleep; pulse 106 and small; the ulcers are all but healed; that on dors. pedis is now quite free from surrounding redness and swelling: hardness of testis quite removed. Omitt. medicamenta.

Dec. 1st. All the ulcers are cicatrized; the old cicatrices of former ulcers have assumed a dark copper colour, and are sunk below the level of the skin, with a sharp, well-defined edge: mouth is less sore; appetite good; no thirst; sleeps well. Ung. Hyd. Fort. ℈ss. bis in die. Pil. Hyd. gr. v. omni nocte.

Dec. 7th. Mouth less sore; some thirst. Repet. med. ut 1st inst.

Dec. 11th. Mouth more affected; some thirst; appetite good. Rep. Ungt. semel in die. Omit. Pil.

Dec. 14th. Mouth very satisfactorily sore, and with more ptyalism than ever. Omit. med.

Dec. 21st. Mouth nearly well; symptoms all removed. I now advised him to take pills of Hyd. Cor. Sub. $\frac{1}{8}$th of a grain three times a-day.

Jan. 13th, 1834. Has taken the pills for eight days; they purged him, especially when he drank strong porter. I shall only add that he has since enjoyed most excellent health, and that he has not had any appearance of even a suspicious symptom.

Here is a case of venereal disease, continuing during a period of three years, in spite of various attempts to cure it whenever it made a fresh attack. The treatment was not on every occasion dependent on the powers of mercurial medicines; for in August, when he was affected with nearly the same symptoms as he laboured under in November, 1833, he began a course of sarsaparilla, and used it with great perseverance and in large quantities, having taken it during August, September, and part of October.

Why the sarsaparilla failed I shall not pretend to determine; but I am fully convinced that the failure of the different mercurial courses was caused by inattention and irregularities on the part of the patient, and misjudgment and mistaken indulgence on the part of the surgeon, who did not insist on the course being conducted under confinement. Let us now attend to the effects of the mercurial course last employed, and which proved a cure for this tedious disease. We find that the report on the seventh day of the treatment is rather unfavourable, for it describes the ulcer on the dorsum pedis as decidedly worse, and that on the calf of the leg as rather worse. We find, too, that on the 15th November, the thirteenth day of the treatment, an improvement is admitted, but is represented as slow, although the action of mercury on the system for the last three days had been manifested by the mercurial tenesmus.

Now, had these occurrences taken place with a practitioner whose mind had been strongly imbued with fears of the mischievous effects of mercury, he probably would have laid aside the mercury altogether on the seventh day of the treatment, regretting that he had ever been tempted to use mercury, or lamenting that he had not begun to use it more sparingly after the fourth day (Nov. 6th), when it appeared to be agreeing so well with the different symptoms. Is it not generally the case that about the seventh or eighth day, when the mercury ordinarily begins to act sensibly on the system, then we see a change apparently for the worse in the condition of the ulcers; this continues for two or three days longer, viz., until the mercurial action comes to be fully established, and then we find a decided improvement take place in the ulcers. Let us not, therefore, determine upon laying aside the use of mercury in cases of venereal ulcers, until we shall have seen what effect this medicine shall have when it has come into full action.

Now I suspect that this timidity in employing mercury, this hurry to lay it aside upon the first appearance of anything like an unfavourable change, has been the reason why it has been so much disused in the practice of many surgeons. Whereas the true rule by which we should guide our practice is this: to desist from the further use of mercury in the *advanced* stage of a mercurial course, as soon as we perceive a decided change for the worse in the ulcers (or rather symptoms). Let us not confound together the bad changes which take place in the onset of mercurial treatment, before this medicine has taken full hold of the system, with those bad changes which we occasionally witness in the advanced stage of the course, when the mercury has for some time had full power over the system.

On the seventh day of the treatment I doubled the dose of the ointment, knowing well that my patient's constitution was one which yielded reluctantly to the power of mercury.

It must strike every reader that in this case the mercury, after ptyalism was produced, was intermitted for much shorter periods than in ordinary cases. This was rendered necessary by the extraordinary quickness with which the action of the mercury subsided in this patient. We know that in ordinary cases, when ptyalism is excited, the action of the mercury continues almost unabated from eight to twelve days. Whereas here the mercury was at no period omitted for a longer period than four days; and indeed during the whole course mercury was employed in larger or smaller quantities, except for the one period of three, and another of four days. I was forced to this by the peculiarity of the patient's constitution, and should not have done so from any desire to accumulate mercury.

N

From the reports of the 22nd and 28th November, we can form a pretty correct notion of the little disturbance occasioned by a very smart action of mercury. On the first of these days, "pulse barely 90, appetite good, sleeps well." On 28th, "pulse 106, and small; appetite failed him yesterday, and this day. Gets five hours of uninterrupted sleep." Surely these slight disturbances of those few functions are not calculated to make a lasting injurious impression on the future general health; particularly when we find, on December 1st, that the "appetite is good; he has no thirst; he sleeps well."

It must be a very crazy system indeed that could not bear these slight disturbances for so short a period with impunity.

Thus, in a period of forty-two days, was this gentleman delivered from a disease which had materially interrupted his pursuits and his enjoyments for a period of three years. If I be asked why I did not at first employ this treatment, I shall find my apology in the impossibility of persuading young men now-a-days to submit to a vigorous use of mercury, under confinement, when they see so many of their acquaintances treated by very mild doses, and very slight action of mercury, used while the patient is debarred of very few of the ordinary amusements and indulgences of life. Besides, so much has been written within the last twenty years in praise of a non-mercurial treatment, and so much in condemnation of mercury, or at least of what has been termed its abuse, that it is scarcely reasonable to expect that patients will submit to the more rigorous treatment, until they have first had sad experience of the inefficacy of milder measures. Mercury, alone, accomplished everything in this case, with the exception of a little laudanum to check the mercurial dysentery; for the ulcers were dressed with spermaceti ointment only.

I shall now only add that this gentleman has continued free from any relapse, or return of any symptoms. In two years after this cure he had occasion to undergo a course of mercury for a primary sore: the cure was effected in the ordinary time, with little more than the ordinary quantity of mercury; nor did any untoward circumstance impede the progress of the cure.

CHAPTER IX.

VENEREAL ERUPTIONS, NODES, AND DISEASES OF THE TESTICLE.

MODERN writers on the venereal disease have bestowed much
pains in observing and arranging, in a nosological order, the
varieties of cutaneous eruptions which form a part of secondary
syphilis. Their labours have been attended with much advantage in the treatment of some of these affections, as we shall
presently remark; and I should think the subject might still
deserve a continuance of that zeal and minute research
which some have bestowed upon it, were I convinced that each
form of these eruptions constituted a distinct species in the
disease. I fear, however, that any superstructure raised upon
this hypothesis will not stand the test of time, as I do not
believe that these eruptions can be considered as characteristic
of distinct and different forms of syphilis.

My reasons for dissenting from others upon this point are
the following:—First, I have not unfrequently observed
varieties of eruption exist together in the same individual; for
example, I have seen small venereal lichen on the face, while
a large form of papular eruption occupied the trunk and the
extremities; sometimes, also, I have found spots of a pustular
character scattered through a general crop of the papular
eruption. Secondly, I have noticed, as a very frequent
occurrence, that when the first eruption has been removed,
either by the use of mercury or by other means, that the
second crop has proved of a different kind; thus, when the
first eruption was of that small pimply kind which resembles
measles, it has been succeeded by a papular eruption, and
this again by a pustular crop. And, thirdly, by injudicious
treatment; for example, by the excessive use of mercury, in
bad habits, any one other form of eruption may be made to
d generate into one which is most obstinate and severe,
namely, that of rupia.

I must declare that, after long and careful observation, I

have not been able to trace particular forms of eruption to particular forms of primary ulcers.

It is hardly necessary for me to repeat, that venereal eruptions, like other secondary symptoms, are often ushered in by a smart degree of fever, and that we almost uniformly observe that the fever which precedes the first eruption runs higher than that which ushers in any subsequent attack. Sometimes the eruption is preceded by lassitude and shifting pains of the limbs, which become more severe in the evenings and during the early parts of the night; while in a few cases the premonitory symptoms, if any, are so slight as not to attract the notice of the patient. By close and repeated examinations we shall find that some spots of the eruption appear to decline, and that fresh spots come out during the first three or four weeks; after this, probably about the sixth week, the general eruption declines so very remarkably as to lead the patient to indulge a sanguine hope that he is about to be relieved from it altogether. But whatever advances it may make towards a cure, these never go so far as to leave the skin free from stains of a pretty deep hue. After a period of four or six weeks of apparent amendment, the patient's hopes will again be damped by the appearance of a fresh crop of eruption, and probably one of a different character, which is generally preceded by an eruptive fever, more or less severe. Not unfrequently, a peculiar, pale, sickly aspect, loss of appetite and of strength, night-sweats, and a pain in some one joint or limb, affect the patient for one or two weeks previously to the second or third crop of eruptions.

In some cases the eruption is very general, occupying almost every point of the skin; in other cases it is very partial, showing itself, for instance, only about the ankle and tendo Achillis. I think we may expect to find, that the eruption which is spread over a wide surface, however thickly it may be set, will prove more tractable than that eruption which is confined to a small portion of the limbs, and which may not exceed the number of twenty or thirty spots.

In long-protracted and obstinate cases of syphilis, confined to the first order of parts, we frequently see this partial eruption as one of the last lingering symptoms; I cannot pretend to say for what length of time any of those eruptions might continue to preserve its identity through successive crops, nor for what length of time any one of them might continue to infest the skin; but I have seen one instance where a single patch of papular eruption, accompanied by other suspicious symptoms, yielded kindly to a course of mercury, at the end of four years from the appearance of the original chancre.

If the practitioner pay close attention to his case, he cannot (and he ought not) often be guilty of error in discriminating venereal from other eruptions; yet such mistakes have occasionally occurred. Thus I have known a medical practitioner treat for measles a young man to whom he was called, while labouring under the fever which ushered in that very small red venereal rash which so much resembles the eruption of measles, and this resemblance was strengthened by the suffused condition of the eyes, which ordinarily accompanies this form of venereal eruption. A still more serious mistake I have known to have been made by an eminent physician, who mistook for papular venereal eruption those papulæ which are often seen on the shoulders and backs of young persons with coarse skins. In this case the subject was a young man with incipient hectic—this the doctor considered as springing from venereal infection; and having subjected the patient to a pretty free use of mercury, he saw him quickly sink, overwhelmed by the rapid and accelerated advance of pulmonary hectic.

If, without trusting to a vain conceit of our own skill, by assuming that we can know by sight all the symptoms of the venereal disease, we merely take the trouble of enquiring carefully into the history of each case, we shall seldom commit any very serious error.

One and the same line of treatment will not prove equally successful in all the various forms of syphilitic eruptions. Experience proves to us that the scaly eruption, the copper-coloured blotch, and the papular eruption, are those which yield most readily, and are most certainly and perfectly cured, by the action of mercury, given in the ordinary doses. But the pustular eruption, and especially when of a larger size

than ordinary, requires a very particular treatment. No one
fact can be more clearly established than this;—that if mercury
be used too largely in cases of pustular eruption, the latter
will quickly degenerate into venereal ecthyma or rupia, or
spreading venereal ulcers. This form of eruption is one of
those venereal affections which, while it is materially aggravated by large doses of mercury, can be certainly and safely
cured by small doses, as I shall endeavour to show in the
chapter on minute doses of mercury. If we watch with care
the progress of the pustular eruption, we must be struck with
the strong disposition which it betrays of running into, or
of being converted into, ulcers. Sometimes the pustules,
spreading widely and still keeping superficial, form ulcers
covered with thin soft yellowish crusts : while in other cases
each pustule, without enlarging much, forms into a deep ulcer
covered with a brown scab, which is depressed below the level
of the skin. Now, in either of these cases, mercury administered in very minute doses, and with extreme caution, will
effect a cure, and that too in a very short time.

It would seem as if the pustular ulcer was closely allied to
the rupia ; for I have in some few instances seen, on the same
person, a soft, white, flat scab on the spots on the face, and on
the bald part of the scalp ; while on the shoulders only scabs
of rupia were to be seen.

The scabs and ulcers of rupia appear to be very little
under the influence of mercury. I have seen this medicine
administered, in cases of this affection, to patients of pretty
vigorous habits ; and although it acted in a most kindly
manner, and produced a full and healthy ptyalism, yet it had
not any effect in causing the scabs of rupia to dry up and fall
off ; nor did it induce, in those ulcers which had been exposed
by the previous removal of the crusts, any disposition to heal ;
the only change induced by it on these was to convert them
into ulcers, which, though florid, presented one uniform
smooth surface, sunk below the level of the skin, and totally
devoid of granulations, which proved very slow and difficult
to heal.

But the administration of mercury to patients afflicted with
rupia is worse than useless in all instances where the patient

is naturally delicate, or has been much reduced and lowered by the previous disease; for in all such it proves almost invariably fatal, by increasing the weakness, and generally by inducing an uncontrollable diarrhœa. It was only in a few very robust men that it could be said not to have proved highly dangerous or fatal.

In treating cases of papular eruption we shall often observe fresh spots of eruption coming out during the early exhibition of mercury. But as soon as the mercury has fully taken possession of the system, fresh spots cease to appear, and the general crop becomes more faint in colour: so far we may rest assured that all is doing well; but if, after this, we observe some fresh spots come out, and some few spots of the original eruption appear covered with a soft scab, around the edge of which a ring of pus appears, then we have most positive proof that the mercurial action has been carried too high; and should we persist in the same line of treatment, using mercury in the same doses, we shall meet with sad disappointment. For the results of such practice will be that these scabs will each of them degenerate into ecthyma or rupia, and that a considerable number of fresh spots of eruption will come out, each of which will be very speedily covered with a scab, and degenerate into rupia. We shall have then substituted for a papular eruption, which is a comparatively mild form of eruption, a very severe one, viz. ecthyma, or one which is most unmanageable, viz. rupia. Nor does the mischief end here; for as soon as the new form of disease is established, we find that the health breaks down; a sharp degree of fever (of the type of hectic) sets in, and this, in no great length of time, is followed by diffused pains of the limbs, and occasionally by swellings of the small bones and small joints of the tarsus and metatarsus, or carpus and metacarpus. But the papular is not the only form of venereal eruption which may be made to degenerate in this manner; I have seen cases of scaly eruption converted into rupia by the injudicious use of mercury. Let it then be a rule of practice to desist from mercury, or to reduce the doses immediately upon perceiving the commencement of such a change; and let us most anxiously examine the eruption frequently, that we may be enabled to discover the first approach of such a change. There is also this additional motive to induce us to watch the changes in the eruption, viz. that the general health does not begin to suffer until the ecthyma has been established; and consequently we cannot derive any indication from the constitutional symptoms, which would enable us to resist those changes in their commencement; and when the fever has set in we have to contend with deranged symptoms, and with broken-down health and an enfeebled system. Should we be so careless as to overlook this change in the eruption, we shall by our injudicious practice convert a very mild form of eruption, unattended by any fever, or at most by one of a very mild and manageable form, into the most obstinate form of eruption, accompanied by a high degree of debilitating fever, and presenting in their combination a disease very tedious and most difficult to manage. We shall recur to this subject when speaking of the treatment of rupia.

There is one sequence of venereal eruptions well deserving of our most anxious study; it is this: a patient whom we imagine has been perfectly cured, by a mercurial course, of some one form of eruption (e. g. the scaly or papular), may, in the course of eight or ten months afterwards, apply to us for advice under the following circumstances: his general appearance may be that of perfectly good health, but he may have a very few spots of eruption; these may be scattered over different parts of the limbs; the entire number may not exceed a dozen; two or three may be found on the fingers of one hand, or on its palm or dorsal surface; perhaps two or three more on the opposite wrist, and two or three in the neighbourhood of the knee or ankle. These are found of the same character in every patient, whatever may have been that of the original eruption : each spot is of a coppery hue, is elevated above the surrounding skin, and is of a horny consistence, or rather of the firmness of an ordinary corn on the toes. Sometimes in the palm of the hand is seen a copper-coloured ring, pale in the centre, with a hard red raised margin, about a quarter of an inch broad. A most important point for our consideration in this case is to decide whether such symptoms are to be considered as true venereal symptoms;

whether the patient should be considered as still labouring under syphilis; and whether the offspring of such a person, whether male or female, would be likely to be infected. Such cases of this peculiar form of eruption as I have had an opportunity of watching for any length of time were either on females too old to have children, or on young men who had as yet remained in a state of celibacy; so that my experience does not enable me to offer any opinion on this subject.

Let not this form of eruption be confounded with that marbling of the surface, or with those broad distinct spots which, as a second and third crop, appear on the trunk, while from eight to twenty such spots appear in the skin of the palms of the hand; for this latter eruption is generally accompanied by pretty extensive ulcerations of the edges of the tongue, and yields readily to the influence of mercury.

I am at a loss to know what kind of treatment is best suited to such cases. I have tried mercury internally, in larger and smaller doses, and pushed it even to ptyalism. These spots have quickly disappeared when the patient became salivated; but even a protracted ptyalism has not prevented them from recurring. They have also disappeared in cases where the mercury was not producing any sensible effect on the system. I am disposed to think that I have seen more benefit, in such cases, from Hydrarg. Calcinatus than from any other preparation of mercury; but I am ready to confess that I doubt whether in any case my plan of treatment was really useful. Indeed, I shall even say that I doubt whether such cases are not cured as effectually, and as well, by the unaided powers of the system, as when treated by any kind of medicine. Sarsaparilla was as little successful as mercury in effecting their removal. I have had opportunities of watching some few of these cases for years together, and I found that this symptom recurred occasionally for three years after the supposed cure of the original eruption.

Venereal Nodes.

Nodes arise only in the later stages of syphilis: the disease may then be said to attack those structures which Mr. Hunter calls "second in order." Such parts of the bones as are most

thinly covered, and are possessed of the most hard and close
texture, are the seats in which this symptom, when purely
venereal, makes its first appearance. No doubt we frequently
meet with nodes on the soft cancellated ends of bones, and on
the small bones of the tarsus and metatarsus, carpus and
metacarpus; but we shall generally find that nodes occur in
these latter situations in patients who have employed more
than one unavailing course of mercury, the venereal disease
being still unsubdued, although much altered by the treatment
and by the attending deterioration of the general health.

A venereal node sometimes forms without much pain, not
more perhaps than what directs the patient's attention to its
seat; but in other cases it is ushered in by severe pain. In
some instances I have observed the tumour to be for a few
days soft and very painful, then it became firm, and at the
same time almost totally free from pain. Those nodes which
are solid at their origin, indeed I might say all nodes, may in
course of time proceed to suppuration, but this is generally a
very slow process, and takes place only in nodes of long
standing.

A node once formed will often remain for months together
apparently unchanged; no increase of size, no discoloration
of integuments, no feel of fluctuation discoverable. In other
cases, however, a node will slowly undergo the process of
chronic suppuration; this I have observed to have occurred
more frequently in nodes on the bones of the cranium than in
other situations. The formation of the pus does not bring
with it any mitigation of the pain, if the node had been previously painful. But in many such the node had ceased to be
painful either long before or immediately on the commencement of this suppurative process.

When the bone has been exposed by art or by the natural
process of ulceration, we see it in some cases exfoliate by a
pretty thick plate; in which case the patient has to submit to
a very protracted ulceration of the integuments. In other
instances the denuded bone is soon perforated with numerous
small holes, through which healthy red granulations arise,
and at length, by their increasing numbers, cover the entire
surface from our view, and in a short time proceed to

cicatrization; thus effecting a cure without our being able to discover any exfoliated bit of bone. Yet that the surface of the bone has been removed we are afterwards assured by seeing the depressed state of the integuments, and feeling the roughness of the surface, which ever after remains.

The bones of adults are subject to some diseases which simulate venereal nodes. One of these is periostitis. I shall not here attempt to describe the local conditions by which we may distinguish the one from the other, because I think this can only be done with certainty and safety by referring to the previous history of the case, and by a close investigation and inquiry for other accompanying venereal symptoms.

When examining a case of this disease, we should most carefully distinguish it from what I would call a "general nodose affection of the bones;" because mercury, which effectually cures the venereal node, exasperates to an intolerable degree this general condition or diathesis. The local condition of the parts can assist us, but in a slight degree; however, I think we may say that a greater number of the long bones are at once engaged in this general nodose disease than we ever meet with in pure syphilis. Again, the form of the swelling differs; for in the venereal node it assumes a rounded figure, whereas in the other disease the tumour assumes a more elongated form. It occupies, for instance, such a length of the anterior surface of the tibia as to give to the bone some resemblance to one which is curved by rickets. Both these bony tumours are attended with severe pain, and in both also this is aggravated at night, especially when the patient becomes warm in bed. This nodose disease, in general, attacks those who have passed the age of puberty, although I have seen a few cases in which it attacked children of both sexes at the age of ten or eleven years, the subjects being in every other respect to all appearance perfectly healthy.

We never can venture to form an opinion as to the real nature of a node until we shall have made the most minute inquiry into the history of the case, and searched most carefully for any other symptom of syphilis; for the accompanying venereal symptoms are often so trifling or so obscure as to be unknown to, or unnoticed by, the patient himself.

The condition and seat of a node afford us some criterion by which we can judge whether it be easily curable or not, and to a certain degree indicate the appropriate line of treatment. The node in the centre or in the hard part of a bone will be found much more easily treated and cured than one on the cancellated structure. The former node will bear, and it also requires, pretty active and full doses of mercury for its cure, while that on the cancellated structure of the bone will require the mercury to be used in moderate doses, administered with much judgment. Very many years ago I heard a most intelligent physician declare that nodes on the bones of the hands or feet were of all nodes the most difficult to cure : a long-continued observation has proved to me the justness of this remark.

In the treatment of pure venereal nodes I believe we may entrust the final cure to the operation of mercury; but until the mercury is brought to act upon the system we must use our best endeavours to mitigate the sufferings of the patient; for this purpose I know not of any remedy more effectual than blistering the part, allowing it to heal, and blistering it again as quickly as we can; indeed the blister sometimes acts like a charm in such cases. Every surgeon must have met with cases of pure nodes which could not be removed; by the mercurial course and blistering they were relieved of all pain, but the swelling remained permanent. Some have proposed the early opening of the tumour, and evacuating all the contained fluid. To this proposal I would object, that in some cases this practice is followed by painful suppuration, and by very copious discharge, and not unfrequently by caries and tedious exfoliation of the bone. It seems to me preferable, in all cases, to try the local effects of blistering and the constitutional influence of mercury; and by means of these endeavour to avert suppuration and ulceration. This rule should be most strictly adhered to in the case of nodes on the forehead, or in any exposed part of the body; for when a node has been of long standing we often find that a sort of chronic suppuration is established, the integuments become thin, and sometimes red; at other times they are reduced to the utmost degree of thinness, and yet may retain the natural colour, so that the surgeon is actually tempted to give vent to the fluid by the puncture of a lancet. Yet if he will but resist the temptation which the very thin state of the skin offers to him to open it, and will still apply repeated blisters, he will have no reason to lament his forbearance. For as soon as the mercury comes to act favourably on the system, he will perceive that the fluid begins to be absorbed, and that this process will finally be terminated by the adhesion of the skin to the surface of the bone. From the depressed position of the skin, and the sunken unequal surface which the bone presents to the touch, after the node is cured, we are convinced that an absorption of the bone has gone on to some depth. Now if we had adopted a different practice,—if we had opened the tumour by even a very small puncture,—the result would, most probably, have been an unhealthy tedious suppuration, perhaps also an exfoliation of bone, and certainly a very unseemly depressed cicatrix.

With respect to the constitutional treatment of nodes on the small bones, or on the spongy parts of long bones, I shall only remark that in general, when they appear, the system is in a state which will be injured by large doses of mercury, and that it requires sound judgment and nice discrimination to employ this medicine in such a manner as to remove the disease; the most dangerous error which we can commit in such cases is to administer it too largely : we are not to omit ths use of other auxiliary means; in the selection of these we must, of course, be determined by the existing circumstances of each case.

It may not be amiss to mention the extraordinary influence which a node on the femur sometimes has upon the condition of the entire limb. The node on this bone is generally seated in its lower third or lower half, and on its anterior surface. Of course it is scarcely perceptible by the eye, but is readily detected by carrying the hand along the front of the bone, and by squeezing it as we descend. Should the disease have existed for any length of time we shall, upon close examination, discover that this limb, through its entire length, is less full an the other which is free from disease : and if we compare the nates we shall be led to suspect that there is morbus

coxæ, so close is the resemblance of the buttock of the affected
side to the condition in which we find it in that disease;
indeed, there is only wanting the painful feeling in the groin
to complete the picture; as not only is the limb more wasted,
and the nates more flat, and its fold more low, but there is
even an apparent elongation of the limb at the knee and
ankle, and in some cases I have seen these characters fully as
striking as in cases of genuine simple morbus coxæ.

I have no doubt that nodes are sometimes excited by the
injudicious use of mercury, or by the irregularities of the
patient.   I have known cases in which mercury having been
largely and repeatedly employed for the cure of other symptoms, and the patient, having been again subjected to a fresh
course of mercury, has complained, even while his system was
decidedly under the influence of this medicine, of a swelling
and tenderness of one or more of the long bones.   The
tubercles of the tibia are frequently the seats of this affection;
when thus attacked they are not seen to become much enlarged,
but are rather soft, and exquisitely tender to the touch; and
not unfrequently the integuments covering them assume a
reddish tint.   Nodes which form from the above cause on
other parts of the long bones are, from their commencement,
very painful, and of different size in different individuals;
but in all cases they are rather soft.   The pain attendant on
all these is more widely spread along the limb than in cases
of purely syphilitic nodes.   Of course patients under such
circumstances are not fit subjects for the use of mercury.
Blistering these tumours will generally procure temporary
relief from pain, while, at the same time, we shall use our
best endeavours to repair the mischief caused in the constitution by the injudicious use of mercury.   And here I must add,
that it has never fallen to my lot to witness an instance of the
use of mercury producing nodes, except when it was administered for the cure of the venereal disease.

Affectiöns of the larger joints are not unfrequent among
patients labouring under secondary syphilis, and more so
while the disease is seated in those parts which Mr. Hunter
has classed as "the first order of parts;" they may also often
be met with when the disease is seated in the parts " second

in order." These affections do not deserve the title of purely syphilitic symptoms ; at least I cannot recollect having seen them, except in cases where the patient had undergone a course of mercury for some form of secondary syphilis, and where, notwithstanding, the disease was not thoroughly subdued.

Effusions into, and distensions of, the synovial membranes and bursæ of the large joints are very frequently met with in patients labouring under secondary syphilis. I cannot venture to say that these never are purely venereal symptoms ; but as far as my memory serves me those I have witnessed might, with very few exceptions, be traced to mercurial courses mismanaged, either on the part of the surgeon or of the patient. In a few cases they may have been caused by over-exertion of the limb. The knee and elbow are the joints most frequently thus affected, the wrist less frequently ; but when it does occur, it presents a more obstinate and intractable case. Although these swellings may not be considered as purely venereal symptoms, yet that they are more or less connected with syphilis would appear from this fact, that when they have resisted (as they occasionally do) blistering and a variety of topical treatment, they readily yield along with the true venereal symptoms to a subsequent judicious use of mercury.

In connection with these affections of the joints, I would just observe that syphilitic patients often complain of pains in the limbs ; when these are described as passing along the greater extent of the limb, and particularly if on inquiry we learn that they, at different times, attack different parts of it, we may be assured that, unless they prove to be the precursors of some fresh eruption, they are not venereal,—that mercury will not relieve them, and that they have been caused by mercury. The only condition which forms an exception to this opinion is the case of a node on the fibula ; for this is frequently attended with a more wide-spreading pain than a node in any other position.

I believe it is unnecessary to notice the incorrectness of that opinion which considers nocturnal exacerbations as pathogno-monic of venereal pains. The pains of gout and rheumatism also obey a similar law.

A very common form of those affections is observed in the elbow joint; the patient has not full power of it, he cannot fully flex it, nor extend it much beyond a right angle; very rarely is there any pain or any tenderness from pressure, when made upon the prominent parts of the joint. Not unfrequently this affection of the joint is attended by tumefaction caused by effusion into some of the adjacent bursæ; but in all these cases we feel the tendon of the biceps extremely rigid, yet not swollen or tender to the touch. As a further proof that this symptom is not purely venereal, we find that it sometimes yields to blisters, often to blisters and to a course of sarsa-parilla. It is not benefited by putting the patient immediately under a second course of mercury, although it is found to yield *pari passu*, with the truly venereal symptoms, to the powers of this medicine when employed with judgment and under favourable circumstances.

A similar, but much more rare affection, is that which engages the knee. The joint in this case, though to a certain degree moveable by the hand of another, can be moved only in a very trifling degree by the patient. No swelling or tenderness necessarily attends, but the hamstring tendons are felt as tense as it is possible to imagine them to be, although the muscles are not in action.

Venereal Affections of the Testicle.

On the subject of Venereal Swelled Testicle I have but very little to offer. It is among the latest symptoms of the disease; it generally takes place slowly, without much pain, and it continues of the same indolent character throughout its entire course. I cannot recollect any case of what I consider purely venereal swelled testis which went on to suppuration, although I have seen this occurrence take place in some where the debilitated and deranged state of health appeared to be the exciting causes of this result. If I were to attempt a descrip-tion of this affection, I should say that the entire of the organ, that is, both the epididymis and the body of the testis, are involved in one common swelling, that the tumour presents to the eye a surface smooth and but little discoloured, and to the feel a firm and uniform consistence, but nothing of a stony hardness, nor is it very weighty in proportion to its bulk; a fulness, but not a hardness, of the chord can also be felt. These I conceive are its leading features. I shall not attempt to discriminate it from the cancerous testicle, nor from that condition of the gland which is by some denominated scrophulous, a term which is often applied with very little reason or accuracy to many anomalous swellings of the testicle. A reference to the history of the case, and the co-existence of some other symptom of a syphilitic nature, can alone enable us to form a correct diagnosis. The syphilitic testicle, in a patient of unbroken health, is one of those symptoms which yields most regularly and uninterruptedly to the action of mercury when judiciously administered; the gland also, to all appearance, regains its healthy structure, and is enabled to resume its function.

But in some instances the venereal affection of the testicle appears under very different characters. When the gland becomes engaged in a patient who is naturally of a bad habit of body, or whose health has been broken down by repeated and mismanaged courses of mercury; one, in short, in whom the local affection may be ascribed fully as much to a reduced state of health as to a venereal taint. In such, I say, the condition of the testicle is very different from that which we may consider a purely venereal swelled testicle; for the swelling in many such cases does not involve in one common mass both the epididymis and the body of the gland; nor does it in general acquire the same magnitude as the purely venereal swelling of the testis. In such cases we not unfrequently find, on a careful examination, one or two points into which the end of the finger appears to sink, as if a fluid or a small cavity existed beneath the skin. But these two affections of the testis differ very widely in another most essential point—namely, as to their treatment; for, in the latter form of the disease, the treatment becomes a matter of great difficulty, and one which requires very nice discrimination and sound judgment; because we have to contend not only with a shattered system, but with such a combination of local and constitutional derangements as frequently cause our patient to sink under accumulated sufferings, the disease of the testicle being the mildest of his complaints.

Such patients may be considered as in the very last stage of the venereal disease, and on the very threshold of being afflicted with those constitutional changes which (from experience) are known to be the prelude of death, and which bring to a close the existence of those who have thus been the unhappy victims of protracted syphilis.

Some few, however, of these cases admit of being cured—that is, life may be saved; but the testicle cannot ever after be considered as restored to its perfectly healthy state, for we shall always be able to discover by the touch a considerable deviation from its healthy feel; we shall find, also, that it is in general rather wasted, and that there is some degree of hardness remaining in the greater portion of it, while a deep depression and a softness may be felt in one or two spots.

Having now glanced at this long catalogue of miseries, to which the venereal disease gives birth, and having also touched upon the appropriate treatment, it is scarcely necessary for me to remind the reader that either the injudicious use of mercury or the misconduct of the patient has a large share in inducing some of these pernicious consequences, and in rendering many of them most unmanageable. How erroneous then must have been that opinion, which for a time took possession of the minds of surgeons, when they imagined that a disease was proved to be not venereal, merely because it was not cured by mercury, or perhaps because it was made worse by this medicine ! By acting on this opinion the surgeon was obliged, in the after-treatment of the case, to have recourse to any, or almost every other medicine, and to every remedial measure, rather than again resort to that which had already done so much mischief. In those protracted cases some surgeons came to the conclusion that the symptoms which they witnessed could not be venereal, because such a lengthened period as two or three years had intervened since the receipt of the original infection, or appearance of the primary disease. If we take the trouble to trace the history of any number of these cases of protracted syphilis, we shall find that they present considerable variety; thus in some it may have continued for five or six years in a very mild degree indeed, while other unhappy sufferers have been subjected to almost every symptom of the disease, many of which too have afflicted them simultaneously, and with peculiar and aggravated severity.

From a vast number of cases of this description which I have witnessed, and of which I have notes, I shall here adduce two only to prove that this disease may exist, in an obvious form, for at least five or six years. The first of these cases shows that the symptoms, few and mild, never completely disappeared; that they did not give rise to any further symptoms; and that, being only kept in check by repeated and insufficient courses of mercury, they continued to exist, and yet did not very materially disturb or injure the general health.

Mr. H., in June, 1830, was treated for a recent chancre by pills of calomel, which he took to rather a large amount, and for a period of six weeks. By this treatment his mouth was not at any time satisfactorily made sore; indeed it was only affected at intervals, and then very slightly, although the mercury was pushed so as frequently to disturb the bowels, to impair his appetite, and to injure his sleep. In the course of eleven months after the healing of the chancre, he was affected with iritis and a few clusters of papular venereal eruption; these symptoms were made to recede, though slowly, by a course of Plummer's pill, which acted in a manner equally unsatisfactory as the plain calomel had done.

In August, 1831, a cluster of these papulæ appeared on the left temple, and he also was affected with a superficial ulceration of the membrane covering the septum nasi, a little way above the anterior edge of the septum. He was then directed to take Hydr. Oxymur. in solution, to the amount of an eighth of a grain every day; and also to combine with it Decoct. Sarsæ c. Extr. Fluid Sarsæ. To the ulcer in the nose he applied Ungt. Hydr. Nitr. diluted with seven parts of lard: this process he continued for three weeks, when he was called away to the country, the symptoms having been nearly removed.

In November, 1832, he again applied for advice, on account

of the same symptoms, which were very much in the same state as in August, 1831. He was again directed the Ungt. Hydr. Nitr. and the Lotio Nigra as applications to the ulcers in the nose, and to take Pil. Hyd. gr. iij. Extr. Conii gr. ij. sing. noct.; this plan he continued without any variation, except that of increasing the dose of the pills, from the middle of November until the 20th of December, when he again went away into the country, his symptoms being a good deal subdued, but obviously not cured.

In April, 1833, he applied to me for the same symptoms, and in precisely the same state in which he exhibited them in November, 1832. He was now directed pills of calomel, antim. tart. and opium, so as to take two grains of calomel every night—the same applications as formerly to the nose. This plan he commenced on the 26th of April, and finished on the 17th of June. Even now the mercury did not act in a favourable manner, and he left town with his symptoms all but well.

January 15th, 1834, he called on me, and declared that he was firmly resolved to use every means, and for any length of time that I might think necessary, for the removal of all traces of this complaint. At that period the symptoms were as follow: the cluster of papular eruption on the temple existed as before; the septum nasi was superficially ulcerated in each nostril; the ulcers were seated not far from the anterior edge of the septum, but not on corresponding parts of it; these ulcers were neither painful nor very sore, nor was there much surrounding inflammation, but they exhibited very distinctly the characters of venereal ulcers.

I now directed for him ℞ Calomelanos ℈i. Pulv. Ipecac. Comp. ʒss. fient pilulæ decem. Sumat unam mane nocteq. The ulcers to be lightly touched, two or three times a day, with Ungt. Hydr. Nitr. Dilut.

January 31st, 1834. Upper gums ulcerated, and sufficiently affected; ptyalism is apparently coming on; he is languid, especially in the evenings. Appetite good, thirst not urgent, some griping. ℞ Inf. Cinchonæ ʒvi. Tinct. Cinchonæ ʒvi. Tinct. Opii Gutt. xxx. Syrupi ʒss. m. sumat ʒi ter in die. Repet. pilulæ.

February 5th. He feels much strengthened since he began the bark mixture; upper gums are much fuller, and more ulcerated; sleeps well. Repet. pil. and mist.

February 11th. Mouth is sufficiently affected; ptyalism is established, but not copiously; ulcers of the nose appear to be perfectly healed. Repet. mist. Sumat pil. unam sing. noct.

February 21st. Mouth is still fully affected; he is more oppressed with languor. He goes to the country to-morrow, being called away by professional business, and promises to keep within doors and take the pills for a fortnight longer.

I have seen this gentleman lately; he continues free from any return of his disease, and enjoys excellent health.

In the history of the following case we shall find the disease passing through its different stages, and yet retaining firm possession of the parts already attacked, while it proceeded to fix itself in some new situation, resisting at various periods the use of all anti-venereal medicines, as well as of various remedial measures—mercury in various forms, sarsaparilla, acids, alkalies, opium, arsenic, sea-bathing, country air, &c.; and yet, at the end of six years, finally cured by mercurial unction, pushed to a smart ptyalism, while the patient was confined to the house.

Mr. A. applied to me in July, 1828, for the cure of a recent chancre. He used mercury under my care for four weeks; although the medicine did not act in a kindly manner, still the ulcer was nearly healed at the end of the month. Circumstances over which he had not control obliged him now to apply to another surgeon, who in a few days advised him to lay aside the mercury. He returned to me again in the course of six weeks, complaining of sore throat and venereal eruption—Psoriasis. I advised Pil. Hydr., which he used for a few weeks, and then went into the country, having derived very little benefit from the course I had directed for him; for in this, as in the first instance, the system received the mercury in a very unkindly manner. After an absence of seven or eight weeks, during which he laid aside all mercurial medicines, he again applied to me with his throat more uneasy and the eruption still out. Mercury was again resorted to, but appeared to disagree with him more decidedly than hitherto. A node formed in the course of a few months, and his left testicle became hard and enlarged. After the complaint in the testicle was established, his general health seemed to improve a good deal; and for some months prior to April, 1829, he complained chiefly of the size of the testis, which was increased by effusion into the tunica vaginalis.

On the 11th of April, 1829, I learned from him that, without any assignable cause, within the last fortnight his health had very much declined; and his symptoms at present are a node on the lower part of the right tibia, from the upper part of which bone another node had been removed about three weeks ago, by the application of two blisters; painful swelling of the left tarsus; the end of the nose is a good deal swollen, and beset with a cluster of pustules, which are covered with white flat scabs; two of the pustules, however, have scales which affect the shape of rupia; a broad flat scab on the left zygoma, testicle enlarged and hardened, and affected with hydrocele; both arches of the palate have white ulcers on their edges; a yellow foul ulcer on the back of the pharynx, high up; pulse 108; skin hot.

I advised a pill, composed of Pulv. Jacobi gr. ij. Calomel gr. i. omni nocte.

April 29th. For some days past he has complained of pain on the inner side of both knees; this is felt most severely when he coughs or sighs; the knees are free from swelling; the mouth, within the last three days, has become very slightly affected; the ulceration of the throat heals in one spot and breaks out in another; pulse still 108; the right nostril feels stuffed, and the membrane covering the lower spongy bone is thickened and ulcerated; some new spots of eruption have appeared. Omittatur Calomel. Pope's Extract of Sarsaparilla.

May 31st, 1829. He is able to walk and jaunt about, has acquired a good appetite, and sleeps well; he has picked off the scabs from the end of the nose, which now presents a very irregular rugged surface; the throat is healed: the nose, internally, is much better, but not yet healed; the testicle is very little improved; pulse still at 96.

July 1st, 1829. Flesh and complexion improved; general

health very good; pulse still 96. He now complains only of weakness in his knees.

November 6th, 1829. About the middle of July his throat again ulcerated; his strength and appetite failed him. A strong solution of Arg. Nitr. (Ʒi. in ʒi.) was daily applied to the ulcers of the throat, and he took small doses of blue pill and cicuta until the beginning of September. No very sensible effect of the mercury on his mouth; yet the throat was nearly healed and his general health re-established. But this happy state did not continue long; for in the latter end of October the right anterior arch of the palate became very much swollen and red, with a whitish sloughy ulcer on its centre, attended by much pain. I now feared that the entire arch would have been destroyed by sloughing; but this evil was averted by the application of various caustics—Solut. Arg. Nit. Mur. Antim., and Solut. Sulph. Cupri; of these, the Mur. Antim. seemed to be the most useful. The edge of the arch, however, and a part of the right tonsil were destroyed.

When his throat was nearly healed, he began to complain of pain passing from the right hip down the thigh; at the same time a considerable effusion into the knee took place. These symptoms were much relieved for five or six days by the exhibition of Sp. Terebinthinæ internally; but after this short period this medicine seemed to have lost its influence altogether. I now gave him Pil. Hyd. gr. iv. Pulv. Jacobi gr. ii. semel in die, and after a week increased them to two pills daily; by this treatment no sensible mercurial action was produced, and yet his symptoms all yielded.

Feb. 8th, 1830. This amendment continued only for twelve or fourteen days; he then complained of renewed soreness of throat and stuffing of his nose. Ulceration now seized on the left anterior arch, and at the same time a deep circular ulcer formed on the back of the pharynx, immediately above the edge of the velum; a small node has arisen on the fibula, a little above the outer ankle; about ten days ago a scab of rupia falling off the right leg exhibited the surface healed, while a larger rupia on the left thigh is daily extending in its circumference.

It were an useless waste of time to describe in detail the

further progress of this case, and the line of treatment adopted ; it will be quite enough to say that I had to contend with symptoms such as have been already described : thus these symptoms occasionally, and with them his general health, improved for a time ; but that the ulcerated throat, ulcers on the surface of the body, and swelling of both testes continued to exist, although the ulcers would heal up in one part, and then seize upon some new spot : and shall merely state that in August, 1832, a node appeared on his forehead, and another on his right thigh. During parts of this long period his general health would improve very much, and this improvement was sometimes apparently produced, but always for a time promoted, by his going into the country or repairing for a time to the sea-coast. Yet such happy intervals were never of a longer duration than four or five weeks. During this amended state of his general health the local symptoms became less severe and less troublesome ; but they did not at any time disappear entirely, even for a very short period.

The treatment was at one time with various internal preparations of mercury, and on two occasions with ten-grain doses of mercurial ointment.   Sarsaparilla, bark, acids, cicuta, alkalies, arsenic, and opium,—each of these would appear for a short time to afford benefit ; but I was compelled to lay them aside when I saw the general health and the local symptoms fall back after each of them had been employed for some time.

In this vacillating condition of amendment and relapses was passed the long period from Feb., 1830, until Nov., 1833.   I now determined to employ mercury in such a manner as to induce a smart degree of ptyalism in a short period, having prevailed on him to employ this process under strict confinement to the house.   The state of his case at that time is here annexed.

Nov. 8th, 1833.   Mr. A. has at least from twenty to thirty ulcers, chiefly on his limbs ; some of these had, about six months previously, commenced in the form of soft round tumours the size of a marble : these came slowly to suppuration, and on opening exposed cavities which were occupied in part by a soft white slough.   This was not confined to the limits of each ulcer, but was seen to stretch beyond the borders of its cavity. Others of these ulcers had begun as pustules, forming on the edge of some old cicatrix, then ulcerating and spreading into sores varying in size from that of a shilling to that of half a crown. From the nose he daily, with great effort, blew down large brown crusts or scabs. Behind the velum palati, and very high up towards the base of the skull, was an ulcer which created very considerable uneasiness, although it did not render deglutition very painful; his face, especially his forehead and cheeks, were thickly studded over with tubercles of a dark copper colour; the upper lip and nose were a good deal swollen and thickened; he had a pretty large node on the head and another on the femur, but these were in an indolent state; both testicles were very much enlarged, quite beyond the ordinary enlargement of diseased testicles; in the tunica vaginalis of the left there was a quantity of fluid; his general health was tolerably good; he was not emaciated, nor had he any degree of hectic; but for the last two months he has complained of a sensation of fulness in the epigastric region, more troublesome after taking food; on examination the liver was found to descend much too low; no trace of jaundice.

Nov. 8th. Sumat Pil. Hyd. gr. iii. Ext. Cicutæ gr. i. ter in die. Ungt. Hyd. Fort. ℥ss. omni nocte.

Nov. 10. No sensible effect from the mercury; he complains that the skin in front of his thigh is extremely tender, although it is perfectly free from any discolouration or swelling; in the centre of this very sensitive skin is a cluster of tubercles, now covered with scabby crusts.

Nov. 12th. Pulse 96; gums swelling; mercurial fœtor pretty strong; some scabs on the thighs are drying, and seem disposed to fall off.

Nov. 14th. The bowels have been dysenterically affected yesterday and this day; has not used any mercury to-day; cheeks are swollen, gums ulcerated, tongue sore to his feelings, but is not ulcerated; says that he has no sense of weakness: the tubercles on his face are very much improved, less swollen or full, and of a less bright colour; the soreness of the skin around the scabs on his thigh, which had been very distressing, is now quite gone. Omitt. Ungt. et. Pil.

Nov. 17th. Mouth as on the 14th; he sleeps well; has a good appetite, but cannot admit solid food; pulse 96; he has lost all feeling of fulness or uneasiness of stomach; the improvement in colour of all those parts of the face which had been occupied by the tubercles is more striking; swelling and thickening of lip and nose are materially reduced; tubercles are smaller in circumference, and much more flat; ulcers of the limbs are all clean and healthy; scabs on the thighs are fast approaching to fall off; they are contracted in circumference and height, their edges dry and detached from the skin; in a word, there is every prospect of a speedy cure of this very tedious disease by the present course of mercury. Repet. Ungt. et Pilulæ Hyd.

Nov. 24th. Mouth still sore; appetite good; pulse 78; skin temperate; sleeps for five or six hours every night; ulcers are all covered with luxuriant granulations, and are healing; scabs all drying, and look as if about to fall off; he can read with pleasure, not only light books, but even those on professional subjects; on two separate days only, since the last report, has he omitted the pills and ointment: the ointment has been always applied to the abdomen, as the numerous ulcers of the limbs prevented its application to them. Pergat.

Dec. 3rd. Has not used any mercury for the last three days; bowels have been purged each day so much as to require opium; the mouth is fully sore; the discoloured spots on the face present a more faded colour, and the appearance of the face is altogether vastly improved; among all the numerous scabs, from one only on the forearm could I press out one small drop of matter.

Dec. 7th. Had a sense of weakness last night when going to bed, but no palpitation then or at any other time; pulse 74; appetite good; no thirst; sleeps well; mouth overflows at night; a crust has come down from the left nostril; both testicles are hard, fully one-third larger than natural; with fluid to the amount of an ounce in each tunica vaginalis. Mist. c. Sulph. Quinæ ter in die. Ungt. Hyd. Fort. ℈i. quotidie testibus affricand.

Dec. 10th. Pulse 76; sleeps well; appetite good, but cannot make use of solids; no thirst: skin cool; countenance good;

mouth quite sore enough; ptyalism reduced; yesterday, immediately after a large discharge from his bowels, felt a weakness for ten minutes, but not the slightest feel of palpitation; scabs have dropt off from every ulcer, except two or three small ones. Omittantur Ungt. Hyd. et Pil. Hyd. Rep. Mist. c. Sulph. Quinæ.

Dec. 14th. Yesterday and this day has been teazed with dysenteric purging, for which he took each day ℞. Tr. Opii. Gutt. xxv.; mouth still very sore; pulse 96; skin hot. ℞. Olei Ricini ʒvi. aq. Cinam. ʒvi. Tinct. Rhei ʒi. Tinct. Opii. Gutt. xxx. M. fiat haust. statim sumendus.

Dec. 26th. General health very good; he feels himself growing stronger; mouth is now well, except around the lower dens sapientiæ of left side, where the soft parts are ulcerated, although free from slough. ℞. Calomelanos ℈i. Piper. Indici gr. vi. fient pilulæ septem. Sumat unam meridie nocteq.

Here let me observe, that I was solicitous to bring his system again under the influence of mercury; because his disease having been of such a very long standing, it seemed natural that it should require a very complete and long-continued influence of mercury to eradicate it effectually. And his constitution had been so much improved by the late ptyalism, that I felt confident it would very safely bear the ptyalism which I was about to inflict on it.

Dec. 31st. The pills on the night of the 27th purged him a good deal, so that he did not take another until the night of the 28th, when he took one only; he has not since that time taken any more, because he found that this fourth pill not only had excited purging, but had also caused very considerable soreness of his mouth; the gums, especially the inner gums in both upper and lower jaws, present one ulcerated edge, and the cheeks at the angles of the jaws are swollen and ulcerated afresh; the edge of the tongue feels to him sore, but it is not ulcerated; with all this his pulse is 76; appetite good; general health very good.

Jan. 9th, 1834. Mouth is still a little sore; pulse quiet; appetite for breakfast is bad,—for dinner is good.

Jan. 30th. Testes, though smaller, are still hardened and

enlarged; all the fluid seems to have been absorbed; no other trace of his former disease remains; general health very good.

Nov. 2nd. Mr. A. called on me this day; he appears in excellent health: can walk twelve or fourteen miles a-day; says that occasionally some crusts are expelled from the nose; whenever this occurs, he uses the Lotio Nigra with immediate benefit; pulse 80; testicles much reduced; the left one is little beyond the natural size; right testis is still much too large.

November 2nd, 1836. Mr. A. continues to enjoy excellent health; very slight induration or enlargement of the testes still continues.

CHAPTER X.

ON THE USE OF MERCURY IN VENEREAL HECTIC FEVER.

I HAVE long been impressed with an opinion that the judicious management of mercury, the. adaptation of the medicine to the general condition of the patient, and to the circumstances and states of his venereal symptoms, afforded the most likely means of enabling us to overcome the disease of syphilis, under all the various forms and in all the various combinations in which it presents itself to our view. In this opinion I feel confirmed by having ascertained that this very valuable medicine may be administered not only with safety, but with certain expectation of effecting a cure under circumstances of venereal symptoms and conditions of the constitution which have hitherto been considered as absolutely prohibiting its use. I shall now proceed to consider this subject in detail.

Should a patient present himself labouring under secondary symptoms, at the same time complaining of total loss of appetite, of urgent thirst, emaciation to such an extreme degree that he is almost exhausted by night sweats, which oblige him to change his night dress two or three times every night; under these circumstances, I believe few surgeons would ask the question, can we venture to exhibit mercury to such a wretched sufferer? I believe the usual answer and the usual practice in such cases would be to postpone the use of mercury, and to endeavour to bring the general health into better order, by means of country air, asses' milk, light nourishing diet, sarsaparilla, acids, &c.; in fact, to use every other means rather than resort to mercury under these apparently disadvantageous conditions; and no doubt many such patients have their health considerably improved by pursuing this restorative plan for a period of two or three months. However, at the expiration of this period, their venereal symptoms are found to be very little, if at all amended, and sometimes they are even in a worse state than when they commenced this plan. I may also add, that in some cases of this extreme emaciation and hectic the restorative plan altogether fails, and these individuals die as if worn out by the long continuance of the hectic : or they are carried off by some of those affections which I have elsewhere described as the sequelæ, or as the last stages of the venereal disease. (Vide History of the Disease.)

I am happy to have it in my power to declare that all this preparation and all this delay are absolutely unnecessary in such cases ; and I most confidently assert that a patient affected with secondary symptoms, even though extremely attenuated, and as it were melting away under the effects of hectic, can with perfect safety and advantage at once commence a course of mercury, by which not only shall his venereal symptoms be removed, but at the same time his general health be re-established.

I shall not, however, attempt to establish this point by reasoning, or to persuade my reader merely by argument or assertion. I hope to convince by adducing facts ; and the first case to which I wish to direct attention is that of Mr. G.

Nov. 21st, 1830. Mr. E. G. says that seven weeks ago an ulcer appeared on the external skin of the penis ; by application of the yellow wash (Lotio Hyd. Cor. Sub. c. Aq. Calcis) and of mercurial ointment this quickly healed. A second ulcer appeared on the prepuce in eight days after the first was healed ; to this he applied Lotio Nigra only : this now appears as a florid granulating ulcer of oblong shape, but with very hard base ; his skin is everywhere covered with a venereal papular eruption ; yesterday, for the first time, he felt his throat sore ; a long superficial foul ulcer runs along almost the entire length of each tonsil ; the gum around the lower right dens sapientiæ is ulcerated ; he complains of pain of the shoulders, pain and tenderness of the sternum, which symptoms distress him more severely at night : on the left side of the neck a large mass of glandular swellings extends from the ear half way down to the clavicle ; under the chin are two or three distinct large lymphatic glands. and about the same number are seen on the right side of the neck ; none of these have the integuments red or discoloured; appetite very bad; bowels free, but not purged; has profuse night sweats, requiring him to change his night-dress three or four times every night: he is much emaciated, and is very pale; pulse 120: he says that every day he thinks some fresh glands become enlarged, and that his strength is daily and rapidly declining, and is now at a very low ebb. Descendat in baln. tepid. R. Pil. Hyd. gr. iii. Extr. Conii gr. ii. fiat pilula, mane nocteq. sumend. Bibat Decoct. Sarsæ c. Extr. Fluid. Sarsæ ad semilibram in die.

Nov. 27th. Pains less: glandular swellings beginning to decline; ulceration about dens sapientiæ exhibits less of venereal character; night sweats appear to have ceased for the two last nights; appetite is now good: he reports himself as being much better in point of strength and general feeling. Repet. Pil. et Decoct. Sarsæ.

Dec. 1st. Sumat Pil. i. sing. noct. Repet. Decoct. Sarsæ.

Dec. 4th. Mouth affected; ulcers of tonsils better, and much less soreness of throat; glandular swellings still more reduced; strength improved; sleeps well; appetite good. Omit. Pilulæ.

Dec. 17th. Mouth well; superficial ulcer of right tonsil improved; ulcer of left tonsil healed; gum still ulcerated, but not at all painful. Sumat Pil. i. mane et nocte. Repet. Decoct. Sarsæ.

Dec. 29th. Gums are swollen; ulcer of the molar gum much better; pulse 100, after he had walked a quarter of a mile.

Jan. 10th, 1831. Mouth slightly affected; pulse 84. Sumat Pil. i. merid. nocteq.

15th. Mouth slightly affected; has been long free from the pains of shoulders and sternum; eruption has long since disappeared; ulcers of tonsils healed, but the anterior arches appear rather too red and slightly swollen; glands of neck reducing more rapidly since he began to take two pills per diem. Repet. Pil. i. bis in die et Decoct. Sarsæ.

Jan. 24th. Has taken three pills daily for the last five days; mouth slightly affected: he is obliged to go to England. Although all his venereal symptoms have disappeared, and the glandular swellings have been entirely removed, and his flesh and strength are now almost completely restored, still I think he has not used mercury enough to secure him against the danger of a relapse.

In the month of February, Mr. G. consulted Sir B. Brodie on account of a sore throat. Sir B. told him that his disease was not venereal, but the consequence of venereal, and ordered a pill of Hyd. Oxyn. gr. ss., to be taken at mid-day with Ext. Sarsæ. He used the medicine rather carelessly. He returned to Dublin in October following, with a decided venereal ulceration of right anterior arch of palate: but his general health was then pretty good, and there was not any return of the swellings of the lymphatic glands of the neck. By the use of Hyd. Calcinatus, combined with opium, and at the same time by the use of Decoct. Sarsæ, he got quite well in the course of six weeks.

Remarks.

I despair of being able accurately to depict the almost deplorable condition of this young man when he came under my care. His venereal symptoms were not, any of them, of a severe character; yet his general health appeared to be completely broken down. When we find him at once affected with loss of appetite, loss of sleep, quick pulse, profuse night sweats, wasting of flesh, loss of strength, enlargement of the cervical lymphatic glands, and every day adding to their number and size, a sure indication that the health was daily and hourly sinking, I think we here have a combination of symptoms calculated to excite our most serious fears for the patient's safety. In fact, so rapid and headlong was the course of his symptoms, that I altogether despaired of being able to check their progress by any restorative treatment, and therefore I determined to rely on small doses of mercury.

I admit that it may be questioned whether as to the sarsaparilla, the cicuta, and the mercury, which was the most useful remedy; or what share each of these had in arresting the progress of these alarming symptoms. But this much at least must be conceded, that in this case the mercury did not bar his recovery, nor did it produce any of those mischiefs which, from recent publications, we might have been led to anticipate. This case I have brought forward to show that mercury may be used (if used in very small doses) while the system is as it were rapidly sinking under the pressure of hectic fever. In fact, we can hardly conceive any man, who is yet able to walk two or three hundred yards, to be more reduced in flesh, colour, and strength. Nor can we imagine a more sharp degree of hectic fever; nor can I recal to mind any instance in which the glandular swellings or the scrophulous enlargement of the lymphatic glands had increased more rapidly. Let this case now stand as an instance of a very reduced state of health, and a very severe degree of hectic fever, not forbidding the cautious use of mercury. As, however, so many other remedies had been used along with the mercury, it is impossible to estimate what share the mercury had in effecting this very salutary change. I therefore feel myself called on to produce the following case as an instance of the safety with which mercury may be used in a very low state of health, and how it will, in such a condition as would almost forbid its use, effect the cure of both primary and secondary symptoms.

John Yeates, a labourer, æt. ann. 30, admitted into No. 11 ward, Dec. 17th, 1835.

On the external prepuce is a painful ulcer; the surrounding integuments are of a dark red colour: there is also a purulent discharge from under the prepuce, which is much swollen; An ulcerated bubo in each groin; through that on the left a lymphatic gland, larger than an almond, rises and projects considerably above the level of the skin; both tonsils are enlarged and of a coppery red colour; on the surface of the right tonsil is a small ulcer: he suffers much pain in swallowing; complains also of pains in the shins, and along the ulna of each arm; the surface of the body is covered with a papular eruption of a large size; appetite is very indifferent; he sleeps badly; pulse 100; says that six weeks ago he first perceived a small ulcer on the glans penis, and a pustule on the prepuce, which inflamed and then ulcerated; very shortly afterwards a bubo appeared in each groin; about ten days ago the gland in the left groin became exposed, and began to protrude; three days prior to his admission he perceived the eruption; thinks he felt soreness in the throat in a

week after he perceived the chancre : he has not taken any
medicine.

From the day of his admission until Dec. 26th he was
directed to take repeated purgatives, generally a combination
of calomel and jalap ; but his bowels proved obstinate, and
these medicines have produced much less of purgative effect
than was wished for.

Dec. 26th. The symptoms being as at the time of his
admission, I directed for him Ungt. Hyd. Fort. gr. x. o. n.

Dec. 29th. Mouth sore, with some ptyalism ; both the
buboes are much improved, presenting clean granulating
surfaces ; the protruded gland, I think, is a little retracted ;
a slight mercurial erythema at the top of his thighs. Mist.
Cinchonæ c. Tr. Opii. Omitt. Ungt.

Jan. 2nd, 1836. Erythema is now gone ; chancre and
buboes looking healthy ; the protruded gland is obviously,
though slowly, retiring ; throat is much better ; eruption very
much faded. Repet. Mist. Cinch. c. Tr. Opii.

Jan. 4th. Repet. Mist. Ungt. Hyd. Fort. gr. x. alt. noct.

Jan. 9th. Pergat.

Jan. 11th. Ungt. Hyd. Fort. gr. x. om. nocte. Repet.
Mist. Cinch. Both tonsils are reduced to their natural size,
but there still remain a few small superficial ulcers on their
surface ; chancre is very nearly healed ; buboes are healing
rapidly ; gland much more retired ; eruption gone, stains of
it are scarcely visible ; prepuce still thickened, and its orifice
much contracted, giving exit to a purulent discharge, for
which I have directed the Lotio Nigra to be injected under the
prepuce.

Jan. 15th. He complains most bitterly of a scalding at the
anus ; the buttocks on each side, where opposed to one another,
are covered with a thick purulent discharge ; on wiping this
away the surface on each side is seen excoriated to twice the
extent of a crown piece ; no actual ulceration ; the ulcers in
both tonsils appear more deep than they were some days ago,
and even seem disposed to open ; the chancre and both buboes
are improving, but now rather more slowly ; mouth is sore.

From the 15th to the 30th of January he had P. Mist.
Cinchonæ c. Opio et Lotio Nigra for the excoriation.

Jan. 19th. Pulse 108, as he stands out of bed; tonsils look better; buboes and chancre have improved rapidly since 16th inst.; he appears much less feverish; excoriation is much less painful, although not yet healed in any one part.

Jan. 26th. Mouth nearly well; throat as before; chancre and buboes perfectly healed; excoriation healed in the more superficial parts; pulse 80 as he lies in bed. Repet. Mist. Cinch. et Lotio Nig.

Jan. 30th. Pulse 96 as he stands out of bed; anus quite well; he feels the throat more sore. Ungt. Hyd. Fort. gr. x. omni nocte.

Feb. 2nd. Repet. Ungt. Sumat Pil. Hyd. gr. v. om. nocte.

Feb. 6th. Repet. Ungt. om. nocte et Pil. Hyd. gr. v. bis in die.

Feb. 9th. Ungt. Ꝺ i. om. nocte et Pil. Hyd. gr. v. ter in die. Mouth is not yet sufficiently sore; ulcers of tonsils are better, and he says he feels his throat better.

Feb. 23rd. Discharged cured; not having used any mercury since the 10th, on which day his mouth was quite sore enough.

Remarks.

In my notes of this case, taken on the 26th December, I find this remark: "I think there is not another surgeon in Dublin who would dare to give mercury to this man in the present circumstances of his case." From which may be inferred how very unpromising was his state at that time. Let us now advert more particularly to his history.

We have here combined, in the same individual, undoubted venereal symptoms (if any symptoms deserve that epithet) of the primary and secondary class; and these too without having suffered any modification from mercury, or indeed from any other plan of treatment. Yet we find that these clear and decided, or unadulterated venereal symptoms of each class, yielded to mercurial treatment, commenced while the system was in a febrile state, and so early as on the twelfth day · ter the eruption began to show itself. So that here the practice was in direct opposition to two rules which I believe

are pretty generally acted on by the profession, *viz.*, 1st, not
to give mercury while there is much febrile excitement;
2ndly, not to give mercury while an eruption is coming out,
and while the eruptive fever continues.   I am perfectly certain
that had I used mercury in the ordinary doses of ʒss. or ʒi.
of Ungt. daily, the result would have been very unfortunate
indeed.   But by apportioning the doses to the state of the
constitution the result has been most satisfactory, and the
patient has been released from his sufferings, and has been
cured of his disease in a period of time shorter by one month
or six weeks than he would have been had he been subjected
to the ordinary routine of practice, and had the mercury been
withheld until the fever had subsided and his general health
to a certain degree been re-established.   Another very remarkable feature in this case is the very short period of time and
the very small quantity of mercury employed in exciting
ptyalism.   Ten grains of mercurial ointment, used for three
successive nights, brought the system as fully and satisfactorily
under the influence of mercury as could be desired.   And the
state of the symptoms, as reported on the 11th of January,
viz. in sixteen days from his commencing the use of mercury,
shows that the effect of this mercurial treatment was as
favourable as is seen in ordinary cases, where mercury is given
in the usual doses to a patient free from fever, and in circumstances deemed most fitted for the use of this medicine.   This
case presents a very striking fact, one for which I candidly
admit that I am not able to account.   We find that ten grains
of ointment, used for four nights, from the 11th to the 15th of
January, had produced that distressing effect, *viz.* a mercurial
excoriation about the anus.   This was only ten grains more
than what had acted so favourably at the beginning of the
course.   But although I do not pretend to account for this
effect of the mercury, I may be allowed to direct attention to
it ; because I look upon this as a most unerring proof that the
mercury is at this time too powerful in its action for the
individual.   Yet surely it will be granted that the doses were
not excessive, nor were they employed for any unreasonable
length of time.   Still as an indication that mercury is now
disagreeing with the system, I look upon this symptom as a

very valuable addition to our rules of practice. Indeed I am of opinion that we cannot too highly prize any effect or test which will point out to us that mercury is disagreeing with the system. It is curious that this excoriation did not appear with the first ptyalism, but at the approach of the second. It is also remarkable that this unkindly action of mercury had more injuriously affected the secondary than the primary symptoms. Had I been using ʒi. or ʒss. of Ungt. Hydr. Fort. instead of ten grains for a dose, then I might be told that the dose was too large. A short time ago I could not have expected such severe effects from so small a dose of mercury.

To those who are anxious to use mercury in the smallest possible quantity, this case must be very gratifying, as the whole quantity employed amounted only to four drachms, and ten grains of ointment, with 110 grains of blue pill.

The venereal nature of this case will not, I believe, be disputed. The state of health at the time that mercury was first employed is described in the report in a very few words; and though the description by no means corresponds with that usually given of hectic fever, yet that he was highly feverish may be inferred from my private remark at that time, viz., " I think there is not another surgeon in Dublin who would dare to give mercury to this man in the present circumstances of his case."

I shall now adduce a case in which the symptoms were considered by me, and by all the hospital attendants, to be secondary venereal symptoms, and which yet did not admit of a verbal description that would satisfy one, who had not an opportunity of personally examining the case, as to their venereal nature. The symptoms were a return of the same ulcers for which he had been in the hospital for two months previous to February last.—Markham, No. 12 ward.

In the history of his case he is described as having an extensive ulcer covering the upper lip and the tip of the nose, which has destroyed part of the septum and part of the right ala nasi; and has also a large ulceration on the fore part of the left thigh, a little above the knee, and a small ulcer on the back of the leg. He had ulcers in all these places, of which he was apparently cured when he left the hospital in February

last, having undergone a full course of mercury. On his
return home he was much exposed to cold, and drank spirits
to very great excess. In two months after leaving the hospital
the ulcer on the nose began to re-appear, and in four months
more that on the thigh also broke out.

From the time of his admission (the 17th) until the 28th of
November he was directed to take the mineral acids. On the
28th he was ordered Ungt. Hydr. Fort. Ɔi. o. n.

On December 1st, ℞ Tinct. Acct. Ferri, ʒi. Hydr. Cor. Sub.
gr. i. M. Sumat Cochl. Minim. bis in die, e Decocti Hordei
Cyatho.

December 5th. His mouth is now very sore; a smart
ptyalism is established. Omittatur Solutio. Curentur Ulcera
Ungt. Hydr. Nitr. ʒi. Dilut c. Axungiä ʒi.

On the 28th of November I made this note:—" This man
was under my care, and was salivated when last in the
hospital. He is now such a wretched-looking creature that I
have been afraid to use mercury with him since his admission;
and I have but little encouragement to do so, as it already
failed to cure him. However, as he must die unless something
decisive be done, I shall venture on the use of mercury; for
die he must under the ordinary treatment, without mercury.
No improvement whatever has taken place since his admission."

On the 13th of December I made the following note:—" It
is remarkable and most extraordinary how very readily
ptyalism has been induced; and it is equally remarkable how
very quickly and materially the ulcer of the lips and nose,
and the swelling of these parts, have improved since the
mercury came into action. The mouth is still smartly
affected, and the checks swollen. The amendment during the
last eight or ten days has been greater than could have been
anticipated."

To finish the history of this case, I shall only add that so
speedy and complete was the healing of the ulcers of the face,
and so decided the improvement of the ulcers of the limbs,
that I did not feel myself justified in ordering any mercury
after that of the 1st instant; we of course ceased at the end
of four days. The ulcers were retarded in healing partly by
the great extent, and partly by being seated on the cicatrix of

former ulcers. He left the hospital on the 27th of February cured.

I cannot readily forget the surprise depicted in the countenance of the students, when they first heard me direct the pupil in charge of my prescription book to write a scruple of ointment every night; indeed it was such as forced me to declare "that if mercury did not cure him, I could not tell what would; and most assuredly, if this man could bear mercury, there was no man who could not." I need not add that the progress of the case (at least as long as under the influence of the mercurial action) was most anxiously watched by all the pupils.

It may not be amiss to state that during (and probably owing to) the mercurial action, a lymphatic gland under the chin inflamed, suppurated, and ulcerated; and it healed spontaneously after the action of the mercury had subsided.

In this case one very remarkable fact should not be passed over. I allude to the cure having been accomplished by such a very minute quantity of mercury; for the entire quantity used amounted only to three scruples of the ointment and one grain of corrosive sublimate; and to use this small quantity seven days were employed.

Some perhaps will lay hold of this one fact as an argument to show that the disease could not have been venereal; because here we have a cure accomplished by a quantity of mercury so infinitely less than what is universally considered as sufficient for the cure of any one symptom in any stage of the venereal disease. But although such objection may be urged with effect, if applied to the cure of a strongly marked and decided, or unsubdued symptom of syphilis, in an unbroken constitution; yet it will not hold good when applied to a case of a feverish, debilitated patient, affected with symptoms which had been subdued and weakened by previous mercurial treatment. In support of this assertion, let me bring forward the following case :—

Mr. O. R., æt. ann. 28, contracted a venereal disease four years ago, and has been for a long time under the care of Mr. ? le for the cure of secondary venereal ulcers; three or four of these, on the face, are most strikingly marked as venereal;

a very extensive and deep ulcer of the calf of the leg is less
strongly marked as syphilitic, but it commenced at the same
time with the ulcers of the face. This ulcer had committed
great havoc; it destroyed no small portion of the inter-
muscular cellular tissue, and had induced a sloughy state of
what remained between the muscles of the calf; so that, even
if the ulcers were perfectly healed, it was obvious that he
must always remain lame of this leg. I had already seen him
in consultation three or four times, at intervals of three or
four weeks. On each of these occasions one principal subject
of our deliberations was whether we should resort to amputation. However, in consultation on the 17th of September,
1833, it was agreed that he should make trial of very minute
doses of mercury: accordingly he was directed to take Pil.
Hydr. gr. v. omn. nocte. Of these he had taken two only in
as many nights, when his mouth became very sore, and a very
small ptyalism ensued. All the ulcers, both of the face and
leg, healed during the salivation; and he has since enjoyed
good health, and is able to go about, wearing a high heel to
his shoe. During the long period that this disease continued
mercury had been tried more than once, and again abandoned,
not having produced any decided benefit.

In this case the ulceration of the leg would have been early
healed, if that object could have been accomplished by applications of different kinds most judiciously chosen, and these
applied with a degree of skill and dexterity not to be surpassed,
and with a zeal and attention which could scarcely be expected
from a very attached friend. Yet the dexterity in applying,
and the judgment in choosing various local applications,
availed but little; nor was the use of mercury more serviceable, so long as the mercury failed to excite salivation; but
the moment that the mercury chanced to be apportioned to
the state of the system, and that it produced salivation, from
that instant all the symptoms changed for the better, and the
deep and extensive ulcers of the leg, and the disfiguring ulcers
of the face, hastened to heal. Here it is remarkable that so
very small a quantity of mercury could have made such a
decided impression on the general system; but it is quite
clear that, in this case at least, this medicine acted not by its

quantity, but by its having luckily taken a right direction, and made what I term a salutary impression on the salivary system.

In each of the consultations our chief objection to the proposed amputation was the apprehension that the stump would be seized with the same kind of bad ulceration. I really thought that the leg was scarcely worth preserving, and that certainly we ought not to endanger the patient's life in the slightest degree from an anxiety to preserve such an useless limb.

I know many will be ready to exclaim, all this does not prove that the ulcers were venereal, for we know that mercury has cured many ulcerated legs where no suspicion of a venereal taint could be entertained. To this I can only answer, that the ulcers on the face presented to me what I considered very strong characters of venereal ulcers, and that they could be traced up to a venereal origin. Waving all arguments on the subject of these ulcers being venereal or not, what I contend for is this—that where we can trace ulcers, or any symptom, to a venereal origin, we may be allowed to suspect that they still partake of the nature of their exciting cause; and that under such circumstances we are justified in resorting to mercury. Had we not in this case been able to trace some connection with venereal infection, I am certain that we should not have dreamed of using mercury for the leg alone, so very different was the ulceration of this from any thing of secondary venereal ulcers. But the ulcers on the face had enough of venereal characters to call our attention, and to rouse our suspicions, as to their real nature.

So very sudden was the improvement that I did not afterwards see the patient until long after all the ulcers were healed, and then he was going about in perfect health, but obliged to use a high heel to his shoe.

A few more such instances would go far to overturn Mr. Hunter's position—" The quantity of mercury to be thrown into the constitution for the cure of any venereal complaint must be proportioned to the violence of the disease."—Page i 3, London edition, 8vo, 1810.

CHAPTER X.

MERCURY DURING VENEREAL ERUPTIONS.

I BELIEVE it is a pretty prevalent opinion at the present day, that when a venereal eruption is coming out we should withhold mercury until the eruption be completed, lest we interfere with or interrupt that process. Another reason for withholding this medicine in such cases is, that the fever which accompanies the eruption is considered as unfriendly to the anti-venereal action of mercury. Now I am convinced, by repeated observations, that this rule not only may be departed from, but that it cannot be followed with advantage to the patient. My observations have been both where the eruptive fever was very slight, and also in cases where the fever was running pretty high preceding the appearance of the first eruption; but I imagine that that line of practice which is found safe and useful in the fever attending the first eruption (which is always of a more severe kind) cannot be dangerous in those slighter forms of fever which occasionally usher in and attend on the second or subsequent crops of eruption.

Some years ago I first adopted this plan of treatment in the case of Mr. P., a young man who had been under my care for four weeks, for cure of a chancre and bubo. From my anxiety to get him quickly cured, I used mercury in doses too great for his system, and accordingly I failed to excite ptyalism, or to cure the local complaints. At the expiration of three weeks the chancre was not quite healed, the bubo was red, painful, and enlarging. Sarsaparilla was now substituted for the mercury. When he had been nearly three weeks on this plan, a pustular eruption, accompanied by an eruptive fever of a hectic type, with profuse sweats, began to appear—some spots of it of a very large size came upon his face. This made him very readily comply with my desire, that he should use small quantities of mercury, notwithstanding his last unfavourable experience of it. I directed Pil. Hyd. gr. iii. to be taken bis in die. I was delighted to see the eruption yield very quickly, the large pustular spots on the face dry and contract, and the crust fall off,—no new spots appeared. And not only did the fever subside, but at the same time the original primary symptoms, which had heretofore proved so obstinate, now got well speedily and permanently. Now I make no doubt that this young man was cured of all his venereal complaints fully six weeks sooner by the course I adopted, than he would have been had I deferred the use of mercury until the eruption had fully developed itself, and had left the system free from fever. Besides, he was saved from some unseemly and permanent cicatrices of the face, which the ulceration of this eruption so constantly leaves after it.

The same plan of treatment I adopted in the following case of a venereal eruption, one very different in form from the foregoing.

July 2nd. Mr. T. informed me that on 7th May he had commenced taking Calomel, gr. iii. bis in die, for cure of a chancre, and continued the use of the medicine until about eleven days ago; during the entire of this time he was travelling through the country in such a manner as did not allow him to guard against cold or wet. He thinks his mouth was made slightly sore in a week after he had begun the use of mercury; and he endeavoured, but in vain, to keep up a slight soreness of it.

At present he is free from mercurial action, and complains much of sore throat; both sides of it are of a colour too highly red, and are swollen. The right tonsil is ulcerated, and a white slough occupies the entire ulcer.

Six days ago he first perceived on his forehead an eruption of small copper-coloured spots, scarcely raised above the surface; these are of the sandy or measles-like eruption. The prepuce, which with him naturally covers the glans, now cannot be brought over it, in consequence of a dark copper-coloured bump, as large as a filbert, which is seated behind the frænum, and occupies the site of the original chancre. He sleeps very badly, although he has not pains of limbs, and he sw ts towards morning. Pulse 108. I directed some James's powder, with nitrate of kali.

July 4th. Throat feels worse, and as if much swelled internally; much pain in swallowing; fauces much inflamed and thickened, especially the uvula; ulcer of right tonsil is more decidedly venereal; snuffles much in speaking. ℞ Pil. Hyd. gr. ii. Pulv. Ant. gr. ii. Rhei gr. ii. Ft. Pil. ii. mane et nocte sumendæ.

July 7th. Symptoms not improved, rather worse; has had gout in one foot and hand since the 5th inst., but not so severe as to confine him to bed. Pil. Hyd. gr. iii. Pulv. Antim. gr. ii. bis in die.

July 10th. Eruption fading, and beginning to desquamate; uvula and velum are less swollen, and have less of inflammatory redness; voice much improved; has a bad taste in his mouth, but it is not brassy; it makes him dislike tea, of which he had been very fond; gums little, if at all, swollen; gout better; pulse 84. Repet. Pilulæ ut 7th inst.

July 14th. Eruption and condition of throat stationary; mouth less affected with brassy taste; no ulceration of gums, but the cheeks exhibit the impression of the posterior molares; gout has seized on right ancle. Sumat. Pil. Hyd. gr. v. ter in die.

July 23rd. Gums not ulcerated; throat less swollen, and ulceration better; eruption less bright. ℞ Calomel Pulv. Antim. Conf. Opiat. a. a. Ɔi. Pil. xii. Sumat unam ter in die.

July 28th. Eruption much faded; skin of forehead much less red; throat less inflamed; voice perfectly re-established; mouth not more affected; appetite good; skin temperate; pulse 84; sleeps for three hours, and awakes in perspiration. Calomel Conf. Opiat. a. a. Ɔi. Fort. Pil. octo. Sumat unam ter in die.

Aug. 3rd. Eruption has quite cleared away from the forehead, leaving only the discoloured pittings; ulcer of throat healed; scarcely any inflammatory redness or thickening; countenance much improved; brassy taste of mouth; gums not ulcerated, but cheeks of a leaden colour, and impressed by posterior molares; has been purged, with some pain; pulse 78; no night sweats. Repet. Pil. ut 28th.

Aug. 9th. Mouth more affected; bowels moved once a day; symptoms have disappeared. Pergat.

Aug. 19th. Gums ulcerated; mouth sufficiently affected; stomach rather weak; no night sweats; lump on prepuce has left only some thickening; for the last four days the skin has come forward and covered the glans. Sumat Pil. unam bis in die. Mist. c. Sulph. Quinæ et Tinct. Cinch. Comp. bis in die.

Aug. 24th. Mouth is very fully and satisfactorily affected; ulceration of gums of dentes sapientiæ, and of the edge of the tongue, corresponding with them; strong mercurial fœtor: spots on the forehead have all left discoloured pits; prepuce lies forward; when retracted, no hardness to be felt; when pushed forward, some small hardness is either felt or suspected. Omitt. Pil. Haust. Purg. et b. tepid.

Dec. 24. In excellent health; a globular hard spot, size of head of a large pin, in site of old lump, is the only vestige that remains of the disease.

Remarks.

In this case I was induced by the importunity of the patient to give mercury in the eruptive fever of secondary symptoms, even while only a very limited quantity of eruption had appeared; for he was obliged to be much abroad, and, of course, was most anxious to get rid of the eruption on his forehead as speedily as possible. I began with very *small doses* in combination with Pulv. Antim., a supposed febrifuge medicine. In addition, we observe that the gout attacked him on the second or third day of using mercury. Yet, with all these disadvantages, on August 7th, when he had only taken twenty grains of Pil. Hyd., there was not only a check given to the symptoms, but even a marked improvement.

Now had I followed the established rule of withholding mercury until the eruptive fever had ceased, what a delay to his recovery, what an injury to his pursuits would he have had to suffer, and possibly, in addition to this, the ulcer of the throat might have created such an aggravation of fever, that it, combined with very inadequate nourishment, would have made it take an unfavourable turn, and begin to slough. T' .s his voice might have been irretrievably injured. Or it is possible that, during the exasperated stage of fever, some spots of the eruption on his face might have ulcerated or have degenerated into rupia, and thus leave him more or less disagreeably marked. All these risks have been avoided, and the entire of the disease has been removed in a period of time little more than would have sufficed to get rid of the eruptive fever only. So that, by following the established practice, the patient would have barely begun the mercurial treatment; for at the time that he would be laying it aside, if treated according to the plan pursued in this case. The duration of the mercurial course might have been a week or ten days shorter, had not the attack of gout prevented me from pushing on with the mercury. In this case, on 23rd July, I ventured to give very full doses of mercury, nearly two grains of calomel three times a day. I was induced to do so because I found that his mouth had not been sufficiently affected by the Pil. Hyd., and I judged it perfectly safe to do so because he had been for many days perfectly free from any febrile disturbance. I think it a satisfactory circumstance that in this case we had other symptoms than eruptions (for we know that these will spontaneously improve after they have been out for a certain time), particularly the hardened purple lump on the prepuce. We observed this to yield in proportion as the action of the mercury on the system was more fully established. The early exhibition of mercury, as here recommended, will therefore be found to possess the following advantages :—

1st. The patient is freed from the venereal disease in a much shorter space of time.

2nd. The constitution is relieved from all the injury which it must sustain by a continuance of the eruptive fever.

3rd. The local symptoms thus arrested at their very onset are prevented from committing any devastation or destruction of important parts, or from leaving after them unsightly cicatrices.

4th. The cure of the venereal disease is effected by a quantity of mercury infinitely smaller than what is ordinarily required for the removal of similar symptoms when allowed to develop themselves fully. This, by some practitioners, will be considered a great advantage.

Mr. D. applied to me with an ulcer on the glans penis,

which I judged to be a chancre, and to be applied to which I directed the Lotio Plumbi Acetatis for a few days. Being confined to my bed by an attack of gout, I did not see him for twenty days; I then found the chancre with so very healthy an appearance that I almost suspected I had been mistaken in supposing it to be a venereal ulcer. I advised him to continue the same lotion. In about three weeks more he called on me; he had now an eruption on his forehead and body of a pustular character; his appetite good; pulse very quick; profuse night sweats. Notwithstanding this febrile state, I advised, May 22nd, Pil. Hyd. gr. iii. et Pulv. Antim. gr. ii. o. n.

May 26th. No visible effect from the pills; the eruption on the forehead is scabbing, and has assumed a darker colour. Sumat Pil. Hyd. gr. x. o. n.

May 30th. Eruption fading; bad taste in the mouth; gums are not sore. Pergat.

June 2nd. Eruption is everywhere fading; some fœtor of breath; mouth not sore; night sweats continue. Sumat Pil. Hyd. gr. x. omne nocte, et grana quinq. mane.

June 13th. Mouth not sore; eruption fading. Sumat Pil. Hyd. gr. x. m. et n.

June 20th. Mouth not affected. ℞ Calomelanos gr. ii. Opii. gr. ss. Ft. Pil. mane noctequ. sumenda.

June 24th. Gums swelled and ulcerated; some ptyalism. Omitt. Pil. per dies x. dein repetantur.

July 3rd. Has been smartly salivated, although he did not take any pills since 24th inst.; eruption completely removed.

The very delicate state of his health during the first month prevented me from increasing the doses of the mercury by larger additions than those above mentioned; but, finding his strength and health so much improved at the end of the month, I then ventured to increase the doses by very large additions, and certainly with the happiest effect. Mr. D. is a young man naturally of a very delicate habit; his strength and flesh were reduced very much at the time I commenced the use of mercury, under all the disadvantages of very accelerated pulse and profuse night sweats.

The changes which the chancre underwent before healing,

and the facility with which it healed under the use of so simple an application as Lotio Acet. Plumbi, should teach us great diffidence in deciding on the nature of primary ulcers.

I think I can scarcely too often repeat what I consider to be the grand rule for the treatment of these apparently deplorable cases, that is, to apportion the doses of the mercury to that weakly state of the system which is thus marked by hectic fever and emaciation. If we neglect this means we shall most certainly fail in our attempts at cure; and I may even add, that we shall very generally render his condition worse by any other plan of treatment. A case which I have lately witnessed renders it extremely probable that the most serious consequences may be quickly induced by a material deviation from this plan. The facts are briefly these:—A young man, by name Mooney, of apparently sound and robust constitution, was received into the hospital, No. 11 ward; he was labouring under a smart degree of eruptive fever, and a crop of small mercurial pimples had made its appearance on the face, and was coming out on the trunk. I thought it a favourable case for trying the plan of employing mercury during the eruptive fever, and while the eruption was still coming out. Accordingly I directed that he should rub Ungt. Hyd. Fort. gr. x. omni nocte; but by some mistake the clinical clerk wrote down Ungt. Hyd. ʒi. omni nocte. I was astonished on seeing the patient again in two days; he was in a most alarming state of fever, not purely inflammatory, although accompanied with very high vascular action, and with very high degree of excitement and great restlessness, and his face full and flushed. This alarming condition made me search most anxiously to ascertain the cause; and on enquiry I discovered the mistake. I need not say that the mercury was laid aside, and the high degree of fever opposed by suitable treatment. As soon as the fever was sufficiently moderated, I made him resume the mercury in doses of Ungt. Hyd. gr. x o. n. By this his mouth became soon, though moderately, affected, and the eruption was removed; so that he was dismissed from the hospital fully one month sooner than he would have been had he been treated according to the established rule of endeavouring to subdue the fever, and allowing the eruption to come out very fully before we ventured upon the mercurial treatment. It is quite unnecessary for me to point out how much more expeditiously this disease may be treated on this plan than it could have been by subjecting the patient either to the non-mercurial treatment or by insisting on a long delay prior to the commencement of the use of mercury.

As this practice is so much at variance with that generally followed, I feel it necessary to produce further evidence of its safety and advantages: and in doing so I shall select two cases of eruption which occurred at the same time, and which were treated on the two different plans. By comparing these together, the superiority of that which I have ventured to recommend will be most clearly established.

On the 17th of September I was called upon to consult on the case of Mr. K., a young man, who had in his boyhood suffered severely from scrophula, which had left a permanent contraction of the right elbow, and numerous cicatrices on his neck. His present complaints are a pustular venereal eruption, pretty numerous on the trunk, with a few on the face; four or five of those on the forehead are disposed to form scabs, and to ulcerate; one on the chin is larger, and incrusted over by the concretion of the discharge among the beard. Pains of the limbs, which had for some days been troublesome, have ceased since the eruption appeared. He complains of pain in the contracted elbow, but this is not attended with swelling, heat, or redness; pulse 72; countenance of a remarkably pale sickly hue.

I saw Mr. K. again on October the 16th, when I found the spots of eruption on the face larger and more disposed to ulcerate; one especially, on the right eyebrow, seemed as if it would not only ulcerate, but spread rapidly; his countenance is still pale and sickly; no night-sweats; the pain of the elbow has almost entirely ceased.

Since my first visit he has been taking Decoct. Sarsæ c. Acid. Nitr. Dil.

His attending surgeon this day agreed to give him Pil. Hydr. gr. iij. et Extr. Conii gr. ij. bis in die.

I have not seen Mr. K., but I have learned that he

bore the mercury very well, and was cured in the ordinary time.

On the 22nd of September Mr. T. called on me, and informed me that for three weeks past he had been troubled with pains of the shoulders, which very much disturbed his sleep; but that within the last week, on the appearance of a rash, these pains had been materially relieved. At present the trunk of the body is thickly covered with a very minute eruption, not unlike small red sand. Some patches of the same eruption, collected in clusters, appear on his face; his countenance is very pale; appetite pretty good; pulse 112.

℞ Pil. Hydr. gr. iij. Extr. Cicutæ gr. ij. Ft. Pil. mane noctcq. sumenda.

September 26th. Eruption out in greater quantity.

October 1st. Colour of eruption less vivid; spots on the face are inclined to desquamate; appetite pretty good; complains of weakness of stomach, which he thinks arises from a nauseous taste in his mouth; no thirst, pulse 108, less languor, sleeps well: no night-sweats; no sensible mercurial effect on his mouth. Repet. Pil. Sumat Sulph. Quinæ gr. ij. ter in die.

October 4th. Eruption on face still more faded in colour; that on the trunk less so; appetite good, sleeps well, and has less languor in the evening. He has now taken twenty-four pills; mouth not sensibly affected; has complained for a few days past of soreness of the right side of the throat, with pain passing into the ear; right palatine arches a little swollen, and of too high a red colour. Repet. Pil. et Affricetur ʒss. Ungt. Hydr. Fort. sing. noct.

October 16th. Eruption has totally disappeared from the face; that on the trunk is of a very faded tint; inner gums of the lower incisors are ulcerated along the edge; no pain of the teeth, or pain from chewing, but painful aching from particles of food resting on the gums; appetite pretty good, sleeps well; pulse 108, when the finger is first applied to the wrist, and 96 when carefully counted for some time. He now says that the pain of his throat has moved from the right to the left side, and to the left ear. On inspection the fauces appear rather swollen, and of a colour too highly bright. No ulcer to be discovered. The cheeks opposite to the last molars are much swollen, and of a leaden colour. Pergat.

October 20th. Eruption completely removed; the gums are universally affected by ulceration of their edges, but without much swelling; he still complains of the soreness of his throat; the swelling and redness of the fauces continue; the edge of the anterior left arch is covered with an exudation of coagulated lymph, and the left tonsil is covered with the same. He feels himself very heavy; pulse 108. Omit. Pilulæ et Ungt. Hydr. Repet. Mist. c. Sulph. Quinæ.

October 23rd. He feels his throat much better, and it is much less swollen and less red; pulse 108. Ungt. Hydr. Fort ʒss. omni nocte. Repet. Mist. c. Sulph. Quinæ.

November 5th. Mercurial action has subsided, although he continued to use the ointment; and for the last four days has also taken Pil. Hydr. gr. v. bis in die. I have now directed for him two grains of Calomel, with three grains of Compound Powder of Ipecacuanha, ter in die. ʒss. Ungt. Hydr. Fort. omni nocte.

November 10th. He has suffered severely the two last days from mercurial dysentery; the gums of both rows of incisors are swollen, and superficially ulcerated; the fauces of a high red colour, not unlike that attendant on scarlatina, with exudation of lymph on both tonsils. The under surface of the tongue is marked on each side by a narrow line of leaden-coloured slough, and a similar spot on the inside of the lower lip; pulse 108, skin hot, appetite very bad. He is continually spitting up fœtid thick saliva, which he says comes from his throat; pulse 108.

The account of this case I shall conclude by saying that the mercurial action was kept up for one week longer, and then the mercury was altogether discontinued.

I have coupled together the history of these two cases, which happened to come under my observation at the same time; and I think it may be useful to compare the progress of the one with the other. This comparison was instituted on the 16th of October; it will be seen that at that time Mr. K. had been for one month affected with the eruption, and in that period no advance towards its cure had been made; all this time was spent in improving the state of his general health, and as it were preparing him for the use of the remedy (mercury) which was to cure this eruption. His attending surgeon was averse to the early use of mercury in this case, although the constitution was at the time so little disturbed; for his pulse was only 72. I must observe that the pain of his elbow was merely owing to an affection of the skin; nor did it appear to his own surgeon, or to me, as connected with the joint. A case in which so little of febrile commotion existed appeared to me as particularly favourable for the use of mercury in the early stage of the eruption. Nor is the loss of time the only disadvantage attending this postponement of the use of mercury; for we find that one spot on his eyebrow threatened to ulcerate and to spread rapidly; now had this taken that course, it must have destroyed the growth of the hair, and left a most unseemly bald spot with a silvery ground.

Mr. T., on the contrary, might be considered, on the 16th of October, as more than half cured, and as perfectly free from any risk of being disfigured by any part of the eruption going on to ulceration; for we find that although he had not commenced the use of mercury until the 22nd of September, yet now the eruption had totally disappeared from his face, and that everywhere it was very much faded. His general health was fully as good as when he entered upon the treatment, although the mercury has now full possession of his system, the pulse being then even somewhat slower.

Nor are we to confine our views to the mere loss of time accruing from the later exhibition of mercury; we should also take into account the ravages that may be committed by the disease during the time we withhold this remedy. Thus in cases of an eruption which has a strong tendency to run into ulceration, such of the spots as form on the face, or on other exposed parts of the body, may produce ulcers which are not only painful and distressing at the time, but which will leave after them cicatrices the most unseemly, on account of the extraordinary silvery whiteness which they assume, and which remains through life. Besides, these cicatrices may form on the eyelids, lips, or nose, in such situations as to render a contraction of these parts absolutely unavoidable.

But I must not omit to mention another circumstance connected with Mr. T.'s case: he was of that peculiar constitution upon which the ordinary doses of mercury cannot be made to act so as to produce a moderate ptyalism. On more occasions than one he had been previously under my care, and also under that of others, for the cure of primary symptoms; but never, in any instance, was a manageable ptyalism induced. On each occasion recourse was finally had to frightfully large doses of mercury, and these had the effect of producing a most profuse and ungovernable ptyalism. By adopting the very small doses of mercury in the early, and in the febrile stage of the eruption, I succeeded in obtaining a moderate and a controllable ptyalism; and, taught by this experience, I succeeded, on a subsequent occasion of chancre in the same individual, in effecting the cure with a moderate ptyalism, by employing mercury in doses considerably below those in ordinary use.

These two febrile states, *viz.* that of hectic and exhaustion, and that of the eruptive fever, require that the mercury should be administered in a manner peculiarly suited to such conditions. Were we to use mercury with these, as we do with venereal patients in general, I believe we should commit most serious mischief.

In these cases we should not commence with a larger dose than ten gr. of Ungt. Hydr. Fort. every morning, or with an equivalent of blue pill; that is about gr. iij. mane nocteque. However inconsiderable such doses may appear, still we shall be gratified to find that on the third or fourth day the mercury is acting in the most favourable and salutary manner on the system; and it but seldom occurs that these effects are deferred to the sixth or seventh day.

I am satisfied, from repeated observation, that we can calculate on the medicine acting in a salutary manner in such cases, with much more certainty than we can when it is administered to those who are in the enjoyment of more robust health, and who are free from fever at the time we ᵉ mmence the course. The benefits derived from this early ...fluence of mercury are particularly striking; for in a few days we have the happiness to witness the sudden change from misery and suffering to comfort and enjoyment. Sleep and appetite return, night-sweats cease, pains are removed; the improved ulcerated throat now allows the patient to swallow with comparative ease; and this, together with the increasing appetite, induce him to take more nourishment.

I do think that under no other circumstances shall we find the secondary symptoms removed so rapidly, and made entirely to disappear (at least for the time) with so small a quantity of mercury. The quantity of ointment employed on such occasions sometimes does not exceed three or four drachms; and not only are the secondary symptoms dispersed, but the general health is proportionably improved, so that the patient rapidly acquires strength and flesh. Those who are anxious to have the least possible quantity of mercury introduced into the system will be pleased with this mode of proceeding.

CHAPTER XI.

Mr. Hunter having clearly shown that in many cases venereal ulcers are kept from healing, and other symptoms from being cured by mercury, in consequence of the supervention of a scrophulous action, surgeons have naturally referred many instances of the obstinacy of venereal symptoms to this cause. This theory has had great influence, and has even reached such a length that the opinion came to be generally entertained by practitioners that scrophulous patients are unfit subjects for the treatment of syphilis by mercury. And it has even been put forward by most of the anti-mercurial writers as one of the great and peculiar advantages attendant on the non-mercurial plan of treatment of syphilis, that scrophulous patients could be cured safely and effectually by that plan, while almost certain death awaited them if their venereal disease were treated by mercury.

Let us, however, appeal to facts and observation, and see whether the dread of scrophula in the management of venereal cases has not been carried too far, and has not deprived many such unfortunate patients of a safe and speedy remedy, instead of subjecting them to the slow, uncertain, and too often unsuccessful line of treatment, by the non-mercurial plan. To decide this question with accuracy, we should first observe that venereal patients present symptoms of scrophula under two very different conditions: first, a patient may have had some scrophulous disease before he contracted syphilis; e.x. gr., a young man labouring under a scrophulous affection of the hip or knee-joint, or a scrophulous ulceration of the cervical lymphatic glands, may become infected with venereal disease ; and, secondly, a young man who in his childhood may have had scrophulous disease, but in whom it has long ceased to appear,—or one in whom though no traces of scrophula having ever been in action may appear, yet may exhibit an indolent enlargement of the cervical glands, at the time that he applies for the cure of some secondary venereal symptoms. On inquiry, we may learn that he had not been subject to any enlargement of these glands until the eruptive fever preceding the secondary symptoms had arisen; that these enlargements took place either at the time that the secondary symptoms made their appearance, or perhaps immediately before them—that is, at a time when the general health became out of order. We therefore may be called on to treat a scrophulous patient in whom the disease of scrophula existed in action for a long time antecedent to the appearance of secondary syphilis, or one in whom the development of the scrophulous symptoms is coeval with, or perhaps induced by, the secondary venereal symptoms. We shall commence with the latter, of which the following are instances :—

Let us first turn our attention to the case of a patient who has not any scrophulous disease in action, and in whom we are not able to discover any trace of the previous action of scrophula. Let us suppose that the eruptive fever which ushers in the secondary symptoms in the first order of parts, either sore throat or eruption, excites at the same time an enlargement of a greater or lesser number of the cervical glands; some of them may be clustered together into an indefined mass, while others remain distinct. This condition of the glands we may observe in some cases does not make its appearance for two or three weeks after the eruption or the sore throat; but sometimes it precedes them by an equal length of time. How are we to treat this complicated case? My answer is, that we are to proceed exactly in the same manner as if the venereal symptoms were uncombined with any such affection. The only difference to be made is, that in this complicated case we should be anxious to bring on a smart degree of ptyalism, and with the least possible delay; but at the same time one that we shall be able to control.

In directing the doses of the mercury we are of course to be guided by the existing state of the general health, and we are to suit the former to the latter. As soon as the mercury

has produced its legitimate effect we must add to the mercurials such other treatment as is best calculated to invigorate
and support the general system, and this the more particularly if the case seems to require a prolonged use of
mercury. I shall now adduce some cases in support of my
assertion, that such scrophulous affections as are ushered in
by the eruptive fever along with secondary venereal symptoms,
not only will bear mercury, but will yield to the action of this
medicine exactly as the syphilitic symptoms themselves yield.

And here I would refer the reader to the case of Mr. G—,
page 206, as a very striking instance of the safety and
advantage with which mercury can be given in cases attended
with this enlargement of the lymphatic glands accompanying
secondary venereal symptoms. In that case the mercury was
commenced under most unpromising circumstances; and yet
we found that so far from these swellings being irritated and
injured, they were, on the contrary, removed by it, and they
appeared to have retired *pari passu* with the truly venereal
symptoms.

In the following case the use of mercury was commenced
under more favourable auspices, the general health not being
much impaired; the doses of the mercury were accordingly
larger, and the result was equally fortunate.

March 14th, 1834. Robert White, æt. 36, was admitted
into the 12th ward; he states that eight months ago he contracted a chancre, which he attempted to cure by taking pills
for three weeks, while he was exposed to the weather; the pills
salivated him.

In December last an eruption, preceded by pain of the head
and limbs, appeared over his entire body: and at the same
time he felt his throat sore. Two months previous to the
eruption, he observed a swelling of one of the lymphatic
glands on the left side of the neck; this swelling did not
ulcerate sooner than three months after it first made its
appearance. For relief of these symptoms he took, for two
months, a mixture given to him at the Talbot Dispensary.
The throat grew better, owing, as he thinks, to the use of a
gargle; the eruption was not improved. At present we
observe on the trunk an eruption of papulæ, of rather a large

size, sparingly scattered ; on the limbs an eruption of similar appearance, but of a smaller size. The eruption on the face is of the small measly kind, and did not come on until a few weeks after that on the trunk and limbs. Mist. Purg. Balu. Tepid.

March 18th. Ung. Hydr. Fort. ʒss. omni nocte, Pil. Hydr. gr. v. omni nocte. I shall not trouble my reader with any more minute details of the practice in this case than to say that during the entire treatment, whenever mercury was employed, the preparations used were those above-mentioned ; and the doses were varied according to existing circumstances. He bore the mercury very well ; ptyalism was induced without difficulty, and kept up with but slight intermissions. He left the hospital on the 12th of May, at which time the ulcer had not entirely healed, although it was reduced to a mere line ; but the skin forming the edges of this line did not adhere to the surface of the granulations which they covered ; and the discoloration in the vicinity of the edges had not entirely disappeared. In this case it was plain that the scrophulous affection of the glands had been excited, either by the unguarded use of some mercurial medicine or by the mere action of the venereal poison acting on a scrophulous habit ; and it is equally obvious that the presence of this scrophulous condition did not prevent the mercury from acting on the system as it acts on the most healthy habits, and effecting a cure of the venereal with the usual readiness and certainty. And surely it is not too much to add that it did not exhibit any unfavourable action on the scrophulous symptoms.

August 2nd, 1834. Michael Bruton, æt. 21, was admitted into the 12th ward. The lymphatic glands on each side of the neck are enlarged and indurated ; particularly on the right side, where three or four clustering together form a large irregular tumour under the angle of the jaw. On the left side the glands affected are fewer in number, and smaller in size ; one of them, however, is soft, and gives the feel of fluctuation. The integuments covering the enlarged glands retain their natural colour ; no pain in the glands ; he complains of night-sweats and of pain in his chest, but he is not emaciated. A fading papular eruption occupies the integuments of the shoulders; and a few scattered papulæ are seen on his forehead, at the root of his hair; he complains of nocturnal pains of the shoulders and arms, and of the right wrist, where a slight enlargement of the bone is felt.

Three months ago he contracted a sore on the dorsum penis, which became as large as a sixpence; for this he used only a few common purges, and it healed in a month after its appearance. The glands of the neck then began to swell, and at the same time the pains attacked his shoulders and arms; and these, in another month, were followed by the eruption. He has not taken any medicine, except a few purgative pills.

August 9th. I put him on the use of mercury, ӡss. Ung. Hydr. Fort. omni nocte, et Pil. Hydr. gr. v. ter in die, and find on August the 30th this report:—"The glands on both sides of the neck have been much reduced by the ptyalism; that on the right side remains solid, that on the left side is soft with a fluid in it, but no redness of the skin, nor surrounding hardness or other sign of active inflammation. From the 16th of August to the 6th of September he discontinued the use of mercury. When resuming it on the 6th of of September, I directed Calomel gr. ij. Opii gr. ¼ twice a day, with the view of quickly reviving the salivation; but in this I did not succeed, for the only effect was a slight ulceration of the gums, accompanied with tumefaction and angina-like redness of the throat, and an increased secretion, but no actual flow of saliva.

On the 25th of October he was dismissed cured. The gland on the left side, which on his admission contained a fluid, had been punctured on the 6th of September, and at the time of his discharge from the hospital still yielded a very small quantity of purulent fluid; and it looked so well that I imagine it must have healed in the course of eight or ten days.

Here is a clear instance of enlargement of the cervical lymphatic glands, arising entirely from a venereal taint; certainly not in any degree excited by mercury, for he had not used any until after his admission into the hospital.

In a note which I made of this case, while the patient was in the hospital, I find it stated—"The result has been, that as soon as the mercury acted fully on the system, the glands

began to decline *pari passu* with the eruption; and that both
remained stationary when the action of the mercury was
allowed to subside, and both finally disappeared under the
renewed action of mercury.

Let me also refer my reader to the case of Mr. G., page
206, where the scrophulous enlargement of the cervical glands
was daily, if not hourly accumulating; and I think that he
will admit that these cases fully bear out my assertions—1st.
That those scrophulous symptoms which come on along with
secondary venereal symptoms, and which are almost a part of
that disease, are all to be cured along with the venereal
symptoms, and by the very same remedies. 2nd. That those
scrophulous affections which had existed prior to the appearance of the secondary symptoms (and which of course are in
no way connected with them) will not be benefited by a
mercurial course, instituted for the cure of secondary venereal
symptoms; nor will they, or the constitution of the patient,
be in anywise injured provided that the mercurial process be
conducted with a reasonable share of skill and judgment.

It may be asked, whence then has arisen the very general
abhorrence of mercury for the treatment of scrophulous
patients when affected with syphilis? I answer, that authors
have, one after another, frightened each other, and the profession at large, on this subject; until at length it became one
of the great boasts of the non-mercurialists to say that their
plan enjoyed this superior advantage—"That by it even
scrophulous subjects, who could not possibly be treated by
mercury for syphilis, could be cured by this plan without any
risk of injuring their general health or of aggravating their
scrophulous symptoms." Yet this strange, this almost
unaccountable aversion to using mercury with scrophulous
patients, does not seem to be felt by the very same practitioners, except in venereal cases. Who of all these will
hesitate to give mercury, even with the view of exciting
ptyalism in a case of iritis, of peritonitis, of synovitis, or of
acute inflammation of any other of those membranous tissues,
in which mercury has been found so eminently useful? Do
we find them very anxious, and very particular, in their
enquiries as to the scrophulous diathesis of a patient labouring

under any of these diseases. Do we observe them hesitate for a moment as to the medicine they should prescribe ? Even in those cases where the existence of scrophula in the habit is obvious, do we observe that they feel any peculiar reluctance in prescribing this powerful remedy ? No, the dangerous nature of these diseases overpowers every other consideration.

I have no doubt that mercury may be administered so as to exasperate already existing scrophulous diseases, or even to excite new forms of it ; I mean when it is given for a length of time, in what has been termed an alterative manner—*viz.* in under-doses, such as will not produce the salutary action of mercury (that is ptyalism), but such as will at length act upon the system as a slow poison ; perhaps, because its influence is only allowed to reach to the nervous system. But I again repeat it, that if mercury be given so as to excite a smart ptyalism in a short time, it will not produce any alarming symptom of scrophula, even though that ptyalism be kept up for such time as may be required for the cure of either primary or secondary symptoms.

We shall next advert to that class of cases in which the patient, who has long been, and who is still, affected with scrophulous disease in action, has the misfortune to become the subject of secondary venereal symptoms. Does not this combination, some will exclaim, forbid the use of mercury ? will not the result of mercurial treatment in such a case be that the venereal symptoms will be aggravated to a dangerous degree ? Such, I imagine, is the generally-received opinion—an opinion, however, which I conceive is rather drawn from theoretical induction than founded on practical observation. For I think I may with confidence appeal to the experience of every surgeon who has had much practice in the venereal disease, and ask him whether he has not treated many scrophulous patients, labouring under chronic abscesses and ulceration of the joints ; and whether he has not even cured many (who had been lamed by morbus coxæ in childhood) of syphilis, by the use of mercury ; and that too without being able to discover what he anxiously watched for, and secretly dreaded—namely, "that the scrophulous complaints had been made worse by the use of mercury." Many, I make no doubt, have felt a secret satisfaction at the result; and some, perhaps, have ascribed their success more to chance, or to some peculiar good fortune which has conducted both surgeon and patient with safety through such apprehended danger.

Out of many cases which have occurred in my own practice I shall adduce a few instances in corroboration of the opinion that such scrophulous habits as those to which I have alluded do safely admit of the judicious use of mercury, for the cure of their venereal disease, without prejudice to any scrophulous symptoms with which they may also happen to be affected.

Joseph Bunbury, æt. 26, of a full habit, was admitted into No. 12 ward, on December 5th, 1834. Eight months ago he had a chancre behind the corona glandis, on the right side of the penis, and immediately began to use mercurial pills. He had taken seventy pills before his mouth became affected. This treatment was continued for two months, and then he was considered to be cured; but in a month after he was attacked by pains in his shoulders and arms, and soon after a papular eruption showed itself all over his face and body, and his throat became sore. He was received into Sir P. Dunn's hospital, was there salivated, and at the end of three weeks was dismissed, the throat being well, the eruption fading, and the pains much relieved, but a small glandular tumour had formed on each side of his neck: these tumours have very much increased in size, and now these tumours are soft; the skin covering the most prominent parts of them is of a purplish red colour, and on each side is a small ulcerated opening, through which a thin pus escapes, the edges of the opening being deeply undermined and thin: he says he had many such tumours in his childhood. His venereal symptoms are several scaly spots in his eyebrows, and red patches on his shoulders, with a slight swelling on each side of the ligamentum patellæ, and severe pains in his shoulders and arms, which prevent him from extending the forearms, and keep him awake at night.

I shall briefly state the treatment, by saying that he was purged, and had a tepid bath, before he began to use mercury.

On December the 9th he was ordered ℥ss. Ung. Hydr. omni

nocte, et Pil. Hydr. gr. v. bis in die. The pills were omitted on the 16th, and the ointment continued until the 23rd. The ointment was resumed on the 30th, and continued until the 10th of January. Again, on the 24th of January, he was directed Ung. Hydr. ƺss. omni nocte, et Pil. Hydr. gr. v. ter in die; this was continued until the 31st, at which time his mouth was again smartly sore. After the 10th of December he took bark, in one form or other, during his stay in the hospital, from which he was dismissed on the 10th of February, perfectly free from any appearance of venereal disease.

Surely this was a fair case for witnessing the effects of mercury in a scrophulous habit; for we may set down this man as scrophulous, not only from the present swellings and ulcers under the ears, but also from the fact that such had affected him very constantly during his childhood. The result of this treatment, in this case, has been that the mercury acted rather kindly than otherwise; that his health did not, in the slightest degree, suffer from a very smart action of mercury, kept up for eight and for twelve days, and even then renewed—for he did not lose flesh during his stay in the hospital. With respect to the scrophulous tumours and ulcers, they were not benefited, nor were they in the slightest degree injured by its use. In a word, the mercury here acted as it does whenever it acts kindly; the scrophulous habit did not seem to have the slightest influence on its mode of acting, nor did the mercury seem to exert any pernicious influence on the scrophulous symptoms or diathesis.

Matthew Connor, a delicate-looking boy, æt. 16, was admitted on 10th June, 1834, into the eye ward, on account of venereal iritis in an advanced stage. He had, on each side of his neck, three or four ulcers covered with yellow scabs. When the scabs were removed by poultices, the ulcers presented all the character of scrophulous ulcers with fungous surfaces. Having satisfied myself as to the nature of the ulcers, I allowed them to scab again; and during the time of his treatment for iritis they remained untouched. For the c... of the iritis, I employed calomel in doses of two grains three times a day, by which a very high degree of ptyalism was induced. During the long period of the ptyalism the scabs remained unaltered in every respect. When the ptyalism had subsided, I again removed the scabs by poultices, and I found the ulcers still retaining the same character and appearance as they had previously to the mercurial course. They did not exhibit any appearance which could lead me to imagine that they had undergone any change during the time of the salivation.

Let us not then be too timid when certain diseases, or conditions of disease, require that we should give mercury to a patient, in whom we see scrophula already in action. The safest rule to adopt in such circumstances is, in my judgment, to give it so that it shall, in six or eight days, produce the degree of ptyalism that we wish for. I need scarcely add, that should it be deemed necessary for the cure of the disease for which mercury is being administered, to keep up a lengthened action of mercury, we should support the system of our scrophulous patient by nourishing diet, by bark, and by opium.

Feb. 1st, 1822. Mr. O. (usher of a school), æt. 26, has been long afflicted with a severe cough, sometimes with hæmoptysis; and is altogether of a very consumptive habit. On Feb. 1st he applied to me for cure of a chancre on the inside of the prepuce.—this is a genuine Hunterian chancre, and had been discovered only a few days previously.

I advised Pil. Hyd. ʒss. Pulv. Ipec. gr. octo fiant pil. octo. Sumat unam sing. noct.; but by mistake of the apothecary he took, according to the label on the pill box, one pill at night and one in the morning.

Feb. 49. He states that the pills gripe him occasionally; gums are very sore, those of upper jaw being smartly ulcerated; surface of chancre clean, hardness much less. Omitt. Pil. per dies ii. dein i. o. n.

Feb. 13th. Mouth more sore; chancre less hard at base; thinks his cough much improved since he began this course.

Feb. 25th. Mouth sore; upper gum severely ulcerated; scarcely any hardness in the site of the chancre. Repet Pil. i. om. nocte.

March 11th. Has not taken any pills since 3rd inst.; mouth is only now well; no hardness in the site of the chancre; cured. Acid. Sulph. Dil. Gutt. xv. ter in die.

Remarks.

I felt quite disappointed when Mr. O. told me that he had been taking two pills a day instead of one, as I had directed, for I was tremblingly alive to the danger of using mercury in almost any quantity with him. However, as he had begun with this dose, I did not like to send him back to a smaller one, fearing that if I did so I should not get the mercury to act favourably hereafter.

This case clearly proves that we may with perfect safety use mercury in small doses, and even excite a smart ptyalism in patients whose lungs must be considered as extremely delicate and tender, and therefore, according to generally received opinions, very ill suited to bear the active use of mercury. Had this been a solitary instance of the safety with which mercury may be given to a patient with hæmoptysis, it might be said that it was an escape and not a cure; but supported as it is by many other instances in which mercury was with safety and advantage given to those labouring under hectic fever in full action, I have no hesitation in pronouncing it a cure. Should this point be established, that mercury can with safety (provided it be with caution) be given to a patient threatened with hæmoptöe, so as to effect a cure of primary venereal symptoms, I say if this point be once established, it will enable practitioners to hold out to such persons the benefit of this medicine.

At the time that Mr. O. applied to me, I had not satisfactory experience of the safety with which mercury may be given to scrophulous patients; and, therefore, I directed very small doses of this medicine, indeed much below what his system could have borne; but by the mistake of the apothecary the dose proved better suited to the existing state of his health, as he was not affected with hectical symptoms. The issue of this case, however, made me reflect more upon this line of practice, and induced me, in subsequent cases of scrophulous patients, to administer mercury with more freedom.

R

Had Mr. O. been a patient in easy circumstances I certainly would not at that period, and with the notions I then entertained of mercury, have preferred a mercurial course. But as his means were extremely limited, depending upon his daily exertions for his daily bread, I knew that he could not procure for himself that ease and those comforts which would be required by a different line of practice. These considerations induced me to direct a mercurial course, knowing that this, if it succeeded, was decidedly the shortest line of treatment, and one which would not materially interfere with his occupations.

The result has shown that mercury may be administered to patients subject to hæmoptöe, provided it be given with caution and judgment; and also that in such the venereal symptoms yield at least as quickly as they do in those of more robust and more healthy constitutions.

It may naturally be asked, if scrophulous patients be found to bear mercury so well as you represent, how has it come to pass that the profession have so generally entertained the contrary opinion, and have been impressed with a notion that mercury would aggravate all the symptoms of scrophula, while, at the same time, the scrophulous diathesis was so opposed to mercury as to prevent it from exercising its salutary influence, and effecting a cure of the venereal disease in such habits? To this I can only answer, that I suppose when men's minds were first instructed, that venereal symptoms in certain habits had resisted the power of mercury, and were afterwards cured by sea-air and sea-bathing, they inferred that this was only attributable to a scrophulous diathesis,—they, of course, became more guarded and reserved in the administration of mercury to patients of this habit. Hence it followed that they not only prescribed this medicine in smaller doses, but whenever they saw it about to produce ptyalism they withheld it for a time. In a word, that they most cautiously watched it for the purpose of preventing it from acting on the salivary system (its natural and salutary action). By this means the constitution of the patient came to be irritated, and teazed as it were, by this long-continued use of mercury in under-doses; the medicine accumulating in the system, and not being allowed its natural action and

natural outlets, it came to act as a poison on the system ; and
thus the scrophulous habit (naturally weak) was thrown into
disorder, and consequently all its diseases, and tendency to
disease, were aggravated. What was considered as an
indulgence to the scrophulous habit ultimately proved to be
the source of the principal mischiefs which have been
observed when mercury was administered to scrophulous
patients.

Let us not, however, consider every instance of enlarged
cervical lymphatic glands occurring in venereal patients,
either simultaneously with or subsequently to the appearance
of secondary symptoms, as excited by the influence of venereal
virus. For we shall meet with some cases in which this
condition of the glands had been produced by the action of the
mercury at the time that ptyalism was commencing. That
the enlargement of the glands is in such instances caused
wholly by the action of mercury is proved by the fact that the
swelling subsides according as the action of the mercury
declines; and that when we again reproduce the ptyalism a
second time, the mercurial action having been allowed to
subside, we shall find that the swellings will return, and that
these can be prevented from increasing in number and size
only by reducing the doses of the mercury, and interposing
brisk purgatives pretty liberally. When we thus know the
fact that ptyalism is occasionally the cause of this enlargement of the cervical lymphatic glands, we can easily guard
against or remedy this occurrence. Whereas, were we to
mistake such for scrophulous enlargement excited by the
venereal virus, we might be disposed to persist in using the
same doses of mercury under an impression that the former
would yield to a little further action of mercury, as some
of the venereal symptoms had already begun to do under
its use.

I have lately seen two cases of this enlargement of the
cervical glands in patients while under treatment for primary
symptoms only, and these afford a still further proof that this
affection was occasioned solely by the action of mercury on
tl salivary glands.

CHAPTER XII.

MINUTE DOSES OF MERCURY IN LATE AND CHRONIC BUBO.

Among those cases of syphilis which can best be treated by very minute doses of mercury, I must mention a rare form of bubo: it is one of an indolent nature; it makes its appearance long after the chancre is healed, perhaps in four or five weeks. The bubo in some cases is single; in others, there is one in each groin: the swelling is not of an acute inflammatory nature, nor is it, during its progress, attended by severe pain, unless it proceed to suppuration, in which case it becomes very painful, particularly if the matter has accumulated largely and has not got vent. These buboes, to which these remarks apply, are very remarkable for the severe and extraordinary derangement of health which attends them. A high degree of febrile excitement exists, and yet the fever partakes very much of the character of hectic fever: a quick pulse, impaired appetite, urgent thirst, languor and lassitude, emaciation, and night-sweats, form its leading features. When, after a tedious suppuration of three or four weeks, the matter gets exit, the pain then diminishes; and by slow degrees the health in some measure improves, but a long time elapses before it is perfectly re-established. In some instances the tumour does not advance to suppuration, but having arrived at a certain size it remains stationary for a short time, and then slowly retires, but only to a certain degree. For whether suppuration shall have taken place or not, the groin is not restored to its natural state, and considerable induration of an indolent nature still remains. Such are the symptoms which attend buboes of this class. I have been consulted on cases in which the disease stood at this point for a very long time; of this the following is an instance :—

October 19th, 1835. Mr. K. six months ago had a chancre on the dorsal aspect of the penis. at the junction of the prepuce with the glans ; for this he was directed, by an eminent

surgeon, to use mercurial frictions : he says that he rubbed in
nearly one pound of mercurial ointment, and with so much
energy that his arms at first became quite muscular ! This
course he continued for six weeks ; during it he lost flesh, and
became unable to read or to attend to any business, but no
affection of the mouth was produced. Three months ago he
again had sexual intercourse, by which the cicatrix of the
original chancre was opened ; this has not since healed.
About this time a bubo formed in each groin ; these were
considered, by another eminent surgeon, as constitutional or
scrophulous buboes, and therefore they were treated without
mercury. During the formation of the buboes his health
suffered materially ; he became emaciated, and had night-
sweats. After this last sexual intercourse the prepuce
became much thickened, especially on its dorsal aspect ; this
thickening affected the integuments of the penis more than
half-way towards the pubes. The ulcer, which hitherto had
been confined to the prepuce, has lately shown a disposition
to slough, and has stretched forwards, so that now it has laid
hold of the corona glandis. To the ulcer Mr. K. had applied
Acid. Nitric. within the last three days. At present the
prepuce and integuments half-way up the penis are red, much
swollen, and thickened ; a hardened and knotted lymphatic
vessel runs along the right side of the penis ; a considerable
hardness and swelling exists in each groin, the integuments
are there also red ; a trifling discharge takes place from a
small opening in each bubo. His health is very much broken
down ; he is emaciated to an extreme degree ; pulse 130 ;
countenance pale and sickly, yet his appetite is good ; he has
no thirst ; has night-sweats ; sleeps until four o'clock a.m.,
from which hour until the time he gets out of bed he is
incessantly teazed by painful erections.

On the most minute enquiry I cannot discover any secondary
symptom. Judging that the original venereal disease was
not cured by the very heavy course of mercury which he had
undergone, although his system suffered so much by it, I felt
no hesitation in recommending mercury, but in very minute
doses, which alone could be borne by him in the present
depressed state of his health. Accordingly I advised Ung.

Hydr. Fort. gr. x. omni nocte; and, at the same time, to drink from half-a-pint to a pint of sarsaparilla broth daily.

October 22nd. The swelling of the prepuce considerably reduced; the ulcer not spreading upon the glans : he says the right bubo is more painful, yet it is scarcely (if at all) more swollen; slept well last night, not having been troubled with erections, but had profuse night-sweats; the pulse is still 130; he relishes the sarsaparilla broth. No sensible effect from the mercury, yet he says that he feels himself stronger.

October 26th. The swelling and hardness of the integuments of the penis much reduced; ulcer with a tolerably clean surface; it discharges healthy pus in sufficient quantity; the right bubo, which had been very painful, is easy, and has now three small openings; the hard chord-like lymphatic vessel along the right side of the penis is reducing. He says his health is in every respect much better: he sleeps well; has no night-sweats; erections are rare, and not attended with pain; pulse reduced to 108; appetite better than when he was in perfect health; some mercurial fœtor indicates a very slight affection of the mouth. Repet. Jusculum Sarsæ et Ung. Hydr. ad gr. x. per duas noctes, et dein augeatur dosis ad Ɔi. sing. nocte.

I was prevented from watching the further effects of this plan of treatment; nor did I receive any account of the patient until the 20th of December, when Mr. K. informed me that when he had rubbed scruple doses of the ointment for three times his mouth became sore, and he continued to use the mercury, so as to keep up the soreness for a fortnight; he then laid aside the mercury. He has taken a great deal of sarsaparilla, and now enjoys tolerable health. The glands, however, in both groins are still too large; the penis is healed, with a broad and deep cicatrix behind the base of the glans.

September, 1836. Mr. K., about the time of the last report, relinquished all further treatment; and yet he has not regained his ordinary state of health, although he has been pretty actively engaged in various pursuits, at times living rather freely. He is still very thin, and suffers from irregular pains in his limbs.

In a word, he is not in perfect health; he is teazed with the above symptoms, not one of which is decidedly venereal; and yet, all together, they excite in my mind a strong suspicion that something of a latent unsubdued syphilitic taint still remains, which obstructs his restoration to perfect health; yet he declines to follow my advice, *viz.* that he should again use mercury, beginning with scruple doses of the ointment.

Such are the leading features of this disease when confined to the primary symptoms. But it does not stop here; for unless it be cured at this stage we shall find that after some time a fresh set of symptoms will arise, which, though not strictly secondary venereal symptoms, yet lead us to suppose that they proceed from a certain modified operation of the venereal disease on the general system.

Thus we sometimes observe that the slight improvement in the general health which takes place on the matter of the buboes being discharged, does not continue unimpaired for many weeks. The patient now, perhaps, suffers from one or more ulcers on some part of the body or limbs. He tells us that each of these began as a small boil or pimple, which he scratched; or that it arose from some very trifling hurt. These ulcers seldom exceed in circumference that of a half-crown piece, more frequently they are of a size intermediate between that and a shilling. When fully established these ulcers exhibit so much of the character of secondary ulcers that almost every surgeon would say they were suspicious-looking ulcers. Generally from two to four of these appear, and they are healed very slowly by means of some active application, for mild ones produce no improvement in them. In other cases no ulcers appear, and the second set of symptoms bear no resemblance to those of the venereal disease; they are such as indicate disturbed health and a broken-down system, and therefore we may expect to find in these an almost endless variety, depending probably upon some constitutional predisposition, or excited by some accidental circumstances; they are such as we could not, by viewing them abstractedly, imagine to be syphilitic, and yet by tracing their history we are led so directly to the previous venereal disease, that we can scarcely doubt their syphilitic origin. The readiness with which they yield to the operation of mercury removes all doubt, and convinces us not only that they arose from a venereal taint, but that they are as quickly and as certainly curable by mercury as the purely secondary venereal symptoms are; and yet these symptoms are among those diseases which altogether forbid the use of mercury under ordinary circumstances. Let us take as an example the following case :—

Dr. G., while pursuing his studies at Glasgow, had an ulcer near the frænum, which destroyed that fold; it had been brought to heal chiefly by repeated applications of Argentum Nitratum. In May following, some weeks after, being in London, a bubo made its appearance in his right groin, and at the same time his health became much deranged, and his strength and flesh were seriously reduced. He applied to one of the most eminent and experienced surgeons in London, who considered his case as one of scrophulous bubo, and accordingly treated him principally with iodine. By his advice he at length removed to Chatham, where, for a time, his general health improved much; but he did not long enjoy this state of good health, for he was then affected with a slight spitting of blood, and cough; some difficulty of breathing quickly followed, and he felt his throat sore. After a little time he became troubled with palpitation of the heart, and along with this a glandular swelling, as large as a plum, took place between the angle of the lower jaw and the chin. Such was the history of his case on 6th Nov., 1835. When he was introduced to me by my friend Mr. White, he had, in addition to a pretty large indurated bubo, the enlargement of the glands under the jaw, palpitation, and quick respiration; he was also affected with extreme quickness of pulse, considerable emaciation, lassitude, and a sickly pallor of countenance. Upon the most mature consideration we agreed to make a trial of minute doses of mercury, in the hope that all his ill health was caused by a latent venereal taint. On Nov. 6th, 1835, we directed for him ℈ss. Ungt. Hyd. Fort. omni nocte.

Nov. 24th. Mr. White informs me that he is getting well very fast; the lump under the jaw has declined; his pectoral complaints are much improved; in every respect his health is much better; he still uses the small doses of mercury; he is now to use doses of a scruple each.

Dec. 18th. He now appears much improved in flesh, and he feels himself in excellent health; he is free from cough and difficulty of breathing; very slight remains of the glandular swelling under the jaw; pulse quiet and natural; he has continued the small doses (℈ i.) of ointment, by which his mouth has been slightly affected; he is ordered to repair to Chatham, and is advised to persevere in the use of the same small doses of mercury for three weeks longer.

The above histories exhibit to our view buboes or swellings of the inguinal glands under circumstances very different from those in which we commonly contemplate venereal bubo. Thus, in the ordinary venereal disease, bubo appears as an acute glandular inflammation; and, when attended by fever, this is of the inflammatory type. These late buboes, on the contrary, exhibit all the marks of a chronic indolent disease; and these are invariably accompanied by a fever of a purely hectic type.

Perhaps a surgeon, in the whole range of diseases in which he may be called on to practise, cannot meet with any instance in which a true diagnosis is a matter of greater difficulty or of greater importance. It is but too obvious that we cannot place any reliance on the present symptoms. Some information may be gained by applying our attention to the history and treatment of the original primary symptoms, and particularly where these had been treated by mercury. We should carefully inquire as to the manner in which this medicine had acted on the system; and in all such cases, if the ptyalism had not been excited at the regular period, and kept up for a fair length of time, then I would strongly incline to the opinion that all the present symptoms arose from a syphilitic source.

On collating together the history of several of these chronic venereal buboes, with that indolent constitutional bubo which I originally described in a paper in the Dublin Hospital Reports, vol. ii. (which paper has been transferred to page 125

of this work), it will be seen that the local conditions of the
groin very closely correspond with each other, and that the
constitutional symptoms of the one cannot be distinguished
from those of the other.   Yet in practice no more serious
error could be made than that of mistaking and of treating
the one for the other.   For were we to treat by the ordinary
mercurial process as venereal the constitutional bubo with its
accompanying symptoms, we should exasperate all the complaints, inflict a severe injury on the health of the patient,
and even endanger his life ; while if we treated a chronic
venereal bubo as a mere constitutional swelling of the inguinal
lymphatic glands, we should only employ an imbecile useless
line of practice, leaving the cause and the root of the disease
untouched, allowing it to pursue its course unopposed, until it
terminated, as it generally will do, in a slow death, except in
some cases, where a peculiar innate strength of constitution
may enable the system to produce the development of some
one or two well-marked symptoms of syphilis.   In confirmation of these assertions we find that one of these cases had
been mistaken by a surgeon of the highest character in the
United Kingdom.   If we indulge a hope that we shall be able
to distinguish the venereal from the indolent constitutional
bubo, we must rest it on our scrutinizing with most scrupulous
minuteness the history of each case.   By so doing we may, in
a few instances, acquire a well-founded confidence in our
diagnosis ; but in the majority of cases we shall find the
history involved in the utmost perplexity and uncertainty.   I
trust, however, that I have now directed the attention to a
perfectly safe and generally an efficient test by which our
doubts may, in a great degree, be dispelled, and by which
our practice may be guided in a safe and satisfactory manner.
It is merely to employ mercury in very small doses, *ex. gr.*
ten grains of the ointment once a day : by employing this
test in cases which are venereal we shall enjoy the satisfaction
of seeing some improvement in the constitutional symptoms,
as also some amendment in the general health in the course
of five or six days, before any mercurial effect can be dis-
cerned ; and those favourable changes will become more and
more manifest, in proportion as the salivary system comes

under the influence of the mercury. Should we have
employed the same test in a case of purely constitutional
indolent bubo, we shall find, on the contrary, that no amendment takes place within the first six or eight days, but that,
even when the mercury acts on the mouth, the symptoms are
rather aggravated.

This plan of employing those minute doses of mercury can,
therefore, lay claim to the double merit of being the most
speedy as well as the safest method of treating all venereal
symptoms, whether primary or secondary, when attended
with hectic fever; and also of affording the most certain and
the only safe criterion by which we can decide in doubtful
cases, which are attended with hectic, whether the disease be
venereal or not.

Were we to employ, as a test, mercury in doses such as are
employed ordinarily in the treatment of the venereal disease,
we should inevitably injure the patient, and at the same time
we should not obtain the smallest insight into the nature of
the disease; for mercury used in such doses, with a patient in
this state of debility and hectic fever, would rather overpower
the system than produce any improvement in the venereal
symptoms.

One word more on the subject of these buboes and their
accompanying symptoms. Here is a modification of symptoms
of syphilis and its consequences, which we are utterly at a
loss to account for; the natural habit and previous state of
health will not explain it; nor do we acquire any real insight
into the cause by referring it to the treatment of the primary
symptoms; for in one of the cases I have related mercury
was used in full doses, and for a very sufficient length of
time, while in the other no mercury at all was employed.
And I recollect other cases in which the patients used only a
few mercurial pills for eight or ten days, without any sensible
mercurial action being induced. If in the case of Dr. G. we
embrace the opinion that the venereal disease had been
modified by the constitution, we shall be fully as much justified
in supposing that in Mr. K.'s case the overpowering doses of
mercury induced the extraordinary deviation from the regular
course of the venereal disease.

CHAPTER XIII.

SYPHILIS IN INFANTS.

PERHAPS there is not, in the entire range of surgical diseases, any one the contemplation of which is more calculated to arrest our attention, or to excite our interest, than Syphilis Infantum. Whether we inquire into the circumstances under which the diseased parent or parents can infect their offspring, or the form in which the disease affects the infant, or the appearance and nature of those diseases which are communicated by the infant to the nurse, or of those communicated to its other attendants, and the further propagation of disease by the nurse to her husband, and perhaps to a large family of children ; I say, in investigating any one of these points, we must be struck with the fact that we find in each a striking deviation from those laws which regulate the venereal disease as communicated by the adult to the adult. Indeed this is so much the case, that some authors have not hesitated altogether to deny that these affections were venereal, while others, admitting the possibility of a venereal disease in infants, have yet made use of those very deviations from the regular laws of syphilis to prove that in particular instances the disease was not venereal, because it did not strictly square with the progress of syphilis in adults.

Before I enter upon the subject of syphilis in infants, I must state a fact which, though I am unable to explain it, has yet been forced upon my observation by more than five or six instances ; namely, that a newly married man, who is himself free from every appearance of syphilis and every other disease, shall yet infect his wife in such a manner that secondary symptoms shall appear in her in a few months after marriage, and these not preceded by any primary symptoms, or by any discharge whatever from the genitals. And although among these secondary symptoms some raised ulcers shall fix upon the external pudenda of the wife, yet the husband shall very rarely suffer in any manner from these. The husband, when

questioned, will admit that at some period within nine months
antecedent to his marriage, he had contracted a venereal
disease; that he had undergone mercurial treatment, by
which all his symptoms were removed and that he had been
declared by his surgeon to have been cured. I am well aware
that such reports cannot always be relied on; and as long as
such cases were confined to mechanics and persons in an
humble rank of life, I could not bring myself to believe
implicitly in their reports; but when I found some occurring
in the higher walks of life, and when the husbands proved to
be men of acknowledged and known veracity, I could no longer
withhold my assent. I now mention one case, of the truth of
which I am perfectly convinced, and which is, besides, one of
the very few instances in which the husband seemed to suffer
from the diseased state of the wife.

January, 1828. In two months after her marriage, Mrs.
K. first noticed certain unnatural appearances on the labia
externa. Having, after another month, obtained an examination, I found a number of white moist raised tubercles on the
internal surface of each labium pudendum. At this time she
was also affected with white superficial ulceration or excoriation
at both angles of the lips. The tonsils appeared with whitened
surfaces, the snail track affection of the mucous membrane, a
deep copperish hue of the skin at the junction of the alæ nasi
with the cheeks. Upon the most minute inquiry I could not
discover any grounds for supposing that Mrs. K. had at any
time been affected with primary symptoms.

At this time Mr. K. had not the slightest trace of any local
or constitutional disease. About ten months before marriage
he had undergone a full and regular course of mercury for
primary symptoms, and was then assured by his surgeon that
he was perfectly cured. Since that period he had remained
free from all appearances or suspicions of venereal disease.

However difficult it was to account for these appearances, I
considered them to be venereal, and accordingly put both
husband and wife under a course of mercury. For although
he did not exhibit the slightest trace of disease, yet it was
uite certain that from him Mrs. K. received her complaints.
By the use of blue pill her mouth was made smartly sore in

the course of three weeks; yet at this period the symptoms on the pudenda were not improved as much as I thought they should have been in that time, and they afterwards appeared to yield rather to a lotion of Sulph. Zinc than to the action of mercury. All the other symptoms were removed solely by the influence of the mercury. This course was continued for eight weeks.

His mouth was with difficulty affected; however, about the middle of the fifth week it became slightly sore, and this effect was continued until the conclusion of the treatment. In the beginning of April they returned to the country, where they both used sarsaparilla for six weeks following. When they had been at home for three months Mrs. K. perceived a few blotches in the skin of the palms of the hands, a few also on the forehead near the roots of the hair, and some over the scalp; on the dorsum of the tongue there now is a small patch destitute of papillæ, and looking as if it had been scalded. Fresh appearances have also taken place on the pudenda. From these she suffers no uneasiness whatever; no discharge. They consist of a number of small raised brown-coloured excrescences covered with a firm cuticle.

Mr. K. now told me that he was strangely affected; for that about six weeks ago he observed some strange appearances on his penis. I found, on the inside of the prepuce, a brown blotch of a rounded form, and a similar one of a much larger size, and of oblong form, raised and with a rough surface on the skin of the penis.

I now commenced another course of mercury, which was carried on for ten weeks, and which proved more satisfactory than the former; for now, in Mrs. K., all the symptoms yielded kindly as the ptyalism proceeded; and Mr. K. was without much difficulty brought under the influence of mercury, and the brown blotches yielded in the same manner as venereal symptoms usually do.

Some authors, with a view to make facts reconcileable to their theories, have supposed that the infant could only be infected by coming in contact during parturition with ulcers in the vagina of the mother. This doctrine is at once overthrown by the following facts:—1st, by having ascertained by observation that no such ulcers were in existence at the period of parturition; and, 2ndly, that many infants have exhibited the symptoms of the disease actually existing at the moment of their birth.

Many of those who admit that the fœtus in utero can be infected suppose that this takes place only when one or both parents have been affected with primary or secondary symptoms at the time of conception, or that the mother has been so affected during utero-gestation; and to this opinion I believe the majority of experienced practitioners, urged by repeated observation, now give their assent, however irreconcileable the fact may be with their theoretical notions of the ordinary mode of propagation of the venereal disease.

But there are cases in which the fœtus in utero has been infected under circumstances so strange and so difficult to explain, that nothing short of actual observation could induce us to allow the fact. The circumstances to which I allude are these:—The father of the child has had primary symptoms six or eight months before his marriage; for these he has been treated by mercury; has undergone a full course of this medicine, under which his symptoms have been removed; and his surgeon has declared himself satisfied with the treatment, and dismissed him as perfectly cured. In six or eight months after this treatment he marries. In the ordinary time his wife becomes pregnant, and carries the child until the seventh or eighth month, when abortion takes place, and this without being preceded by any of those circumstances which in ordinary contribute to its occurrence. The same fatality attends on the second and third, and perhaps fourth, pregnancy, in spite of every attention paid to the directions of her accoucheur. At length suspicion arises in the mind of the accoucheur; he examines the product of the next abortion, and finds that the cuticle is loose, and that it readily peels off in patches of greater or less extent; thus is explained what the midwife had termed a putrid state of the child; he may find too that the nails are not formed, and in general that the child appears as if it had been badly nourished. It should be ¹ re remarked that sometimes the child is produced alive, in such a weakly but emaciated state that it does not survive

more than a few hours; and such often bear unequivocal
marks of the venereal disease.   Until these repeated abortions
shall have attracted the attention of the accoucheur there
has not been any one circumstance which could have raised
suspicion as to the true cause of them; for both parents
continue, all this time, to live in the enjoyment of perfect
health; no trace of disease is to be discovered in either.
When the husband is questioned he candidly avows that he
had before marriage been affected with primary symptoms;
that he had been (as he thought) cured of them; and that
having allowed six or eight months to pass over before his
marriage, without perceiving any sign of a return of the
disease, he had concluded that it had been perfectly eradicated
from his system.   On further inquiry it is ascertained that
his wife had never complained of any sensations which might
lead even to a suspicion of her having had primary symptoms;
nor has any appearance taken place in her which can even
bear a resemblance to secondary symptoms.   In a word, both
parents are pronounced (after the most minute investigation)
to be in the enjoyment of perfect health.

In some cases we may discover equivocal appearances of
disease in the father, yet so faintly resembling syphilitic symptoms that we could not think of considering them as venereal,
unless our suspicions came to be strengthened by some col-
lateral circumstances.   Of this kind is one of those cases
related by the late Dr. Beatty, of this city, in his paper
entitled—"On a Species of Premature Labour," published in
the fourth volume of 'The Transactions of the Association of
the King and Queen's College of Physicians.'

Dr. Beatty, having concluded the history of one case of this
kind, says, "Several similar cases occurred to me from that
time, with similar success, which I shall pass over, as they
rest only on my own experience, and shall therefore confine
myself to a very few, in some of which I was assisted by Mr.
Colles and Mr. Todd, in their capacity of surgeons.   In my
case-book, to which I have referred, I find that, in August,
1812, I attended the wife of a staymaker, who was delivered
of a putrid child, in the seventh or eighth month, which she
said was the third that she had born dead.   I discovered so

much of venereal affection, as to recommend that they should put themselves under the care of some experienced surgeon for the use of mercury. They applied to Mr. Colles: and, when she was pregnant in the following year, Mr. Colles told me that they had not continued a sufficient time under his direction to satisfy him that they were cured of their venereal complaints; which I found to be the case in July, 1813, when I delivered her again of a putrid child, in the eighth month. I then declared that I never would attend her again, until Mr. Colles told me that he was satisfied with the result of the mercury used. They again returned to him; and fully attending to his directions, in October, 1814, I again attended her, when she bore a living girl, at the full period of gestation. She has had several living children since."

I can recollect that the symptoms exhibited by the husband were of such a very doubtful nature that I should not have thought of treating them as venereal were it not for the communication made to me by Dr. Beatty with respect to the abortions of his wife.

Although in this case there was enough in the appearance to allow us to connect them with a preceding venereal complaint, yet Dr. Beatty's experience had informed him that similar unfortunate results might have taken place where no visible symptom of venereal disease could be discovered in either parent, for his next case, which was treated by him and the late lamented Mr. Todd, was of this description.

In a few instances the child comes to its full time or nearly so, but is born in such a weakly state that it ceases to live in a few hours. This infant exhibits unequivocal marks of venereal disease, in the copper-coloured eruption which is always to be seen about the anus and genital organs, and which is often spread over the entire surface of the body. The countenance of such a child generally exhibits somewhat the appearance of extreme old age.

But the manner in which I believe the syphilis of infants more generally makes its appearance is this:—the child is born, to all appearance, healthy and well-nourished, and continues apparently in good health for a period varying from eight days to as many weeks; then a number of copper-

coloured spots appear about the anus and genitals, on the
inside of the thighs, corresponding to the external genital
organs; these sometimes spread along the groins, and
degenerate into ulcers. The voice of the child is now observed
to change; it has a peculiarly hoarse cry. Superficial ulcers
next appear at the angles of the mouth; these sometimes
affect more particularly the mucous membrane of the mouth;
sometimes they occupy rather the external skin, and when
this is the case we find the cracks or fissures made to bleed
by the stretching of the skin whenever the child cries. The
tongue, palate, and throat also are affected with those white
and very superficial ulcerations which authors have denominated aphthous. The nose too begins to be more or less
obstructed; a sharp thin discharge flows from the nostrils,
and seems to irritate (if it does not excoriate) the adjoining
skin; this discharge occasionally dries, and forms a hardened
crust at the opening of the nostrils, which, by obstructing
the breathing, seems to add considerably to the sufferings of
the child. If the disease be allowed to proceed we find ulcers
or fissures in various folds of the skin, e.c. gr. between the
chin and throat, on the back of the neck, in the folds on the
back of the thighs—in a word, in any or every one of the
folds of the skin. The voice of the child soon adds another
symptom characteristic of the disease; the cry of the child
becomes hoarse and raucous to an extreme degree. The child
then begins to emaciate, the skin becomes flabby, and unless
active treatment be employed, the unfortunate sufferer dies
extenuated and exhausted. Very frequently a number of the
lymphatic glands become enlarged, which older authors have
improperly termed buboes. The glands of the neck, and those
on the occiput, suffer in this manner, especially when the
eruption forms a crusty covering on the scalp; indeed everywhere on the surface are the lymphatic glands occasionally
liable to be thus affected. But there is nothing in this
enlargement to be compared to the inflammatory enlargement
of these glands in the buboes of adults; on the contrary, they
remain perfectly free from the condition of active inflammation, although some of them often undergo a sort of slow
suppuration and ulceration, similar to those processes in

scrophulous glands. These changes in the glands, when they do occur, take place only in the last stages of the disease.

It deserves to be mentioned that some infected infants (in addition to the other symptoms) have a muco-purulent discharge from the eyes. This is in such quantity as scarcely to overflow the eyelids, and is accompanied with only a very slight redness of the tunica conjunctiva. It bears no resemblance to the purulent ophthalmia of infants, which latter (I need scarcely add) is not a symptom of syphilis. My limited observation does not furnish me with a single instance of purulent ophthalmia, in conjunction with truly syphilitic symptoms.

The unfortunate victims of this disease generally perish apparently worn out by marasmus. In a few some spots of the eruption, or some of the glandular swellings, degenerate into ulcers, which, becoming irritable and painful, add severely to the sufferings, and hasten the death of the patient. But in no instance do we oberve those destructive processes of ulceration by which portions of the throat or nose are carried away; nor do we see those frightful affections of the bones which too often attend the other symptoms of secondary syphilis in adults.

Let it not be supposed that we are to meet with all the above symptoms affecting every infected child; scarcely ever are they all combined, nor indeed is there any regular order in which they necessarily appear. In some cases the first indication of the disease is a hoarseness or a snuffling; and in such the throat or the nose exhibits the first traces of the disease. More frequently the eruption about the anus and genitals is first in appearing; but this remarkable phenomenon we may expect invariably to witness, *viz.* that when the throat is first affected we may shortly expect to see the affection of the anus, and *vice versâ*.

I shall only further add, that sometimes the copper-coloured eruption is most remarkable at the fold between the lip and chin, or below the chin, and on the fore part of the neck; a coppery hue, along the joining of the lip and nose, is also o' n a very characteristic symptom of this infection.

There is still another manner in which the infant may

receive the infection, *viz.* by sucking a nurse affected with secondary symptoms of syphilis; but I am in doubt whether the diseased nurse could infect the child unless she had ulceration of the nipple; I cannot at this moment recollect an instance. The following case, which has been very lately under my care, is an instance of the infection being communicated to the infant, the nurse having, at the same time, an ulcerated nipple.

Esther Warren, æt 29, admitted into the 9th ward, 23rd of February, 1836, under Mr. Colles.

In September last this woman, then nursing the child at present at her breast, and having more milk than it could use, was induced to suckle another infant, and continued to do so until it died, at the end of three weeks. The manner of its death and the state of its body, as related by the woman, leave no doubt that this infant had been affected with syphilis. Previous to her reception of this child, she and her own infant (then four months old) were in perfect health; about the time of the death of the strange infant a sore appeared on her breast, near to the nipple; and not long afterwards an eruption occurred over various parts of her body, preceded by the customary febrile symptoms. Not being aware of the nature of her illness she did not apply for medical advice, and she has remained up to the present time without treatment, the eruption fading in one place and then reappearing in another. In the meantime her health began to decline; her appetite failed her; she was troubled with night perspirations; and, from a previous state of health and strength, she soon became weakly and emaciated. Her throat, also, has latterly become affected; it presents a deep red appearance, but there is no ulceration to be seen; the angles of her mouth are also affected by fissures: at this period of her complaint, seeing that matters were going on from bad to worse, she determined on coming into the hospital. She has had but one other child, now two years of age, stout and healthy. The infant, now ten months old, is not at all emaciated, but of a pale and rather sickly appearance. Several parts of its body are covered by a copper-coloured eruption, slightly raised and smooth; this particularly affects the parts of generation and the neck; it became affected nearly at the same time with the mother.

Mercury was administered to both the mother and child, and they were dismissed cured in the course of some weeks.

I shall now mention the outlines of a case which proves that the dry nurse (if labouring under syphilis) may infect an infant committed to her care.

My friend Mr. Cusack says:—"The child of A. B., a year and a half old, was brought to me with small superficial ulcerations about the nates; the dry nurse, whom I examined, was at the time perfectly free from any disease. Before the expiration of three weeks the nurse again called on me, and showed me an inflamed spot, about the size of a sixpence, just over the dorsum of the metacarpal bone of the thumb. From the previous affection of the child I was induced to suppose it venereal, and accordingly received her into the hospital. I placed her under the influence of mercury, but this did not prevent the occurrence of secondary symptoms, for she subsequently suffered from pains of her limbs, general copper-coloured eruption, ulcers in her throat, and iritis. During her residence in the hospital, the cook from the same family was admitted, labouring under very similar symptoms. On inquiry, I ascertained that the cook had originally infected the child, and the child had contaminated the dry nurse." I shall extract from the hospital registry an account of these two cases.

Mary Margrane, æt. 36, dry nurse, married, and mother of several healthy children, admitted into No. 9 ward, on the 21st of January, 1834, under the care of Mr. Cusack.

About four months ago a pimple appeared on the dorsum of the metacarpal bone of the thumb, which, being frequently rubbed off, was converted into a circular ulcer, which was accompanied with a slight eruption on the fore arm. She did not suffer any constitutional disturbance. When she had the pimple on the hand she was in the habit of washing a delicate child, affected for some months previously with sore eyes and an u'cer on the buttock. The child had been weaned for five onths. The ulcer on the buttock of the child was healed by the use of black wash. On the 15th of November, 1833,

Margrane was admitted for the first time into the hospital; she had then a circular sore on the hand and a slight eruption on the fore arm. On the 25th of November she was ordered Pil. Hyd. gr. v. bis die, which she continued to take till the 9th of December, when, being apparently quite well, she left the hospital, but without being salivated. In about a month afterwards a scaly eruption appeared on the face, unpreceded by constitutional symptoms.

January 21st. She was admitted a second time; and the eruption, which was of a scaly kind, had not extended to any other part. N.B.—Was examined by Mr. Cusack, and no appearance of ulcer on the genital organs. R. Hyd. Oxym. gr. ¼, c. Decoct. Sarsæ.

March 10th. Has continued the same medicines to the present time; is now free from the eruption, but still there is some slight discoloration; is made an external patient, and to continue medicine. N.B.—She continues in good health.

Mary O'Donnell, æt. 30, cook in the same family with Margrane, admitted into No. 8, on the 25th of January, 1834, under the care of Mr. Cusack. She always enjoyed good health until the end of October last, when an eruption came out all over the body, preceded by pains in the bones, palpitations of the heart, loss of appetite, nausea, and thirst. Says she had previously no ulcer in, or discharge from, the vagina; but the child, mentioned in the preceding case, was usually left in her bed during the day. She got some cathartic pills and woods from a dispensary, during the use of which the general symptoms subsided and the eruption was fading away, when, having caught cold, a fresh eruption came out, about last Christmas; she has taken no medicine since.

January 25th. Scaly eruption, of the same character as Margrane's. all over the body; pains in the bones occasionally at night; tongue clean, pulse regular, appetite good. Decoct. Guaiaci, Pil. Purg.

February 3rd. No fresh spots have come out; the eruption has ceased to desquamate. R. Hyd. Oxym. gr. ¼, c. Decoct. Sarsæ.

February 24th. Eruption gradually declining; left the hospital, and was desired to continue her medicine.

Of course we cannot say whether the infant was infected by the primary or the secondary symptoms of the cook.

I had an opportunity of seeing these two women when in the hospital, and I considered their diseases to be most unequivocally syphilis.

Here then are two clear examples to establish the opinion that secondary symptoms are capable of propagating the venereal disease; for, in this case, no suspicion whatever arose in the minds of any of the medical attendants that the disease of the child had been derived from the parents; indeed, the advanced age of the child at the time that it first exhibited any signs of the disease was quite sufficient to remove all doubt or suspicion on that head.

In cases of so much danger and so much delicacy it is a matter of the first importance to form a correct diagnosis. No man would be so rash as to be decided in his judgment by any single symptom, however strongly it may be marked. Every circumstance which can throw any light upon the subject must be taken into the account, *viz.* the history of the father's health previously to his marriage, the frequent abortions, the appearances of the expelled fœtus (if these can be discovered), the present symptoms, their succession and combination.

We sometimes meet with an infant having numerous spots of moist button-scurvy about the anus, genitals, and inside of the thighs. Should other members, also, of the family, exhibit similar symptoms, we are not in much danger of mistaking these for syphilis; because a slight attention to the history of each case, and to the marked difference in the more grown subjects, between button-scurvy and syphilis, will at once enable us to decide; but where the infant is the only member of the family so affected, we are liable to commit a mistake if we form a hasty opinion. A little attention will enable us to decide correctly; for although the anus and external organs of generation may present appearances pretty closely resembling those of syphilis, yet we shall be able to distinguish these from the latter by observing that they are
ised above the surface, that they are most distinct and distant from each other; and while those in syphilis begin

with a number of blotches pretty thickly set, these enlarge a little, and then ulcerate, without being at any time raised above the sound skin.

An eruption of a few copper-coloured spots is not unfrequently seen about the anus and genitals of infants, otherwise perfectly healthy—these we must not hastily pronounce to be venereal; they are sometimes caused by inattention to cleanliness, and sometimes by a disordered state of the alimentary canal. But these spots remain the single symptom; this disease is not found to make progress—it does not show itself in the mouth or throat, or on other parts of the body, and it soon yields to persevering cleanliness and a regulated condition of the alimentary canal.

The anus of infants is subject to fissure, which, beginning on the skin, extends in upon the mucous membrane of the rectum. We know that such is not venereal when we find it remain a single and solitary symptom; this yields to black wash and ordinary local treatment.

A superficial ulcer is sometimes found on the rugous skin of the anus; generally with one part of the edge rather deep, the rest of the edge thin. This ulcer is not very painful; it is indolent, and sometimes continues for many weeks or months. the general health of the child remaining unimpaired. Strict attention to keeping this ulcer constantly covered with some stimulating application will effect its cure in a very moderate time.

It is asserted by many authors that this disease in the infant at the breast may be most safely and very certainly cured by subjecting the nurse to a full and regular course of mercury. This, no doubt, is true in many instances; but not a few cases are reported to have required, in addition, that the child itself should be subjected to mercurial treatment. In my opinion the cure, if not more certain, will be more speedy by subjecting both nurse and infant to the use of mercury. Indeed, in some instances it is absolutely necessary to give mercury to the infant, as well as to the nurse, as in the following case :—

Eliza Redmond, æt. 35, an unhealthy-looking woman, of spare make and dark complexion, with her child (aged five months and a half), admitted into No. 7 ward, under Mr. Colles, September 3rd, 1836, gives the following history of herself :—

She lives in the country, and was married in November, 1832 ; she states that in seven weeks afterwards she perceived a sore on the right labium pudendum, which spread to the size of a sixpence, accompanied with ardor urinæ, and a discharge per vaginam, great heat, and swelling of the parts. For this she took seven pills ; her mouth became sore, and she discontinued the medicine. She was not aware of her husband having any sore, but he also took pills, and was similarly affected. On salivation being produced in her, the sore healed, and all her other symptoms disappeared, except a slight discharge. In two months from the date of the sore healing, an eruption broke out on both herself and her husband ; this was attended with pains in the large joints, particularly the shoulders and ankles. The eruption continued, off and on, for five or six weeks, but finally disappeared without her taking any medicine, but leaving her much broken in health. The husband's symptoms went off similarly.

In June, 1833, she states that a sore appeared on the same place, and of the same character with the first, attended by gonorrhœa, &c., &c. She again had recourse to the pills ; took six, and, being exposed to much cold, a profuse salivation was produced. The husband also took pills. Her sore healed, and the discharge gradually went off; but she complained of general ill health, with loss of appetite, and occasional pains at night. No eruption or sore throat followed this attack, in either herself or husband. With these symptoms she went on to the period of her becoming pregnant (in the summer of 1835), the first time since her marriage. About the eleventh week she miscarried, caused, she supposes, by fright. There was no fœtor, or marks of putrescence of the fœtus.

On the 20th of March, 1836, she was confined of this child, a girl, which she suckled herself; healthy to all appearance, and it remained so till the ninth week, when she perceived a number of copper-coloured spots, the size of from a pea to a sixpence, on the neck, face, and body of the child ; these symptoms were attended with fits of screeching and a remarkable alteration in the voice. In twelve days from this, the child getting worse, she applied for advice—got pills, seven of which affected her mouth. The child at this time improved a little ; it got no internal medicine, but had medicated baths. Shortly after this the child's mouth became sore at the angles and along the red margin of the lips, appearing first in the form of whitish blisters, running into one another, and afterwards extending to the inside of the mouth ; the spots gradually disappearing, the voice and mouth remaining much the same. The child sucked well.

June, 1836. About three months ago sores began to appear on the pudendum of the infant, cleft of the nates, and the angles of the body generally, coming on in the latter places in the form of fissures. The child's health became much worse ; it was much wasted and constantly crying.

August. Three weeks ago she came to town, and, under the care of a surgeon, took ten pills, and eight rubbings, herself ; the child received no treatment, and evidently became worse ; the mercury did not affect the mother.

September 3rd. Came into the hospital ; child is extremely emaciated and weak, has a pale sickly look, its eyes dull and heavy. takes no notice of anything, is constantly crying—in a word, is in a deplorable and almost hopeless state ; its voice is of that peculiar croaking hoarseness characteristic of its state, and its whole aspect bespeaking disease ; its lips, and the mucous membrane of its mouth and tongue, are covered with white spots, and small superficial ulcers interspersed ; at the angles of the mouth are superficial ulcers with white margins ; at the cleft of the nates, opposite the front of the coxæ, and not engaging the anus, there are two ulcers ; their edges are red and well defined ; in the centre of each there is a small spot, about the size of a pea, standing up like an island, unaffected by the disease. The rest of the sore is covered over with a layer of greyish white tenacious matter, without any appearance of pus ; the whole of the ulcer lies beneath the surrounding integument. There is an ulcer of similar character, but not so marked, across the throat ; also in each hand, and the marks of others in the folds of the skin of the thighs ; one on the pudenda is healed ; there are also slightly discoloured marks of the spots on different parts of the body, but no appearance of desquamation.

The mother has no symptoms of disease, except slight wandering pains and general weakness ; her nipples are perfectly sound, and have not been at any time sore. She was ordered, on September the 3rd, Calomel gr. ij. opii gr. ¼, bis in die ; and the child, Pil. Hydr. gr. ij. in Syrupi ʒi. bis in die ; and black wash to all the sores.

September 8th. The mercury has disagreed with the mother, and caused a purging ; it was discontinued, and she had an oil-draught. The child is greatly improved since its admission ; it is more lively, takes more notice, its eyes have lost the dull look, its voice is much improved, it cries less, and when it does it is much stronger than before. The sores are all better, and those about the anus are losing the elevated margin, and have thrown off the white tenacious coating ; are clean red, and the surface is coming on a level with the skin : that on the neck is nearly healed ; the mouth is cleaner—the white coating has peeled of, the ulcers are healed, leaving the tongue redder than before. It sucks and sleeps well ; ordered to continue the Pil. Hydr. and the wash. The mercury has not been resumed with the mother : she complains of weakness, and loss of rest and appetite ; mouth slightly sore. Ordered a gargle, and cardiac mixture.

September 11th. Mother much the same ; child continues to improve ; the mouth better, and the sore contracting and healing both from the edges and the central sound spot.

She is not aware of her husband having any symptoms of disease since June, 1833.

This case pretty clearly proved that mercurial treatment of the mother was not sufficient to cure the infant ; for while the mother was treated by a surgeon in Dublin with mercury (no medicine being, at the same time, administered to the child), the latter was not at all benefitted ; indeed, its state was all but hopeless on its admission into the hospital. Again, when the child was put upon mercury it rapidly improved, although at this time the mercury had not made any salutary impression on the system of the mother ; indeed it had acted on her, at this time, rather as a poison. Vide Report of September 8th.

The ulcer near the anus of the infant resembled, in a degree, the secondary ulcer of an adult when it is about to heal; for we often find the latter to heal in the centre, while the edges show no disposition to do so. I merely notice this fact, because I think we cannot be too particular in noticing every fact in those diseases which are not well understood; and I imagine that we are as yet but little acquainted with the syphilis of infants.

It is unnecessary to enter into particular rules as to the treatment of the nurse : this must be conducted as in ordinary cases of secondary syphilis.

For the child we may employ either the Ung. Hydr. in very minute doses, Gr. v., or Gr. iij. semel in die ; or Pil. Hydr. gr. ij. may be given daily, suspended in mucilage ; or we may give Calomel, gr. ¼, combined with sugar, once or twice a day.

It is quite in vain that we look to the state of the mouth and gums as an index to point out the action of the mercury on the child's system. No man, in the present day, requires to be told that mercury never does produce ptyalism, or swelling and ulceration of the gums in infants. The morbid affection of the mouth and throat may be attended with profuse flow of saliva, before any mercury is employed; but this excessive flow is actually found to decline in proportion as the medicine acts in a kindly manner on the system. We have no better proof of the wholesome and efficient action of mercury on the system of the child than our witnessing an improvement in the condition of the symptoms, especially of those which are not treated by any topical applications. When the mercury begins to act the bowels may be deranged, or even dysenterically affected ; but I have not yet seen any instance of mercurial erythema in an infant. In one case, which I was treating with frictions of mercurial ointment, on the second day of visible amendment of the symptoms, the child was seized with a convulsive fit, in which it expired. I ascribed the death, in this case, to the first impulse of mercury upon a highly sensitive system, for I could not discover any other probable cause of it.

The amendment, once begun, proceeds with greater rapidity than we observe in cases of similar symptoms in the adult.

Here, as in adults, the disappearance of symptoms is not always the cure of the disease; for it occasionally happens that the symptoms return after an interval of one or two months, and require us to resort again to the use of mercury.

In some cases we see one symptom linger for weeks, and even months, after all the rest have disappeared; and this, although such symptoms may enjoy the additional advantage of a topical application. This remnant of a solitary symptom would seem to be merely a local affection; for it may be removed by changing the topical remedy for another possessed of similar qualities, though perhaps not of superior activity. I believe that under these circumstances we need not resume the use of mercury.

From the commencement of the treatment we employ topical remedies for the relief of the external ulcerations, wherever seated; those around the genitals and anus imperatively demand them, as they add so severely to the sufferings of the patient. The black wash will be found of inestimable value; it soothes the irritability of the ulcers, and improves their appearance and condition, even before the mercury can have acted on the system. When a larger spot of ulceration at the anus proves obstinate, and seems stationary, even after all the other symptoms have disappeared, we should then apply some other topical means—perhaps the Ung. Hyd. Nitr. one part, with eight parts of lard, will then be a good substitute for the black wash. I am not acquainted with any remedy which can be with advantage applied to the ulcers of the lips, palate, and throat; the very act of applying anything to these irritates the ulcers, and causes them to bleed, so that I have imagined they did fully as much harm as good; and, on that account, I have long since relinquished their use altogether. I am not aware of any local treatment for the swellings of the enlarged lymphatic glands; they yield readily to the constitutional influence of mercury. The slight purulent discharge of the eyes is benefitted by a mildly astringent collyrium.

Should the parents, from ignorance of the nature of the disease, have wasted time in unavailing efforts to cure it, and have allowed the child to be so far reduced that a diarrhœa has set in, and the countenance has assumed that confirmed appearance of old age, we must consider the case as in the last degree dangerous ; however, by using the ordinary means to check this discharge, while at the same time we use our best efforts to bring the system under the influence of mercury, we afford some chance of life to the unhappy little sufferer.

The Appearance and Nature of those Diseases which are communicated by the Infant to the Nurse.

In some days, or at least in a very few weeks, after the nurse has observed the venereal symptoms appear on the child, her sufferings commence. She is first affected with what she terms a sore nipple. On inquiry it will be found that one or two pimples, or pustules, have appeared near the nipple ; these soon degenerate into an ulcer, which presents the characters of a secondary rather than of a primary venereal ulcer ; this becomes exquisitely sensitive. A slight enlargement and tenderness of some of the axillary glands quickly follow, but these glands do not betray any strong tendency to run into suppuration ; on the contrary, I must say that I have not seen one single instance in which this occurred. In two or three weeks more the nurse complains of sore throat, or of an eruption ; and not unfrequently these two symptoms appear almost simultaneously. An eye accustomed to view the secondary symptoms of syphilis does not discover any difference between these and the venereal sore throat and eruption which follow a primary ulcer in the adult. The pudenda of the infected nurse are very generally beset with small raised ulcers which discharge copiously ; these are to be seen perhaps in every case where a general eruption of the skin exists, but not unfrequently they may be found accompanying the superficial white ulceration of the throat where no general eruption exists, and where the skin remains free from a general venereal eruption. I have occasionally seen venereal iritis also attack the nurse. What other symptoms might arise, or what course the disease would follow, if longer unattended to, I cannot pretend to say, because I have scarcely ever seen any case in which the course of symptoms might not have been disturbed and deranged by the exhibition of mercury. But many of those infected nurses to whom mercury was administered became affected with very obstinate ulcerations of the throat, and with pains of the bones and joints. And in the greater number of these the disease appeared very obstinate and intractable. Indeed we have instances recorded in which the nurse has suffered the loss of the bones of the nose, and a few in which the disease terminated fatally. I am, however, disposed to think that these severe symptoms of the disease might be traced to the injudicious and excessive use of mercury employed for the cure of the early symptoms, without due consideration of the nature of those symptoms and of the reduced state of the system, which at this time is often debilitated by nursing. I am inclined to think that in most of these cases we should commence with under-doses of mercury. The following fact appears to me very deserving of notice :—I have never seen or heard of a single instance in which a syphilitic infant (although its mouth be ulcerated), suckled by its own mother, had produced ulceration of her breasts ; whereas very few instances have occurred where a syphilitic infant had not infected a strange hired wet nurse, and who had been previously in good health.

It is a curious fact that I never witnessed nor ever heard of an instance in which a child deriving the infection of syphilis from its parents has caused an ulceration in the breast of its mother. The following case when received into the hospital appeared to form an exception, but on closer inquiry this proved not to be the case. The particulars are transcribed from the hospital registry.

Anne Cullen and child, admitted into No. 9 ward, on the 19th of March, 1834, under the care of Mr. Cusack.

The child is about three months old. The extremities, the parts of generation, the anus, and the chin, are covered with copper-coloured spots slightly raised and smooth. The child is pale but not emaciated, its bowels are rather relaxed. At birth it was healthy and well-looking, and continued so until it was about a month old, at which time the body became co·red with those spots, which were of a brighter red colour than they are at present. The mother of this infant was admitted a patient into the hospital about four years ago for syphilis. The only symptoms she had then were pains in the bones; but about six months before that, while suckling her own child, when it was half a year old, she got a sore on the breast, near the nipple, which was soon after followed by an eruption. She stated that other women occasionally suckled her child. In a short time after the sore had appeared on her breast, she observed that the child's mouth was sore; an eruption afterwards came out all over its body; two other children, and a girl that attended them, became similarly affected—namely, with eruption and sore throat. They were all admitted at that time into the hospital; were all salivated, and discharged cured. After this she had two children; the first about two years since. When this child was a month old an eruption came upon its body, and in two or three weeks afterwards it died. The other child, now three months old, is (with herself) a patient in the hospital. She herself says that she never enjoyed better health than at present: her husband also is perfectly healthy.

Remarks.

In this case there was a very strong presumption that the infant was infected by some of those women who were allowed occasionally to suckle her child; that, from the child, she and all the other members of the family also were infected, except the husband, who was probably infected by the mother. I say there is a presumption that this is the manner in which the infection was conveyed, because as yet I have not seen any instance in which an infant infected by the mother communicated a venereal ulcer of the nipple to her. This case shows that parents, who are not at all conscious of any derangement of health, may yet produce a child which shall, in a few weeks, exhibit the genuine characters of venereal disease, necessarily derived from one or both parents.

With respect to the treatment of the nurse and child, I believe that very little difference of opinion is to be found among authors or practitioners. All agree that it is desirable (even though the child alone show signs of the disease) to bring the nurse under the influence of mercury; and many

assert, that this alone is quite sufficient to effect a cure of the
disease in the child. To the latter part of this statement I
have seen some exceptions, and I am certain that I have seen
a good many cases in which the child appeared to be daily
growing worse until mercury was daily administered to
it. Of this much I feel assured, that the cure of the infant
will be more speedy if it be subjected to the use of mercury
at the same time with its nurse. I am also disposed to think
that in the treatment of the nurse we shall succeed better with
small doses of mercury than with large ones. I am disposed
to ascribe to over anxiety of the surgeon, and consequently to
too vigorous treatment, the frequent failures, and still more
frequent tedious cures, of nurses infected by syphilitic infants.

The disease, as it appears in the syphilitic infant and the
infected nurse, seems to differ very little, if at all, from the
secondary symptoms which result from this disease of adults,
as contracted by impure sexual intercourse. But if we proceed
to trace further the consequences of this infection in the nurse
and in the infant, we shall discover some striking particulars
in which the diseases propagated by them materially differ
from the ordinary forms of syphilis.

With respect to the nurse, we find that very frequently her
husband becomes affected with ulcerations on the genitals ;
and these, in a short time, are attended by superficial ulcerations of the throat and mouth. If we have an opportunity of
examining the nurse, at the time her husband first complained
of ulceration of the genitals, we shall find a greater or smaller
number of small raspberry-like, moist, raised excrescences, or,
as some term them, ulcers on the genitals, and insides of the
top of the thighs of the nurse, and this while there is no
eruption on the general surface of the skin. Every one conversant with the appearances of the venereal disease in females,
knows that when they are the subject of general venereal
eruption their genitals are beset with them, and that all those
on the genitals, or on the skin in contact with them, assume
that raised moist condition above-mentioned. But in the
instance of infected nurses we shall sometimes find that
that condition of the genitals is an attendant on the affection
of the throat, without the intervention of a general eruption

T

of the skin. And here we must remark, that while spots of ordinary venereal eruption so generally fix upon the pudenda of the female, and there degenerate into those moist raised ulcers above mentioned, we see very few spots of venereal eruption on the male organs of generation, and those which are found there retain the characters of the general eruption, and indeed, we may add, are usually of the scaly nature.

The ulcers which are seen on the genitals of the husband very closely resemble those on the pudenda of the woman ; but they are not so much raised, do not yield as much discharge, and give us the idea that they are in a less moist structure or tissue. In a short time after the appearance of these ulcers on the genitals of the husband, he begins to complain of soreness of the throat, or of the lips and tongue. On inspecting these parts, we do not find the truly syphilitic ulcer ; we see generally the palatine arches pretty extensively covered with a very superficial ulceration, with broken patches of whitish lymph on and around the really ulcerated spots ; a similar appearance, but more in the form of pretty large circular superficial ulcers, may be seen on the inside of the cheeks and lips. The tongue, if affected on the sides, presents appearances similar to those on the cheeks ; but if the dorsum be affected we observe a smooth patch, perfectly bald and polished, as if the papillæ had been carried off from this spot. The common occurrence of ulcerations in the throat of the nurse being attended with ulceration of the pudenda, and, on the other hand, the ulcerations of the genitals of the husband being so constantly attended with affections of the throat, seem to point out some sympathetic connection between these parts which have hitherto been overlooked. But when we come to examine more closely the circumstances under which this apparent sympathetic connection takes place, we shall find reason to call in question this solution. For we find that such sympathy between the throat and parts in the vicinity of the anus never takes place when the former is affected with any other form of ulceration, except that which is distinguished by the milky-white state of the mucous membrane of the throat ; so that, after the most strict investigation, we are obliged to adopt the conclusion that this apparent connection is caused rather by something peculiar in the morbid condition of the part affected than in any natural or morbid sympathy between those remote parts, the seats of co-existent disease. And, in corroboration of this opinion, I observe that in every instance of such superficial white ulceration of the throat, I always enquire, and generally discover the corresponding affection of the anus; while, on the other hand, when the patient complains of the soreness of the anus, and that we there find the superficially ulcerated or excoriated state of this part, I immediately enquire for the corresponding affection of the throat; and this latter I have not unfrequently discovered when the patient was not conscious of its existence, so little inconvenience was he suffering from it. I readily admit that we may meet with instances where only one of these parts are thus affected at the time of our examination; but if we closely attend to the progress of the case, we shall seldom (if ever) find any which, at one period or other, either in its earlier or its later stages, has not had the other accompanying it.

I must here remark that my observation does not supply me with an instance of the husband having contracted the disease by sexual intercourse, except when ulcers had formed on the pudenda of the nurse.

Having traced the propagation of this disease from the parents to the infant, from the infant to the nurse, and from the nurse to her husband, let us now attend to the manner in which it may be made to contaminate every other member of the same family. If we suppose, as frequently happens, that the child is suckled in the house of the wet nurse, and that she has a numerous family of children, to one of her daughters (more particularly) is assigned the care of dressing and carrying about this infant; this is the child which first suffers from the contamination. The whole family being completely igno-rant of the nature of this disease, this girl sleeps as usual with the rest of her brothers and sisters; for among the lower orders of Irish that family is considered to be in comfortable circumstances which can afford a separate bed for the parents, while all their numerous progeny are huddled together in another bed; the discharge from the ulcers about the anus and vagina coming in contact with one of her brothers or sisters will, produce a similar ulcer on their persons, and in this manner, obviously, a number of the family contract the disease.

The readiness with which this disease is communicated by contact cannot be exceeded in this property by any other disease with which I am acquainted; I look upon it as equally infectious with the itch itself. Another manner in which this disease is made to spread through the family is by the use of the same spoon, and drinking out of the same vessel with another of the family, to whose mouth the disease may have spread. Those who are acquainted with the very scanty furniture of an Irish cabin will readily comprehend with what facility and rapidity the disease can be propagated in this manner; but to others it may be necessary to say that the family are quite satisfied with the possession of one single spoon, and the stock of cups and cans is nearly as scanty. Exposed thus to the double risk of contracting this infection when sleeping or taking nourishment, we cannot be surprised at finding three or four, in a family of six children, all at the same time infected. I shall adduce two cases to show the readiness with which the disease is communicated.

August 5th, 1834. Mr. D., a respectable mechanic, and his wife, had (about two years before) been under my care, on account of secondary venereal symptoms, of which I supposed them both to have been cured, although Mr. D. had not confined himself to the house while using mercury. Previously to this treatment Mrs. D. had many miscarriages, but I could not learn any particulars of the appearance of the abortions. She now applied to me on account of her child, which is five months old: when two months old an eruption appeared, of large copper-coloured blotches on the thighs and in the folds of the hams; the latter have degenerated into ulcers. Around the arms the skin is infected with venereal ulcers, from which, however, the child does not seem to suffer much uneasiness. On the lips and at the angles of the mouth there are white superficial ulcers; no affection whatever of the nose. Both parents appear to be in the enjoyment of perfect health, and the child has all the appearance of a fine thriving child; no symptom of disease in the mother. The mother says that the eruption on the child has been improved by washes, and some internal medicines, which she got from her apothecary in the country. The child does not seem to be fretful, or suffering any pain. I advised Hydr. c. Creta for the infant, and directed both parents to commence a course of Pil. Hydr.

As the parents lived thirty miles from Dublin I had but few opportunities of witnessing the progress of the disease. I did not see the child again until the 28th of August, 1834, when the symptoms of the infant appeared much relieved. The mother was sensibly (if not fully) under the influence of mercury; the father very slightly affected.

They now introduced to me another patient, a young country girl, about thirteen years of age; she is their servant, and has had very much the care of the child. She has a superficial white ulceration of the throat, stretching low down on the right arch and tonsil; the body and limbs covered with an eruption of copper-coloured spots, very thickly set; on the legs these spots are smaller, and of a much darker hue; on the thighs they are larger, and more distinctly of the copper hue; she has no soreness or ulceration about the anus or pudenda. Sumat Pil. Hydr. gr. iv. mane nocteque.

Neither eruption, ulcer, nor any other symptom, is discoverable in the mother.

October 2nd. The eruption on the child is scarcely visible; ulcers of the mouth healed; the skin of each buttock, close to the anus, is superficially ulcerated, presenting pretty strong characters of venereal ulcers.

The servant girl is affected by the mercury; eruption faded a good deal; throat free from ulcers, but those spots which had been ulcerated retain too high a colour. She now complains of a superficial but venereal-looking ulcer, which occupies the adjacent sides of the third and fourth toes.

December 12th. The child is now free from every symptom except the ulcers at the anus; these appear very little improved, but they do not give much uneasiness; the marks of the eruption are still of too high a colour. Lotio nigra ulceribus apud anum. Sumat Hydr. c. Creta gr. ij. alternis noct.

The servant girl appears quite free from disease; the ulcers of the throat, and that of the toes, have healed; both she and the mother of the child have had their mouths smartly affected with mercury for the last six or seven weeks.

The father's system has been, in a slighter degree, under the influence of mercury for ten weeks past; I have therefore concluded that the parents and servant are well, and have desired the mercury to be discontinued.

May 16th, 1835. The father informs me that his wife and the servant maid are both well, and have not had any return of the symptoms.

The child was weaned a month ago; it appears in excellent general health, and the ulcer on one side of the anus is healed, the other is still very little improved. ℞. Ung. Zinci ʒss. Ung. Hydr. Nitr. ʒss. ft. Ung. ulceri applicand. Repet. Pulv. Hydr. c. Creta alt. noct.

June 30th. The ulcer of the anus was healed in three weeks by the use of the ointment. The child is in excellent general health, and is now a year and a quarter old. Omitt. medicamenta.

February 13th, 1836. The child is in excellent health; no return of any symptom.

The mother was delivered of a fine child six weeks ago; the infant continues very healthy.

In this case it is perfectly obvious that the servant girl received the infection from the infant; it would seem to have first affected the mouth and throat. The ulcer between the toes was so strongly marked as a superficial venereal ulcer that I had no hesitation in considering it as part of the same disease; and here let it be observed that the mother had not in any degree, or in any way, suffered from the disease of the child; she had not received any infection from her own child, yet the disease was capable of infecting the servant girl. This child's health was in all probability upheld, and the venereal complaints kept at bay, by the medicines which were given to it before the parents brought it to town; and it is not very unreasonable to suppose that mercury formed one ingredient in its medicines, when we consider how very generally Hydr. c. Creta is given to children. I do not, however, suppose that the apothecary was aware of the real nature of the case, as he did not even hint any suspicion of this sort to the parents.

The ulcer near the anus, when it appears in a late stage of

syphilis of infants, is generally very slow in improving, and requires that the internal use of mercury should be assisted by some stimulating mercurial application locally. Indeed I have some doubts whether this ulcer could be healed by the mere action of mercurial medicines on the general system.

Michael Swain, æt. 60, admitted into No. 3 ward, under Mr. Colles's care, November 9th, 1829. He states that about ten weeks since a fissure formed in the left angle of the mouth, and at the same time the inside of the lip became excoriated; since that time the disease has been increasing. At present there is a fissure, with indurated edges, in the part before-mentioned; the gums are soft and red; the lower lip, the posterior part of the palate, and the pharynx, are studded over with superficial ulcerations. Between the root of the tongue and the arches of the palate, on the right side, are fissures resembling that in the corner of the mouth; all these parts are of a bright red colour, spongy-looking, studded with whitish eminences, and harder than natural; the parts so affected are but slightly painful; the secretion of saliva is profuse, and that fluid is turbid.

A few days since sores of the same description formed around the verge of the anus.

He asserts that he had not exposed himself to venereal infection by any illicit intercourse; but believes that a woman and child, lodging in his family, had the disease, and he and his children used the same spoons with them. He has taken some pills given him by an apothecary, and used a lotion for the mouth, but he is ignorant of the nature of these remedies.

November 10th. ℞. Decocti Hordei ʒvi. Spt. Terebinth ʒi. Vitell Ovi. Ft. Gargarisma.

14. Repet. Gargarisma. Utatur lotio nigra ulceribus circa anum.

21st. Ulcers round the anus are nearly well. Repet. Garg. et lotio.

27th. Excoriations on the inside of the lips and on the palate are more healed; the back of the pharynx and side of the tongue are much improved; those around the anus are quite well. Repet. Gargarisma.

December 5th. Nearly well; slight redness of the soft palate and pharynx alone existing at present, with very superficial ulcers at the bases of the palatine arches. Repet. Gargarisma.

26th. Discharged; free from all complaints.

Dorah Swain, æt. 10, and Margaret, æt. 8, daughters of Michael Swaine, admitted into No. 9 ward on the 18th of November, 1829. Dorah has numerous excoriations on the covering of the hard palate; the uvula and tonsils in the same state, the latter organs being much enlarged; on the dorsum of the tongue, near the base, are two large circular patches, elevated, and much resembling condylomatous ex- crescences, as they appear near the anus: the pharynx also is excoriated, and covered with viscid matter; the parts in the immediate vicinity of the sores are of a colour too highly red; the diseased parts are not very painful; deglutition is very slightly affected.

There are several condylomatous excrescences on the inside of the labia pudendi. This girl was attacked with sore throat about one week after her father. Ordered Lotio Nigra.

Margaret Swaine has large superficial excoriations engaging the posterior part of the dorsum of the tongue, in the centre of which is a slight furrow; the gum surrounding the last molar tooth in the upper jaw of the right side, and all the soft palate, tonsils, and pharynx, present an uniform red and excoriated surface; there are also several ulcerated spots on the inside of the cheeks: she has also condylomatous excres- cences on the inner surface of the labia pudenda.

Both these children were exposed to the disease in the same manner as the father; they also used the same remedies.— Extracted from the case books of Stevens's hospital.

I shall now give the history of this disease spreading through an entire family, all of whom were previously healthy, and to whom the infection was brought by a servant girl, who was dry-nursing a syphilitic infant in another family, the infant having died of the disease. The servant girl was received into Stevens's hospital; and the history of her case and reports of the treatment, are as follow:—

Margaret O'Reilley, æt. 23, an unmarried country girl, admitted on the 18th of December, 1836, into No. 9 ward, under Mr. Colles. On the right side of each lip, near the

median line, and within the red border, exists an aphthous-
looking sore, with a whitish surface, irregular edge, and indurated base, a little raised above the level of the surrounding
parts ; that on the upper is about twice the size of a split pea,
while that on the lower is something larger ; they are placed
directly opposite each other. At the angle of the mouth is
the raised and soft cicatrix of a former ulcer ; on the left side
of the tongue, near its apex, there is an oval-shaped ulcer,
about the size of a sixpence, possessing the same characters as
those on the lips, but the surface appears more elevated, and
the upper edge more regular and defined ; she only feels pain
in these ulcers while eating, and is not annoyed by a constant
flow of saliva into the mouth. The dorsum and remaining
sides of the tongue are fissured irregularly where former sores
had existed, but which are now healed. Beneath the chin
four or five glands can be felt enlarged, firm, but not very
hard ; the central one is a little painful when pressed on.
Two or three small depressed bluish cicatrices mark the
situation of former openings there. She complains of occasional pains in the shoulder, which come on accompanied
with some stiffness, generally towards evening ; is sometimes
affected with pain in the head, and with pyrosis ; her appetite
is good, sleeps well at night, and general health unimpaired ;
pulse 86.

She states that six months since, while nurse-tending a
child which had a sore mouth and some spots of eruption on
the body, and the nurse of which had a sore on the nipple,
she became affected with a sore at the angle of the mouth and
inside of the cheek of the left side ; in a fortnight after it
began to affect the tongue on the same side, and she has continued to be affected on different parts of the mouth at different
periods ; the sores healing up in one place and breaking out
in another, but never being completely free from disease since
its first commencement. The present sores have existed for
the last six weeks, and have never been touched with any
caustic applications, which the others had frequently applied
to them. The pains in her shoulder came on about three or
four weeks ago, accompanied with some soreness of the throat,
which she ascribed to cold ; she had also some slight soreness

about the anus; the pains were so violent as to prevent her working, but they had remitted a few days before her admission.

The swellings under the chin have been there these two years past, and did not subside after suppuration had occurred, nor are they larger since the present disease made its appearance. It is worthy of remark that, seven or eight weeks after the commencement of the sores, she went to the house of one of her relatives (Hogarty), and having remained there for more than a month several of the family caught the disease from her.

December 20th. The sore on the tongue is much smaller since the last report, and several red patches are appearing through its white surface; that on the lower lip has spread in a small patch near the gum.

26th. The sore on the upper lip has almost healed; that on the lower is smaller, except where the new spot had appeared, as mentioned on the 24th. The ulceration on the tongue is now distinctly striated red and white; had the pains in her shoulder last night.

March 10th.   Ung. Hydr. ℥ss. et Pil. Hydr. gr. v.

19th.   Repet. Remedia.

22nd.   Repet. Remedia.

26th.   Bal. Tep. Ung. Hydr. ℥ss.

29th   Repet. Omnia.

April 2nd.   Ung. Hydr. ℥ss.

5th.   Pergat.

9th.   Pergat.

12th.   Olei Ricini ℥ss. c. Tinct. Opii. gtt. xx. Pulv. Dover. gr. x.

16th.   Vesicatorium Nuchæ.   Pulv. Dover. gr. x.

19th.   Bal. Tep.   Mist. Purg.

26th.   Nitratis Potassæ gr. x. ter in die.

30th.   Mist. Purg.

May 3rd.   Bal. Tep.   Mist. Purg.

14th.   Gargarisma Astr.

Discharged on the 22nd of May cured.

In this case a variety of remedies were used before recourse was had to mercury; and she passed through a severe and tedious fever, shortly after her admission into the hospital.

If we attend carefully to the history and to the course of this case, we cannot but be convinced of the greater danger of infection which those run who are exposed to contact with that person who is only one degree removed from the infected child.

Hogarty and Family.

December 15th, 1835. The disease was most positively introduced into this family by the servant girl, Margaret O'Reilley, who was hired to dry-nurse the youngest child while weaning. The mother asserts that this girl when she came to live with her had (in addition to the symptoms mentioned in O'Reilley's case) sores on the pudenda, similar to those on the inside of the lips.

The infant is reported to have taken the infection first in its mouth; after some few weeks it appeared on the genitals. Next the mother took it; in her it appeared first in the mouth and then in the genitals; she says that in the latter situation it appeared like small buttons. The genitals are now free from any appearance of it, but she has a patch of superficial ulceration, accompanied with a slight swelling, on the inside of the lower lip, a superficial ulcer at each angle of the mouth, and two white ulcers, or rather blisters, under the apex of the tongue.

The daughter, twelve years old, had it first on her mouth, where now only a too high state of redness is to be seen. Her mother said that it was on the pudenda of the daughter, but on examination I could only discover on the external labium one raised spot, exactly resembling the cicatrix formed by button scurvy.

The father says it first appeared on him as a sore on the penis; this has healed without leaving any mark: it next appeared at the anus : at present there is on each side of the groove leading to the anus two small circular copper-coloured spots; the surfaces of these are constantly moist, yet one only of them can be said to be ulcerated; on the under-surface of the tongue, near to its apex, is a small white raised spot; on the right anterior arch of the palate is a superficial white u!.eration, or rather rugose patch ; a long white ulcer is seen on the inner surface of the right cheek.

I directed each of the females to take Pil. Hydr. gr. v. om. nocte, and the father to take a similar dose night and morning.

February 8th, 1836. None of the family used the medicine for a longer period than three or four days. They have now come into town (a distance of seven or eight miles), and their present symptoms are—

The father: the small ulcers on each side of the anus are as described on the 15th of December last; on the right tonsil is a distinct ulcer, not deep; on the left anterior arch is a pretty broad excoriation; a superficial ulcer on the red border of the lip.

Wife: her symptoms have scarcely altered since December; so little that the difference could not be described by words.

The daughter has now no trace of it in the mouth, but she has a slight swelling, with superficial ulceration, on the labia pudendi and on the preputium of the clitoris.

One spot on the mother's lip is exactly the same as the spot on the lip of the servant girl O'Reilley.

I rubbed Sulph. Cupri to the diseased spots in the mouth of the mother, and gave a wash of Acet. Cupri to be applied to the ulcers at the anus of the husband and the pudenda of the daughter.

September 22nd, 1836. Father—general health very good; his mouth is well, except one white spot on the right side of the tongue.

The mother says that she had an ulcer on the inside of the lower lip for five months previously to her confinement, but that since that occurrence she has been well, and now nurses her infant. The infant is three months old and perfectly free from any symptom of the disease.

The daughter died a fortnight ago of scarlatina; I am informed by her mother that she had one sore spot in her mouth at the time she was seized with scarlatina.

The child, which was the first of the family infected, was brought to me this day. It is now three years old, and, apparently, in robust health. It was one year and four months old when it was first infected, the disease commencing in its mouth and throat.

I have had an opportunity of seeing the members of this

family to-day, not because they wished to consult me on account of this extraordinary disease with which so many of its members had been afflicted, but that I should prescribe for the father, who had received a very slight wound of the cornea from a splinter of stone which struck it while he was at his work of a stone-mason. The family living at the distance of seven or eight miles from Dublin, and being poor, could not be induced to come into Dublin more frequently than above represented; of course I had no opportunity of watching the effects of remedies, and I wished rather to watch the progress of the disease undisturbed by treatment than to direct any particular plan which I well knew would not (and, indeed, could not) be followed with punctuality by persons in their condition. Their sufferings were not so severe as to oblige them to seek for admission into an hospital. The small quantity of mercury which any one of them used was so very inconsiderable that it may be left out of the account altogether.

The history of this disease, propagated by an infected infant, would be incomplete if I did not add that sometimes one or more of the soft raised ulcers will be found in the axillæ, and occasionally, though more rarely, in the folds of the groin.

That this disease is derived from lues venerea, I think has been clearly shown; and yet it does not in every respect resemble the lues of adults, but possesses some characters peculiar to itself. And, first, it appears to be infinitely more contagious; for very rarely do any escape who are for a short time exposed to its infection; this is not the case with the syphilis of adults, for we know that very frequently men escape, though exposed to infection under circumstances calculated to render the parts highly susceptible, and to facilitate the introduction of the poison. And here I would remark, that in the very remote parts of Ireland the poor people are so strongly impressed with the notion of the very infectious nature of the venereal disease, that if they be told that a stranger whom they had lodged in their house for a night had this disease, they would instantly burn the straw seat upon which he had been sitting. I may be told by some that men may contract syphilis by sitting in a public privy; to this I can only answer that I have never witnessed a single instance; nor did the late Mr. Obre, who had been for many years most extensively engaged in treating the venereal disease; for on asking him if he believed that the disease was propagated in this manner, he shrewdly answered, that it sometimes was the manner in which *married* men contracted it, but *unmarried* men never caught it in this manner.

Secondly. The symptoms of this disease bear a most exact resemblance to each other, in every individual affected by it; neither age nor sex, nor difference of temperament, seem to make it differ in its characters, even though it may have existed for some months. Now we know that syphilis in adults is subject to very considerable varieties, modified perhaps by peculiarity of constitutions; *e.r. gr.*, the various forms of eruption, the great varieties in venereal ulcers of the throat, &c., &c.

Thirdly. This disease, in the third remove from the original syphilitic infection, appears to be permanently fixed to the parts it first seizes, and to be of a much milder nature than the corresponding symptoms of pure syphilis. Thus we do not find that in the throat it ever assumes that destructive form of ulceration which too often attends venereal sore throat of adults. I have never yet seen an instance of loss of substance caused by this disease, even though it may have existed for more than twelve months.

Again: I have never seen this disease produce ozæna or iritis; I have not known it in any case affect the bones or joints; nor, in men, affect the testes.

Fourthly. This disease appears to become less virulent as it becomes farther removed from the fountain head; of this we become assured, by observing the greater facility with which the cases of more remote infection are cured. Thus we find that when the disease originates with the infant, the use of mercury is absolutely necessary for the cure of the infant; unless this be employed, the child dies. The nurse infected by this child must undergo a course of mercury for the cure of her symptoms; the husband also must undergo a mercurial course, to effect a cure of his symptoms. Such of the attendants as contract the disease by immediate and direct contact with the ulcers of the child will also require mercury for their cure. But such other members of the family as receive the infection from the attendant thus affected, these may all dispense with the use of mercury. These positions are many of them borne out by the history of Hogarty's family; for the servant-girl, O'Reilley, who was infected directly by the diseased infant which died, could not be cured without mercury. The length of time that she remained in the hospital, and the number and variety of other medicines which she in vain employed, are sufficient to establish this fact. Nor was the disease removed, even temporarily, by a very smart attack of fever, which continued for three weeks and upwards. Yet neither Hogarty's child, which was contaminated by this servant-girl, nor his wife, who received the infection from her own infant, required mercury to cure them: the infant appeared to have been cured by a medical practitioner in their neighbourhood, who employed washes of different kinds, and a few powders internally. The disease of the mother seems to have been eradicated by parturition, although it had resisted various local applications irregularly employed, and had continued to afflict her for a period of many weeks. The disease of the father may be considered as cured, as he had only one small white spot on the tongue.

One fact well deserving our attention is this: that a child born of a mother who is without any obvious venereal symptoms, and which, without being exposed to any infection subsequent to its birth, shows this disease when a few weeks old; this child will infect the most healthy nurse, whether she suckle it or merely handle and dress it; and yet this child is never known to infect its own mother, even though she suckle it while it has venereal ulcers of the lips and tongue.

The foregoing chapter will be admitted to be one of the most valuable as regards original observation in Colles's work. In it he—

1st. "Establishes the opinion that secondary symptoms are capable of propagating the venereal disease."

2ndly. While showing that a child may receive the infection of syphilis " by sucking a nurse affected with secondary symptoms of syphilis, he cannot recollect an instance in which the diseased nurse infected the child, *unless she had an ulceration of the nipple.*"

3rdly. It is in this chapter that he more than once very clearly announces the fact which is in truth the basis of the important generalization which Mr. Jonathan Hutchinson has very well named " Colles's law."

His observations go to show that those who have once suffered from syphilis have more or less immunity from the disease afterwards. But more than this, he adduces facts which lead to the belief that when a mother acquires syphilis by blood contagion from the fœtus, she gains an immunity without suffering herself from any severe form of disease.

As regards the former of these propositions we have the following remarkable pronouncement :—

He states a fact which has been forced on his observation by more than five or six instances—namely, that a newly married man, admitting that he has had syphilis within nine months before, yet who is himself free from every appearance of syphilis and every other disease, shall yet infect his wife in such a manner that secondary symptoms shall appear in her in a few months after marriage, and these not preceded by any primary symptoms. And although among these secondary symptoms some raised ulcers shall fix on the external pudenda of the wife, *yet the husband shall very rarely suffer in any manner from these.*

As regards the immunity of the mother, he says :—" The following fact appears to me very deserving of notice : I have never seen or heard of a single instance in which a syphilitic infant (although its mouth be ulcerated), suckled by its own mother, had produced ulceration of her breasts ; whereas very few instances have occurred where a syphilitic infant had not infected a strange hired wet-nurse, and who had been previously in good health.

" It is a curious fact that I have never witnessed nor ever heard of an instance in which a child, deriving the infection of syphilis from its parents, has caused an ulceration in the breast of its mother."

The concluding paragraph of this remarkable chapter reiterates the same statement even more strongly, with the addition of the fact, the importance of which is so justly dwelt on by Mr. Jonathan Hutchinson, that the mother may all the while be *without any obvious venereal symptoms.*

In order to give their true value to these observations of Colles, we must carry ourselves back to the period at which he wrote. We must remember that Hunter had denied the contagious character of secondary lesions, that Ricord still in a great degree upheld the doctrine of Hunter on this subject, and that the results of the unjustifiable inoculations of Wallace were as yet unpublished or little known.

That neither the husband who had had syphilis got it back again from his wife, nor the mother from her own child, although the wet-nurse became infected by the same child, seem justly to be regarded as the great fundamental observations on which is based the dogma that as a rule syphilis " is not doubled "—is not taken a second time by the same individual.

Lastly. There is reason to believe that a mother may become diseased by infection from the fœtus without the intervention of any chancre, and without the appearance of any symptom recognisable as secondary syphilis. Yet such a mother may give birth to a child, itself infected with syphilis, capable of infecting the wet-nurse, its own mother having gained an exemption from the possibility of infection.

I am not acquainted with any observer who has so fully laid hold of the importance of these facts as Mr. Jonathan Hutchinson.

I may be allowed, therefore, to give his views in his own words:—

" If," he says, " Colles's law be one which has no exceptions, it follows that all women who bear syphilitic children contract syphilis; for how else can they obtain immunity? And, since it is notorious that women under such circumstances scarcely ever show secondary symptoms, it follows, further, that we have here a form of syphilis which is protective, but which is unattended by any cutaneous outbreak. Thus, syphilis acquired by blood-contagion from the fœtus would appear to be, for the mother, a parallel with vaccination in regard to smallpox: she gains immunity without suffering from any severe form of disease. The botanist will at once suggest that probably in both cases the explanation is to be found in heteromorphism or alternation of generations on the part of the fungus. To him the clinicist might aptly rejoin, that really one might almost have expected it; for, when the mother gets syphilis from the fœtus, she obtains it from fluids in which the plant-life is evidently under some very special restraint; for in the fœtus itself, as a rule, no development of it takes place during the nine months of intra-uterine life. We have only to suppose that the same condition of the yeast which existed in the fœtus is perpetuated in the mother, and the thing is done. We had no right to suppose that infection by inoculation of solids (or chancre-contagion) would be exactly the same in its results as direct imbibition by the blood.

" As we may call vaccination developed yet protective variola, so we may consider syphilis, derived from the fœtus, as an undeveloped yet protective form of that disease; and we have here another most interesting point of analogy between syphilis and the exanthems.

" Before, however, we accept as probable such a possibility as that just hinted at, it is desirable to look at the facts with the utmost incredulity. Let us doubt unsparingly at every stage of the reasoning. First, is Colles's law true? I cannot see any escape from the conclusion that it is. It was announced in 1837, and has received, I believe, the assent of every authority who has written on the subject since. It has attracted attention both at home and abroad, and I am not aware that a single exception to it has been recorded. We have all of us seen chancres on the nipples of wet-nurses. They are, indeed, not very infrequent. We have, however, none of us seen such on those of the mothers of infected children. Let us remember that it is very unusual to put a syphilitic infant out to wet-nurse,—a thing which no prudent surgeon would ever permit,—and that, probably, for one so nursed a hundred are suckled by their mothers ; and we shall appreciate the weight with which this entire absence of proof that mothers ever suffer bears. It amounts, I think, to all but proof that they are absolutely insusceptible. It is as strong in that direction as is the rarity of smallpox within short intervals after successful vaccination. We must remember also that these mothers of syphilitic infants not only suckle them, but handle them, dress their sores, and in various ways through long periods expose themselves to risk. If it be granted that it is proved that these mothers—a very numerous class—have really in some way had syphilis and acquired immunity, I do not think there can be much dispute as to the next fact, that they do not during preg-nancy show any of the usual symptoms of the disease in its secondary stage. This is a matter of everyday experience. There remains, how-ever, the possibility that the syphilis may have been gone through prior to pregnancy ; and I am well aware that the few remaining writers who teach that syphilis can be inherited only from the mother will hail this confirmation of Colles's law as a strong support for their opinions. Here, however, again I must appeal to everyday experience. Is it not the fact that women bear syphilitic children without having ever themselves, either before or during pregnancy, had any symptoms, either primary or secondary, of that disease ? If this happened only once or twice, we might reasonably doubt the histories given us. But it is not so ; it is in hundreds of cases ; and few, I think, of much experience, can doubt that, as a rule, syphilis is inherited from the father ; and that the mother never shows any external signs of the malady."

CHAPTER XVI.

THERE is no class of complaints in which we may not occasionally meet with instances in which the distinctive characters of the disease are but faintly marked, and in which the symptoms of some other disease are more or less closely imitated. The venereal disease does not form an exception to this assertion: on the contrary, it is among those which are most subject to this irregularity; and this remark applies to each symptom separately, as well as to the combination and to the order of succession of its symptoms.

In the present state of our knowledge I think it is impossible to lay down any rules by which our judgment can be correctly guided in deciding on the nature of those simulating cases; for surgeons now no longer acquiesce in the opinion of Mr. Hunter, page 439: "The venereal matter, when taken into the constitution, produces an irritation which is capable of being continued independent of a continuance of absorption, and the constitution has no power of relief; therefore a lues venerea continues to increase. This circumstance is perhaps one of the best distinguishing marks of the lues venerea, for in its ulcers and blotches it is often imitated by other diseases, which, not having this property, will therefore heal, and break out again in some other part. Diseases in which this happens show themselves not to be venereal; however, we are not to conclude, because they do not heal of themselves, and give way only to mercury, that therefore they are venereal, although this circumstance, joined to others, gives a strong suspicion of their being such." Nor do surgeons agree with Mr. Abernethy, in admitting that those simulating diseases are not venereal, "because they got well without mercury." On the contrary, I should think that there are w surgeons of the present day, who have read the statements published by several military surgeons of Great Britain, of the appearance and progress of syphilis as it appeared in the late Peninsular war, who do not believe that the symptoms of syphilis occasionally disappear without the use of mercury; and there is no surgeon extensively engaged in this branch of the profession who has not in his own practice repeatedly observed that the venereal disease, both in its primary and secondary forms, will sometimes be made worse by the use of mercury, and yet at some future period this very same case will yield to and be cured by mercury, and by mercury alone. We now no longer entertain the opinion that mercury is a test of the syphilitic or the non-syphilitic nature of any particular case. I think that, in deciding on primary ulcers, there is one source of doubt and difficulty very deserving of attention, and yet it is one which I believe has been very generally overlooked; I mean the discharge from those excrescences on the genitals of females which attend on venereal eruptions, and on some forms of venereal sore throat. This discharge is capable of producing an ulcer on the genitals of the male; and this ulcer, though in appearance and character very unlike the true chancre, is yet capable of contaminating the general system, and of inducing secondary venereal symptoms. The confusion which may arise from our overlooking this fact does not stop here; for by watching the progress of the secondary symptoms consequent on this form of primary ulcer, we shall soon discover that these also differ, in many material points, from the corresponding secondary symptoms which follow an Hunterian or a true venereal chancre. By referring to the chapter on the Syphilis of Infants we shall see these positions clearly illustrated and amply proved.

The next circumstance which tends to embarrass our judgments, in deciding as to the true nature of primary symptoms, is the variety in the effects of mercury, according to the manner in which it has been employed, and according to the influence which it has exerted on the constitution. Thus the appearances exhibited by a chancre or bubo, when mercury has been administered in doses so large as to have induced a general febrile irritation, will be very different from those changes which a legitimate ptyalism will effect. And even the appearance produced by this injudicious exhibition of mercury will be different, according as this medicine has been early discontinued or has been obstinately persevered in for a length of time. I believe few will hesitate to admit that inordinate and unsuitably large doses of mercury alter the appearance of primary symptoms, and often derange the succession of the secondary. But I fancy it is not so generally known that moderate and suitable doses of mercury may be mismanaged, so as to be productive of less serious mischief to the system in general, but of equal or even greater perplexity to the surgeon who is to decide on the nature of local symptoms. I mean when these moderate doses have been so employed that the patient was made to desist from their use the moment even a slight approach to ptyalism was perceptible, and was desired to resume them again when all tendency to ptyalism had subsided; when .by repeatedly resuming and relinquishing mercury, without allowing it ever to act on the salivary organs, the general system comes to be irritated into a sharp degree of fever, and to be acted on by mercury as if by a poison. I imagine that many surgeons indulge the hope that moderate doses of mercury can do no harm; but they do not seem aware that when employed in this very timid manner it may alter materially, though not cure, the symptoms; and may as completely, though more slowly, break down the constitution, as when it is exhibited in excessive doses and for a shorter period. This error in the administration of mercury I look upon as a very fertile source of those derangements or irregularities in the appearance and in the order of succession of the various symptoms of the venereal disease.

Now, while I admit that some ulcers of the genitals and some affections of the inguinal glands are occasionally met with, which, though not venereal, yet strongly resemble the primary symptoms of that disease; and while I freely declare that I have repeatedly met with a combination of symptoms which were with difficulty to be distinguished from secondary venereal symptoms, yet I cannot believe that the number of those diseases which resemble or simulate syphilis is so great as many authors would lead us to suppose. And, indeed, I must say that the attempt to form a new class of diseases, designated " Pseudo Syphilis " or " Cachexia Syphilitica," has only tended to embarrass our practice and to divert our attention from the real causes of those irregularities. Instead, therefore, of any attempt to point out the distinctive characters whereby we are to discriminate between the true syphilitic and the pseudo syphilitic case, I should rather recommend to apply ourselves sedulously to search out the natural history of the venereal disease, and to become acquainted with all those (probably minute) circumstances which influence the effect of mercury on the general system, and on the particular symptoms of the venereal disease. In the meantime, I do not apprehend that much mischief is likely to arise from our occasionally treating as venereal another disease resembling it, provided that the mercury be given with due caution, and regulated by sound judgment. Mr. Abernethy himself makes the following admission :—" The effect of exciting a mercurial affection of the constitution in diseases resembling syphilis is, as far as my observation enables me to determine, very various. It sometimes cures them very suddenly, and very differently from the gradual amendment which it produces in truly venereal diseases. Sometimes, however, these diseases yield more slowly to its operation, and are cured permanently. Sometimes the diseases recur in the same parts, after a severe course of mercury ; sometimes mercury merely checks the disease, and can scarcely be said to cure it ; in which case it seems important to support the strength of the constitution, and to keep up that mercurial effect which controls the disease, and can be borne without material derangement of the constitution for a great length of time. Sometimes, also, the use of mercury aggravates these diseases."

Now it appears to me that, unless in some few instances which involve the moral character of the patient, it is not material by what designation we mark the disease, provided that we do but cure it. The patient will not be the less grateful for the favour; though his disease may have been misnamed, he will be quite satisfied if it have not been mismanaged. I suspect that a frequent cause of our failures in the treatment of such cases,—whether those broken down cases of syphilis with their distinctive characters very much impaired, or those cases of general cachexia which assume some characters of syphilis,—I say the cause of our failing to cure such has been an over-anxiety to push the mercury to such an extent as to afford a security against a return of symptoms which had previously eluded the action of this medicine. If we lay it down as a rule of practice in such cases to use mercury in very moderate doses, suited to the condition of the general health, and to support the strength of the patient during the influence of a mild ptyalism, I imagine that our success in these hitherto deplorable cases will be most gratifying; and, assuredly, by such a line of conduct we shall avoid all those mischiefs and dangers which have followed in such cases from the too large use of mercury.

From the records of the hospital I shall extract the following case, which is headed " Cachexia Syphiloidea " by the clinical clerk, a young but a very intelligent surgeon.

Anthony Brady, æt. 32, a shoemaker, greatly emaciated, admitted into No. 3 ward, 12th of December, 1835.

He states that he was first subject to the influence of mercury eleven years ago for the cure of a gonorrhœa; and again, three months afterwards, for a bubo, unaccompanied by any sore on the penis.

Ten years since he contracted an ulcer on the penis, together with a gonorrhœa, for which he was salivated: the sore had scarcely healed when he was attacked with pains in the larger joints, followed by emaciation, febrile symptoms, and an eruption of pustules, which degenerated into small ulcers covered with scabs. He was now admitted, for the first time, into a London hospital, where he again used mercury; the spots of eruption healed; their white and smooth cicatrices are still evident on the face and arms. He left the hospital of his own accord two months from the time of his admission, and returned again with a second eruption of the same kind, and more debilitated than ever; one of the spots in each ham assumed the phagædenic character, became larger than any of the others, and were very painful; but he got rid of all these symptoms in nine weeks, without his mouth having been affected by mercury. Pains in his joints were now the only symptoms remaining, which continued to affect him for a year and a half; when, after exposure to cold, he was attacked with iritis of the left eye, was admitted into the hospital, and again salivated. About the same period the right testicle began to enlarge, became very hard, but not painful: the treatment not seeming to produce any good effect, he was recommended to have it removed; but before he was strong enough to bear the operation, an abscess formed in its substance, which was opened, and a fungus soon after protruded, but was removed by caustic applications and alterative doses of mercury. He was discharged, to go to the country, three months from the date of his admission, the testicle remaining enlarged, and the situation of the fungus not quite healed. The ulcer, however, soon after cicatrized, and the testicle finally wasted away to its present size. He returned again in six months afterwards, with pains in his bones, and nodes on his forehead and shins, and two spots of phagædenic venereal eruption on the hip and thigh of the left side; he also had sore throat and ozœna. The node on the forehead was opened; no bit of bone ever came away from it, though a depression exists there, marking the loss of substance. He had exfoliations from both tibiæ, and lost several of the nasal bones and the greater part of the septum.

His mouth was made sore twice while he remained in the hospital, from which he was discharged one year and nine months from the time of his admission; and the ulcers on his legs, where the bone had exfoliated, were not yet healed.

It is now six years since that time. Debility, emaciation, night-sweats, pains, and loss of rest, were then his principal symptoms; for which latter he was obliged to take laudanum to procure sleep, and which he has continued to do ever since. When he went to the country he became somewhat stronger, and gained flesh: the pains, however, continued; his knees became stiff, so that they were kept constantly in the semi-flexed position, and a year afterwards he was admitted into another hospital, where, after remaining in bed for three months, and using some medicine which did not affect his mouth, he was again able to move about. The ulcer on the right leg was also healed, but broke out again immediately on getting up; some swelling, which had existed in the right knee, had likewise disappeared. From this time, though not quite free from the pains, he was able to move about, and work at his trade, until about six months ago, when he got cold, and the pains were increased, which have ever since kept him almost constantly confined to bed. He accidentally hurt the left testicle about four months ago, which became swollen and painful soon after; the pain subsided, but the enlargement continued to increase.

His knees have again become stiff these last three months, and his body more emaciated than ever.

The left testicle and epidydimis, which can no longer be felt distinct, are enlarged to the size and shape of a turkey's egg, even and uniform on the surface; they have a firm and solid feel posteriorly, while anteriorly the swelling is tense and elastic; there is some effusion into the tunica vaginalis, but not in sufficient quantity to render the tumour transparent. He has not the least pain in the testis, but he feels a sensation of weakness in his loins when it is allowed to hang. The spermatic chord is very slightly enlarged; the vas deferens possesses its natural size; the scrotum moves freely over the testicle, and is not discoloured. The opposite testicle is only half its original size.

The liver appears to be enlarged, but from the tense state of his abdominal muscles it is impossible to examine it accurately with the hand. There is also an enlargement of the glands of the groin, but none of those of the axilla or neck.

He has pains in his shoulders and knees, which latter are kept permanently semiflexed, and cannot be extended; the left one is very tender to the touch, with some effusion into the joint. The ulcers on the legs, where the bone exfoliated, present a very irregular appearance; that on the right leg is situated at its upper and outer part, sinks between the bones, its margins are uneven and of a purplish colour, and its surface covered in some places with a whitish matter that cannot be wiped off. That on the opposite limb is lower down, smaller, and deeper, and a probe, passed to its bottom, detects a portion of the bone to be in a carious state. A soft node exists above the outer ankle of this side. The legs are

attenuated to the last degree, not the slightest appearance of
the calf; they certainly answer to the description of being
only skin and bone.

He complains of languor, debility, and constant lowness of
spirits; his sleep is broken and disturbed, scarcely ever more
than two hours at a time, though to procure it he is obliged
to take half an ounce of laudanum every night; and, when
omitted, his condition is most miserable and the pains are
dreadfully severe. His appetite is bad, tongue clean, and
bowels regular; pulse 76, soft, and weak. He complains of
an occasional fluttering about his heart and coldness of his
extremities, especially if he did not take his usual dose of
opium; the sounds of the heart are however perfectly healthy.
No perspirations at night.

Soon after his admission he was put on hydriodate of
potash, which was stopped in consequence of its disagreeing
with the stomach and bowels. Since that time he has been
taking sarsaparilla and quinine.

February 13th, 1836. Appetite has improved, but he has
not gained flesh or strength since admission; the pains in his
shoulders have disappeared, but are still present in the knee-
joints at night; the fluid effused into the right one has been
absorbed.

The dose of tincture of opium has been diminished to three
drachms every night. Pergat cum Quinæ Sulph. Vini Rubri
ʒiv. o. d.

February 22nd. No further change; the soft node has
disappeared. Ung. Hyd. gr. x. o. n. genibus affricanda.

28th. Has rubbed in five times; pains in the knees are
somewhat better, and he is able to extend them a little.
Pulse 80; other symptoms continue. Pergat.

March 6th. His mouth is beginning to be affected by the
mercury; the pains in his knees have disappeared, and he is
able to stretch them further than at the last report. The
ulcer on the left leg has assumed the healing aspect, and its
surface is clean and granular; the one on the opposite limb
still presents a whitish surface. The effusion into the tunica
vaginalis is less, and the testicle is smaller; appetite improved; bowels regular; pulse 84. The dose of tincture of

opium has been diminished to two drachms every night. Pergat.

8th. Mouth sore. Omitt. Ung. Hyd.

10th. The salivation continues; he feels much stronger; he is now able to extend the knees better, but not yet perfectly; the fluttering at his heart has disappeared; the swelling of the testicle has decreased; he sleeps better at night; pulse 84. Simple dressing to ulcers.

15th. The ulcer on the right leg is almost healed; that on the left has extended, and its surface is whitish; discharge thin, and in some quantity; he sometimes feels stings of pain in it. His appetite is now good; pulse 88; strength increasing. Tinct. Opii ʒss. o. n.

18th. The sore on the left leg has improved in appearance, is red and granular, and discharges healthy pus. He looks much better, and is beginning to gain flesh. Pergat et Lotio Nigra Ulceri, et Tinct. Opii ʒiss. om nocte.

25th. Sulph. Quinæ gr. ij. ter in die, Tinct. Opii ʒj. o. n.

April 2nd. He has gained flesh and become much stronger since last report; his appetite is very good; he has no pains, and he sleeps almost as well, after taking ʒss. of tincture of opium at night, as he did before when his usual dose was half an ounce, but is very easily awakened; tongue clean; pulse 84.

The ulceration on the left leg is beginning to fill up with granulations, and is now not larger than a sixpence; the discharge from it is diminished; the one on the opposite limb has healed.

He is not yet able fully to extend the legs, nor can he bear on them from weakness; he feels no pain in the knees, but is often annoyed at night with starting in them, extending down to the soles of the feet. He now walks about the ward with the help of a walking-stick.

The left testicle is now only a little larger than natural; but remains hardened, though the induration is not so great as on admission. Lint. Camph. genibus affricand. Sulph. Quin. gr. ij. ter in die.

This man felt himself so much improved that he quitted the hospital about the middle of April, having continued to improve daily.

In my opinion we often meet with cases of cachectic patients whose health is much more quickly restored by the use of very small doses of blue pill, or some other mild mercurial, combined with such other medicines as the circumstances of their case may require. The mischief occasioned hitherto by giving mercury in such cases seems to me to have arisen from its having been employed in the ordinary doses, instead of doses extremely minute.

CHAPTER XV.

ON THE NON-MERCURIAL TREATMENT OF SYPHILIS.

I SHALL not, with my limited experience of this plan of treatment, attempt to enter into a detailed account of its application to particular symptoms; I shall only offer a few general remarks. When this plan first attracted the notice of the surgeons of Great Britain both my colleagues and myself adopted it in our hospital. In general we confined this treatment to men who had not used any mercury; but as in Dublin it is extremely difficult to meet with venereal cases in hospital who have not applied to some apothecary, or received medicines at a dispensary, we could not adopt it as the general practice of the institution. However, we tried it until we all became convinced of this fact—that it was not suited to patients who were obliged to earn their bread by labour; for we saw that after they had left the hospital and got into employment they generally found themselves weak, and unequal to their usual labour; and often, at the end of two or three months, they returned emaciated, pale, and enfeebled, in consequence of the hectical form of fever which was about to usher in a new series of venereal symptoms. Their stay in the hospital also proved in general very protracted; so that they then became impatient of this treatment, especially when they saw others with similar symptoms in the same ward have their complaints more quickly cured by the use of mercury. In private practice, also, I employed it for a time; but not finding it superior in point of quickness of cure, or of security against a relapse, and observing that these relapses were more frequently reiterated, in a short time I ceased to employ it, except at the express solicitation of the patient. But I had many opportunities of witnessing the results of the practice of some of my brethren who adopted it more fully. Of course I could not know much of those cases in which this practice was successful; but in many of those who suffered from secondary symptoms, and from relapses of the different series of secondary symptoms, I had melancholy proof that this treatment was too often unequal to remove syphilis. No doubt fewer of the non-mercurial patients complained of affections of the bones than those who had been ineffectually treated by mercury; but I saw instances of closed pupil and opaque lens, produced by iritis which had been neglected, not having been considered as venereal symptoms. I have seen many cases where the soft parts of the throat had suffered severe mutilations : and, above all, I had too many opportunities of watching the very slow and silent, but sure inroads which the often-repeated attacks of secondary symptoms made on the constitutions of the patients ; of witnessing this phenomenon—that the venereal disease, from year to year, showed itself with less striking characteristics, while other diseases appeared to spring up ; so that, for some months before the death of such patients, it would require a close examination to discover the one or two slightly-marked symptoms of syphilis which remained : and also required close research to trace the symptoms of apparently the last and fatal disease to its true source, the infection of syphilis. But on tracing the state of health, from the primary ulcer down to the final and fatal disease, I could clearly see that at no period was the unhappy sufferer altogether free from the venereal disease ; so that both the patients and their friends, in many instances, lost sight of the original syphilitic disease, and referred the death to some other apparent cause, such as ascites, or some disease of the lungs. Nor is it to be wondered at that non-professional persons should form such an erroneous opinion, seeing that the course of some of these fatal cases had occupied a period of four or five years between the appearance of the primary ulcer and the fatal event. In a word, after the experience of one year's full trial of the non-mercurial plan, we have since, in our hospital practice, only employed it rarely, and generally in very mild slight cases of primary symptoms.

But we must acknowledge that the profession is indebted to those who have lately introduced the non-mercurial plan of treatment ; for we have now not only acquired a second line of treatment for venereal cases, but, what is of the highest value, we have been released from an inveterate and deep-rooted error—from an unfounded conviction that the venereal disease could not be cured by the innate powers of the system, unless aided by mercury. I need not add, that all the opinions and practices consequent on this prejudice have been subverted.

I think it must, however, be admitted, that the non-mercurial plan has not fully answered the expectations of its early admirers. All will allow that many cases of syphilis have resisted this treatment, and were afterwards cured by mercury. Let us revert to the history of its introduction. It was introduced at a time when (in Great Britain) mercury was rather in disrepute; and when in the Peninsular army the venereal disease, treated by mercury, was making frightful havoc among the soldiery. It was tried on a scale beyond that of any other medical experiment, and under circumstances the most favourable; the patients being subjected to military discipline and restraint. In civil life it would naturally obtain a preference from both patient and surgeon; the former would be relieved from all the horrors of a mercurial course, the latter find in it a line of practice simple, plain, and safe; one that required not any extensive observation of the venereal disease, or any nice and accurate judgment in the employment of the remedies. Yet it has not superseded the mercurial plan of treatment; on the contrary, it seems now, after a trial for twenty years, to have fewer and less warm admirers. This may possibly be fairly accounted for by observing that this plan subjects the patient, especially when the symptoms are of an inflammatory character, to the most rigid quiet, even continued confinement to bed, to repeated bleedings both general and local, to the long-continued use of nauseating medicines, to frequent purging and very low diet. So that, during the treatment, the patient is subject to all this severe discipline, and, after it, suffers a proportional degree of weakness; and, to all this, I am disposed to add relapses more frequently repeated, although less severe than those which follow after the mercurial treatment.

Still I trust that this line of treatment will not be hastily given up. And here I may be allowed to express a sincere hope that those surgeons who continue to employ it will direct their attention to the following points, *viz.*, for what symptoms or combination of symptoms, or conditions of the general health, is this line of practice best fitted? whether it be preferable in warm weather, or in what seasons it is most likely to succeed? But, above all, I trust that they will turn their attention to the natural history of the venereal disease, and thus furnish the profession with what is so great a desideratum, and one so difficult to attain, as long as the mercurial treatment alone is employed. Let not the spirit of enquiry be biassed by any rivalry between the two plans of treatment. Let not the non-mercurialists try to advance their cause by depreciating the mercurial treatment. Neither plan is entitled to the praise of *cito, tute, et jucunde*. Further observation is required to improve each and give them some claim to the above character.

While the non-mercurial plan appears to have rather retrograded of late in public estimation, I am yet aware that equally strong objections may be urged against the mercurial treatment; for it may be said that this medicine has been employed in the cure of syphilis more universally than any other medicine for ages past, and yet at this day it must be admitted that many cases have resisted its use,—and still further, that often where it did not cure, it considerably aggravated the sufferings of the patients; and this, although mercury has obtained the title of a specific for the venereal disease. But I think much may be said in vindication of mercury.

Hitherto, and indeed at this very day, surgeons have been very unreasonable in their expectations of the powers of mercury. No doubt many weak minds, misled by a name, have thought, if they administered mercury, which is considered a specific, they must cure the venereal disease. Yet it is strange that men have dealt more justly with other specifics: thus Peruvian bark is looked upon as a specific for the cure of intermittent fever; and it will, if judiciously administered, succeed in a great majority of cases of this disease; but still all must allow that this specific may not only not cure, but may aggravate this disease, and even induce others of a more dangerous nature—if, for instance, it be given when the stomach and bowels are loaded; the same unfortunate result will also follow its use if it be given at an improper period, *ex. gr.*, during the paroxysm. Or, again, if it be directed in doses unfit in point of quantity, it must be acknowledged that it has failed; thus if given in very small doses it will prove unequal to arrest the progress of the disease; or if given in excessive doses it will derange the stomach, so that this organ can no longer retain it. I need not add that this specific must also fail if it be administered when the constitution has been much broken down, and when a material derangement of structure, in any important viscus, has been induced by the long continuance of the intermittent fever. Now, when we come to recollect that the intermittent fever is a disease pursuing a single and regular course, we must see how much more easy it is to lay down plain and simple rules for its treatment than we could do for a disease so complicated as syphilis is. Again, although Peruvian bark, is in intermittent fever and some few other diseases, an useful and effective remedy, yet we know that it is not possessed of such active powers over the animal body as mercury is; and, therefore, that when mismanaged it will not be productive of as much mischief. Let us not impute to mercury more evils than those which really belong to it; let us draw a distinction between the judicious use of this powerful medicine and the mal-administration of it. When we recollect how very universal the venereal disease is, how very numerous its symptoms in every one of the stages of this disease, how strangely these symptoms are modified by the habit of the patient, or by accidental circumstances—*ex. gr.*, by inflammation; when we call to mind the astonishing powers of this medicine over the animal economy; and when, in addition to all these considerations, we reflect upon the immense number of injudicious, ignorant, uneducated persons who fearlessly and constantly venture to undertake the treatment of syphilis by mercury, surely the wonder should be that it has not done it :nitely more mischief in such hands. Surely mercury is not chargeable with all the mischief done by the ignorant who presumed to direct so powerful an engine. We see every man who has the slightest connection with medicine, from the hospital and dispensary porter up to the presidents of the College of Physicians and of Surgeons, prescribing mercury for the cure of lues venerea.

In my opinion nothing can more clearly establish the claims of mercury to be considered as a *specific* for this disease than the frequent instances of cures made by it in the hands of those who must employ it at random, and very frequently on patients as careless and incautious as the prescriber is ignorant and pretending. It is wonderful that an engine so very powerful could have been so long wielded by the hands of the inexperienced, the injudicious, the uneducated and unprincipled, and yet should not have committed much more havoc than it has done. Mercury is either a valuable medicine or a dangerous poison. I trust that hereafter the profession, at least, will no longer be misled by a name, and suppose, because the remedy they employ is entitled a specific for syphilis, that on that account they have only to exhibit it in certain doses, and that they must thereby effect a safe and permanent cure. Let the actions of this medicine on the system at large, and on the various parts of it, be diligently studied; let the very earliest indications of its agreeing or disagreeing with the system be investigated and made known: let the doses and forms of this medicine be suited to the state of the symptoms and to the conditions of the general system; and then I am convinced that this medicine will rank high among those which restore to man the invaluable blessing of health. Should mercury, when administered according to the above suggestions, be afterwards subjected to the same obloquy which it has lately experienced, I would then think that Mr. Hunter was correct in saying, " Nothing can show more the ungrateful and unsettled mind of man than his treatment of this medicine."

PART II.

ON THE USE OF MERCURY IN AFFECTIONS OF THE NERVOUS SYSTEM.

EVERY surgeon who has been engaged in the practice of his profession during the last twenty years cannot fail to have remarked the following fact, *viz.*, that during that period mercury has been applied much more frequently to the cure of certain diseases than it formerly was, although its powers over these had not been previously acknowledged. How to account for the increasing partiality to this medicine for the cure of other diseases, while its use in venereal complaints, for which it had so long been considered a specific, has within the same period of time been by many practitioners objected to, and by some totally abandoned, is a problem not easily to be solved. This, which is almost a paradox, I cannot attribute to mere fashion or caprice: were I to offer any conjecture upon this point, I should say that in the treatment of the Walcheren fever this medicine evinced such salutary effects and such superior efficacy to any other of the various remedial measures that were resorted to during the prevalence of that severe, and too often fatal, epidemic, that some surgeons became disposed (almost unconsciously) to ascribe to it some very peculiar and very superior powers, and were thus prompted to employ it in the treatment of those diseases in which it had not hitherto been used, at least to any extent, and in which the various remedies in ordinary use had been generally found ineffectual.

It is not my intention at present to treat of the application of mercury to those acute diseases in which it now is, and long has been, freely and successfully employed; for example, in iritis, and in acute inflammations of various membranes, peritonitis, pleuritis, synovitis, &c. The concurrent experience of all practitioners has so fully justified the high reputation of this medicine in these acute complaints that I shall not dwell upon this point, nor attempt to enforce by any examples a truth which is now so generally admitted.

I now merely wish to report the result of my own experience as to the efficacy of mercury in some classes of disease, in which, as far as I know, it has hitherto been but seldom, and even then but sparingly, employed. I allude to certain derangements of the brain and nervous system, sometimes accompanied with more or less of paralysis of the voluntary muscles. In these diseases I consider mercury, when actively, and at the same time judiciously, administered, to be a most invaluable medicine.

This position, I think, I can best establish by a brief statement of a few cases which I have selected out of several that have fallen under my own immediate observation.

CASE 1.—In the year 1810 I was requested by a respectable apothecary in this city to visit a young gentleman, a near relative of his, who had been attacked about three months previously with hemiplegia. I attended him every fourth day during three weeks; his complaints were becoming rapidly worse, and indeed his condition was most alarming. He was reduced to a state of great debility; he could not stand; he passed the entire day in the recumbent posture on the sofa. He burst into tears whenever he was spoken to; his urine and fœces passed involuntarily and unconsciously. I scarcely thought he could survive many days, unless some successful effort were made for his relief.

I suggested to his relative a trial of calomel, to be administered so as quickly to induce ptyalism. This plan was put in execution, and in four days he was in a state of salivation. In less than a week from this he was able to walk with the assistance only of one crutch. The bladder and rectum regained their retentive faculties. In the course of a fortnight he considerably improved in every respect; his intellects and strength of mind were perfectly restored.

I may add that he still lives and enjoys perfect health and happiness in the midst of a young family. The hemiplegia, however, still, and I conclude will remain to the end of life.

CASE 2.—In April, 1834, I was called into the country to a near and very valued relative who was seized with hemiplegia,

a gentleman about sixty years of age, and of a very full habit
of body. He had been always very temperate in regard
to drink, but he had indulged very fully a remarkably good
appetite, and he used but little exercise. For some weeks
previous to this attack he had the usual premonitory symptoms : before I had seen him he had been bled largely and
purged freely. I recommended the use of calomel, with the
view of exciting ptyalism ; in three days this effect was
accomplished, and it was maintained for nearly three weeks ;
by this time the powers of the limbs were gradually restored ;
and, at present, the recovery is so far complete that no eye
could discover the slightest paralytic affection of the leg. He
can walk with as firm and quick a step as ever. The
upper extremity also seems to have regained its wonted
power ; he can employ it in any vigorous exertion or laborious
occupation as effectively, and with as little fatigue, as
the other arm ; but still one defect remains in it, namely, he
is not always certain of catching or of holding a small object,
such as a pin or needle. However, as far as regards all the
ordinary business of life, both limbs may be pronounced recovered and well. No paralytic affection of the face remains ;
no distortion or twitch can be observed in laughing, or in the
expression of any emotion of the mind ; his intellects too are
perfectly clear, as some late productions from his pen on
political subjects abundantly testify.

CASE 3.—December 19th, 1833. I this day saw (in consultation with Mr. Corr, a surgeon in this city) Mr. H., who had
been within a few days attacked with paraplegia. Mr. H.
was a young man of intemperate habits. The paralysis was
recent, and was rapidly increasing ; he could not stand
steadily for any time without support ; nor could he walk even
from one side of his bed to the other without the greatest
difficulty, and without clinging to the bed-posts.

Mr. Corr and I agreed to try the effects of mercury, given
so as to induce a speedy salivation. We therefore directed
him one grain and a half of calomel every fourth hour.

December 24th. Mr. Corr informs me that Mr. H. is now
in a sort of salivation ; and that he is so much improved as
to be able to walk through his bed-room without assistance, or

without laying hold of any object within his reach. He was now directed a blister to the nape of the neck, and the ptyalism to be maintained.

In the course of a short time Mr. H. recovered the perfect use of his limbs without any other remedies, except occasional purgatives.

He is now in excellent health, and a remarkably active man, and even practises certain feats of agility, such as, holding his hands behind his back, jumping upon a sofa and down again in rapid succession, and without turning his person round.

Case 4.—The following case was in Stevens's Hospital, and excited so much interest and attention that I feel certain the pupils of that day cannot readily forget its particulars.

—— Mooney, a young woman, admitted during the winter of 1835, labouring under almost total paralysis of every limb. She was scarcely able to stand, and totally unable to walk; her upper extremities also were in the same useless and helpless condition. She states that a few weeks previous to her admission into this hospital she had been in the Whitworth institution for a severe fever; that she left that hospital in a very weakly condition, and in a very delicate state of health; she then went into uncomfortable lodgings, where she was necessarily exposed to many privations, and where she found not only her general health to decline, but she also experienced a sense of numbness, and a peculiar weakness and loss of power in all her limbs; these sensations have daily increased to the present date of her admission.

I immediately resolved to try the active mercurial treatment in this case; and directed calomel in such a manner as to induce a speedy and a smart degree of ptyalism. The effect of this treatment in this case was most striking: ptyalism was induced in the course of three days; and, coincident with it, there was observed a great improvement in all the symptoms: the limbs, in particular, were benefitted; the sense of weakness and loss of power in and control over them were speedily diminished; her strength returned, so that she was soon enabled to walk, feebly and unsteadily no doubt.

The ptyalism was maintained, steadily and prudently, for some time. Her head was directed to be shaved, and rubbed with tartar-emetic ointment: this application she, from her anxiety to recover, used too freely, and for too long a time; in consequence of which, sloughs formed in different parts of the scalp, and the ulcers caused by these were slow and difficult to heal. When the ptyalism had been continued for about a fortnight, it was allowed to subside; and although all paralytic symptoms had disappeared, yet as some peculiar debility as to particular motions, or in some particular muscles, still remained, strychnine was prescribed. Under this medicine, and afterwards tonics and generous diet, she was gradually restored to strength, and was discharged in perfect health.

CASE 5.—October 3rd, 1836. Mrs. B., of Rathmines, æt. 50. For nearly the last two years this woman has suffered many severe family afflictions, and considerable loss of property; in consequence of which, as she thinks, she has become subject to what she terms " great confusion in the back of the head," which of late has extended to the right side of the head also. She has latterly avoided all society, and has sought for solitude. At the same time she observed a failure of memory, which, within the last six weeks, has increased considerably, so that now she cannot find words to express her ideas. If she chance to lay a key, or any other thing, out of her hand, she cannot, in a minute after, recollect where she had placed it. She is unable to read, as the attempt instantly brings on the "confusion in her head;" if she attempt to recollect any thing, it all ends in the same confusion; nor can she even attempt to do any needlework, as this would be followed by the same distressing sensation. Her temper has become extremely peevish and irritable; she suffers from a constant sickness of stomach, like sea-sickness. When she attempts to walk she staggers, as she says, in consequence of a dizziness in her head, yet she can walk in a dark room; nor is she alarmed by looking down from a height. Her appetite is good; the bowels are very costive; she sleeps very heavily; there is no emaciation; pulse 96.

I determined, in this case, to try the effect of ptyalism; and having premised a strong purging draught, directed " Calomel gr. iij. bis in die."

October 7th. She has had a slight attack of mercurial dysentery yesterday, with some soreness of the mouth and gums; she describes what she terms "the confusion" in the back of the head as being much less; there is a decided improvement in her memory; she can now much more readily, and more constantly, find the words to express her ideas. She has some sickness of stomach, but different from that kind of sickness she has so long suffered.

October 10th. She can now read and attend to figures, and can even cast up an account, which she could not previously attempt to do; but still she feels she would become confused if she attended to them beyond a very short time. Habeat " Mist. c. Quin. Sulph. gr. ij. ter in die."

October 18th. Whenever she stoops, or turns about her head suddenly, she feels a sense of confusion, and then a pain in the head. Her sleep is less heavy, and much more refreshing; she can now read as much as twenty pages of a book at once—she takes an interest in it, and can recollect what she has read. Says she feels as if some great weight had been lifted off her; ptyalism was still maintained by occasional doses of mercury. Habeat Ung. Ant. Tart. Vertici Capitis.

October 24th. Pustules have been produced by eleven applications of ointment. She can now stoop, and look up suddenly, without any unpleasant sensations; her memory and spirits are improving; she can now attend to her household affairs, and can recollect what she has to do.

November 3rd. She now reads with interest, and recollects what she has read a week before; her temper is still very irritable, especially if hurried; her spirits are much better in the latter part of the day; in the morning she feels very nervous: jolting of the car makes her head still feel a little giddy.

November 13th. She feels much improved in every respect; she becomes fidgetty and uneasy at 10 p.m., before she goes to bed; this is the principal nervous uneasiness she now experiences. She feels her temper much improved since the scalp has healed.

November 16th. She walked from Rathmines to my house, upwards of a mile, this day, and feels no inconvenience, except a very slight giddiness. Her sleep is now refreshing and natural; her temper much improved; she now does not suffer from confusion when she is hurried.

I observed that during the entire treatment her bowels required very active purgatives, the uneasy feelings in her head uniformly becoming aggravated by costiveness.

CASE 6.—On May 12th, 1836, I received from my friend Mr. Pierce, surgeon of the King's County Infirmary, a letter relative to the state of health of Mr. Elcot, which was handed to me by the patient himself. In this Mr. Pierce mentions that this man's habits of living had been irregular, his occupation as a farmer and cattle-dealer having obliged him to attend fairs at great distances from home; and there, from exposure to wet and cold, Mr. E. said he could not avoid often resorting to the use of ardent spirits, as he fasted during the entire day, from an early hour in the morning till late at night. He applied to Mr. Pierce on account of anasarca, accompanied with diseased liver. By the use of calomel, digitalis, and squills, pushed so far as to induce a slight ptyalism, these complaints were removed. In a very few weeks afterwards he again applied to Mr. Pierce, with the following symptoms: "an unsteady gait when he walked into the room: a considerable inclination of his body towards the left side; and, in progression, a deviation towards the left from the course he wished, or which I desired him to take; along with this, a most striking alteration in his countenance. The left eye seemed as if sunken in the orbit, with ptosis of the lid; a languid and vacant cast of countenance; added to which was an obvious inattention to and neglect of personal cleanliness; he was inclined to remain always in bed; his urine was passed frequently, with difficulty, and in small quantities."

Such was the state of this man when he called upon me with Mr. Pierce's letter, some days after it had been written; and then I observed that his intellects had in some degree suffered; his memory was much impaired, and he seemed almost perfectly indifferent to everything around him. I feared he was likely to fall into a state of fatuity.

I advised that mercury should again be applied, and pushed so as quickly to induce a smart ptyalism; that then the top of the head should be rubbed with tartar-emetic ointment; and on the subsiding of the ptyalism that he should take strychnine.

In August, 1836, I had an opportunity of learning, by a letter from Mr. Pierce, the result of this plan, and of again seeing Mr. Elcot. From Mr. Pierce's letter I learned that ptyalism had been induced in the course of seven or eight days; that the symptoms appeared to show some improvement as soon as ptyalism was established; that on the subsiding of the ptyalism he was directed to take one-twelfth of a grain of strychnine three times a day, and to have tartar-emetic ointment rubbed on the top of the head; and that the effects of the strychnine were most satisfactory, a most decided improvement having taken place in his limbs, his intellects, and countenance.

I must say that I seldom enjoyed more pleasure than in seeing Mr. E. walk into my study, and finding him give such a satisfactory account of the improvement in his health. Still he admitted that his urinary complaints distressed him a good deal; that he frequently suffered from incontinence of urine, although he was able to pass half-a-pint of urine at a time.

I also learned from his sister that he was still inclined to remain in bed all day, and that he had a great aversion to walk or to use any kind of exercise. On examining his abdomen, I felt the bladder distended, and as hard as a foot-ball immediately above the pubes : I drew off the urine ; and repeating the operation again, after an interval of three days, I found that the bladder had now regained the power of emptying itself completely. His sister remarked a very decided improvement in his habits from the day I first drew off the urine; that he walked out every day since, and remained for a considerable time in the open air each day.

The three succeeding weeks he passed at the sea-shore, near Dublin, and he improved considerably in every respect.

Since his return to the country I find his improvement has been equally progressive, as in a letter from Mr. Pierce, of the

11th of September last, he states, " I should have written to you before relative to Mr. Elcot, who, I am happy to say, continues to improve in health and strength. He is, I am credibly informed, this very day attending the fair of Banaghar, which is a distance of seventeen miles from his residence."

CASE 7.—John Brady, æt. 30, a publican, admitted into No. 3 ward on the 26th of October, 1833, under the care of Mr. Colles.

His right lower extremity is paralysed; the left is very weak, and bends under him: he has a sensation of coldness in it, and a stinging pain shooting up from the instep and foot. The urine flows from him involuntarily when in bed; but whenever he sits up, and tries to make water, he experiences great pain just before doing so—a sensation as if the bladder was over distended; and he has to force very much: there is a white powdery sediment in the urine. He has also a feeling of weakness and pain in the small of the back. His appetite is pretty good; he has no thirst; he perspires very much at night; pulse 100.

He states that on the 1st of September last, after exposure to cold and wet, by sitting up for seven successive nights to protect his tent in Donnybrook fair, he was attacked with a profuse diarrhœa; after which he had a sensation of being stung with nettles in the lower part of the body and in the thighs; then a numbness in the right leg, which was growing weak, so much so that on the 15th of September, three weeks from the commencement of the attack, he was obliged to use a stick in walking; and he felt a slight pain shooting from the foot up the right leg; also a weakness in the loins, and pain, which prevented him from stooping. He had also a difficulty and delay in making water, which gradually increased, and amounted to a complete retention of urine on the 20th of September, when the water was drawn off for the first time. He was now received into Sir P. Dunn's Hospital, when he was given some pills, during the use of which he was enabled to make water; he was also acupunctured, but without benefit. On the 15th of October he got incontinence of urine; and about a week before admission into this hospital he had completely lost the use of the right leg.

When admitted into Stevens's Hospital the incontinence of urine continued unabated; he had lost the power of both lower extremities, so that he was unable to stand at the bed-side unless he were held up by a strong man; but he said that the right leg was weaker than the left. He has a sensation of coldness, and stinging pain shooting up the limbs from the feet; he also complains of weakness and pain across the loins. His appetite is pretty good; has no thirst; he sweats profusely at night; pulse 100.

November 1st. ℞. Calomel gr. iss. Pulv. Doveri; Pulv. Jacobi, aa. gr. ij. ft. Pil. ter die.

6th. He is now able to retain his water, but has still difficulty and forcing in making it. ℞. Cal. gr. ij. Pulv. Jacobi; Pulv. Ipecac. aa. gr. ij. ft. Pil. ter die.

9th. Has retention of urine: the left leg does not now bend under him when he rests upon it. Rep. Pilula.

12th. He has no longer retention of urine; the power of motion is returning in the right leg; his gums are sore, and salivation is established. Omitte Pilulas.

26th. His urine has gradually become clear, and he has now no difficulty in making it. He can walk without assistance. Discharged.

September 29th, 1836. This man has continued to enjoy very good health, although he still follows the business of a publican, and drinks rather too freely by his own admission. He called on me this day, and walked into my study with a firm step, not exhibiting the slightest trace of any paralytic defect.

CASE 8.—My dear Sir.—I have great pleasure in sending you the case of Mr. ——. Mr. —— is now about 30, of a fair florid complexion, with light hair. In childhood he was very subject to severe attacks of croup: during boyhood, and until he entered college, he was seldom free from headache, for which Dr. Cheyne was consulted, and which were only relieved by repeated cuppings on the back of the neck and leechings. During his college life he enjoyed good health, and a perfect immunity from head complaint; he became very full and corpulent, and was remarkably strong and athletic. Some time after he obtained his degree he had frequent attacks of slight fever, with headache, which generally confined him to bed for three or four days, and always ended with profuse perspiration; still his general health was good, and his habit full and plethoric. In the September of 1832 he was suddenly attacked, while walking, with a severe epileptic fit; this was his first: the only treatment he then received was a general bleeding, aperient medicine, and a more abstemious regimen, with regular and constant exercise. He went on well for some months; but in February, 1833, he had another epileptic seizure: after which, by the advice of Mr. Crampton and Dr. Marsh, a seton was put in his neck, and he was ordered the nitrate of silver, which he took for eight weeks, to the amount of two grains three times each day; regimen and exercise were more strictly observed, and he was particularly enjoined to sleep very little. After this he continued without any return of absolute fit for twelve months; but he became more subject to those attacks of feverish cold, attended always with headache, and, he remarked, he found it very difficult to keep his bowels open. The seton was then taken out of his neck, and there was no particular symptom to create alarm until the winter of 1834, when, after being exposed to fatigue and cold, he began to complain of his limbs becoming numbed and stiff; he found a difficulty in getting upstairs: he could not run or leap: his right leg was weaker than the left; he felt, when on horseback, that his legs, from his knees down, were numbed, and that he had no power over them; for this he was ordered friction, warm baths, blisters to the sacrum, cupping on the spinal canal, and an issue in the nape of the neck, and electricity, from all of which he derived little or no benefit; his limbs became weaker, he found a difficulty of passing water, and an inability of emptying his bladder, and his headaches came on at shorter intervals.

In the summer of 1834 he had another epileptic fit; and in the autumn of that year his symptoms were an evident drag in the right leg, and great weakness of both; inability to walk more, or so much as, one mile; occasional loss of vision, amounting, in some instances, to total blindness for days; slight squinting; complains of a constant pain and weight in the vertex; occasionally of suddenly losing the use of his right

hand and arm, and recovering it again after some time ; mic-
turition more difficult.   At this time it was agreed, in
consultation with Mr. Crampton, Dr. Marsh, and Dr. Colles,
to put him fully under the influence of mercury ; for, although
he had often taken it partially as a purgative, he had never
taken it so as to affect his system.   He commenced, early in
December, taking two grains of calomel, with one of Dover's
powder, three times each day, and continued it for three weeks
before it had produced active ptyalism : from that period he
lost all headache, never complained of any defect of vision,
his spirits (which before had been gloomy and desponding,
with a great dislike to reading, study, or any exercise of his
memory whatever) became after that very good ; had less
inclination to sleep in the evening, was fond of reading, found
his memory again quite restored, and he began the study of
German, preparatory to his travelling in the ensuing summer.
His limbs still continued weak, and he had not lost the drag
in the right leg.

In the summer of 1836 he went up the Rhine, and found
great improvement in his limbs by the baths at Baden; and
since ttha, he writes to say, he has been into Italy, by Swit-
zerland, and has travelled seven hundred miles, over bad
roads, most of which he has walked.

Such, dear Sir, is his case : exhibiting, I should think, in
the strongest light, the beneficial effects of mercury in cerebral
disease.

From an impartial consideration of these cases I think it
may be fairly deduced that mercury is entitled to be classed
amongst our most powerful and valuable therapeutic agents,
in those morbid affections of the nervous system which admit
of any remedy or relief from the healing art ; I say admit of,
for it must still be conceded that of all complaints that come
under the notice of the practitioner none are more obscure in
their nature and uncertain in their issue, and generally none
are more obstinate in resisting every remedy that is applied
for their relief, than those in which the functions of the nerves
and muscles are deranged.

I shall refrain from offering any theory, or attempt at

explanation, of the *modus operandi* of mercury in this class of diseases; partly because we are totally unable to do so in reference to other diseases in which its influence is still more marked and obvious, and in which its power over disease is almost certain and unerring, or specific; but, principally, I abstain from offering any theoretical observations whatsoever, because the class of diseases to which I have alluded are, as regards their pathology, involved in deep obscurity. Even at the present day we possess but little exact knowledge of the structure or function of the various parts that compose the nervous system; how then can we venture to theorize as to the nature or the cause of its diseases? How are we to discriminate between a functional derangement in its mysterious powers, or an organic lesion in its delicate and (as yet) unexplored tissue.

No doubt, in many instances, the impaired function may depend on some organic change in the nervous fibre, or in some of the effects of inflammation in its investing membranes, and a suitable treatment be hereby indicated; but I fear that in a vast majority of cases no such connection can be ascertained as that between cause and effect; and that we must remain contented to pursue a mode of practice, which, though it be empirical, inasmuch as it rests on no just reasoning, is yet sanctioned by analogy and fortified by experience.

I shall not trouble the reader with any detailed account of the benefits derived from mercury in cases where derangements of the stomach are dependent on organic diseases of that or of some neighbouring viscus.

The following case of gastrodynia, which was one of the most severe I had ever witnessed, shows the great benefits to be derived from the use of this medicine, in a case where the most strict and repeated examination could not detect any organic affection of any of the abdominal viscera. The subject of it was a farmer, æt. 34, in rather comfortable circumstances, of full habit and florid countenance, with every appearance of perfect health. He was received into the hospital on the 24th of September, 1836: he stated that he was a man of temperate habits, but, from his occupation, was much exposed to wet and cold. This disease commenced

about twelve years ago. Whenever he used any green
vegetable food for dinner, he was attacked on the following
morning with a peculiar squeezing pain in the stomach, as if
(to use his own words) the stomach was squeezed between
two rollers; this pain gradually increased for an hour, when
it at length became almost intolerable; he then vomited, and
threw up sometimes a greenish thin fluid, of very acid taste;
at other times a whitish fluid, like the white of an unboiled
egg. The difference in the quality of the fluid vomited did
not depend upon any particular kind of food. After the
vomiting, he felt quite relieved and perfectly well. As the
disease advanced, he found that salted meats and eggs produced the same effects; and that the pain came on much
earlier after the meal: at this time he remarked that change
of air agreed with him, and that wet and cold greatly aggravated his sufferings. He next found that when the pain
seized him he was not able, by a natural effort, to discharge
his stomach; and with the view of relieving himself from the
pain, he would then excite vomiting by thrusting his finger far
down into his throat: this expedient after some time failed him,
so that he could not, even by these means, provoke vomiting.
He now discovered that he could still excite this act, if he
first drank large quantities of boiled milk, so as to over-
distend the stomach, and then introduced the finger in the
manner before mentioned. The milk thrown up was completely changed into a curd. He says that the quantity of
milk he used on these occasions generally amounted to two
gallons; and this he always had boiled before he used it.
Within the last two years the pain has increased greatly in
severity, and comes on after any kind of food; but he thinks
it more severe after eating vegetables or eggs. Pressure on
the stomach procures some slight mitigation of the pain; and
he has sometimes found relief by lying on his back upon the
cold pavement.

He handed me the following communication from Mr.
Cane, a very intelligent surgeon, practising in the city of
Kilkenny :—

"Sept. 20, 1836.

"Sir,—The bearer, Mr. Walshe, was a patient of mine about twelve months since for gastrodynia of the severest character I have witnessed. Since then I had nearly lost sight of him, until this day, when he applied to me to state in writing what I had then done for him, as he wished to consult you, for he is still suffering as severely as ever. I have no notes of his case, but have a tolerable recollection of his very great sufferings, and the remedies applied.

"He was cupped repeatedly, rubbed with the tart. ant. ointment, and blistered over the epigastric region; he got calomel and blue pill occasionally, and the carbonate of soda in inf. quassiæ daily. Cupping gave temporary relief; the other remedies none. He was then put upon sedative treatment; he got the acet. morphiæ, and had a belladonna plaster over the stomach: these failed; he afterwards took the muriated, and then the acetated tincture of iron daily, with gentle aperients at night; all which failing, prussic acid was tried, and this with as little good effect; creosote increased his sufferings. At length he grew tired of medicine, and the restrictive diet which accompanied its use, and withdrew from my care. Since then, he says, his sufferings have been rather increasing; and he expresses the agony he now endures to be so great that he desires death.

I have the honour to remain, Sir,

ROBERT CANE.

To A. Colles, Esq., M.D., &c."

September 24, 1836. I ordered for him ℞. Aq. Calcis ℥vi. Magnesiæ ℥ss. Spt. Amm. Arom. ℨij. m. Sumat ℥i. ter in die. Alternis noctibus Sumat Pil. Alöeticas duas.

October 4th. He passed the three first nights, after commencing this medicine, perfectly free from pain; but has since passed three or four nights of most severe suffering. ℞. Pil. Hyd. gr. iv. Ext. Hyosciami gr. i. fiat Pilula ter in die sumenda. Repet. Mist. c. Aq. Calcis, &c.

October 9th. Repetantur Pilulæ et Mist. ℞. Tinct. Acet. Opii Gutt. xx. Aq. Font. ℥vi. m. ft. haustus urgente dolore sumendus.

October 18th.   His mouth has been smartly affected the two last days.   Omitt. Pilulæ; Repet. Alia.

October 28th.   He has not had any return of pain for the last ten days; appetite very good; mouth is yet slightly sore. He imagines himself perfectly cured, as he has had such a long immunity from pain.   He is most anxious to return to his farming business, and cannot be prevailed on to remain longer in the hospital.

Being anxious to learn whether this man continued free from this severe affliction, I wrote to Mr. Cane, who was so kind as to favour me with the following communication.

" November 18, 1836.

" Dear Sir,—I have seen Walshe, and he has furnished me with the following particulars :—The night he left the hospital, and the following day, when he travelled outside upon the coach to Kilkenny, he felt some slight return; since leaving Dublin he has had no attack *by day*; but is attacked, *very slightly*, however, about twice a week, but at *night only*.   He has had vomiting but once, and that after having eaten of stewed goose.   He complains of some uneasiness in the region of the stomach, occurring after even slight exercise ; and, also, that his nights are restless.   He has a clean tongue, good countenance, pulse 78, and regular ; has no tenderness upon pressure over the epigastric region.   On the whole he is, he says, *well, comparatively to what he was*, and that the pain he now suffers ocasionally is quite trifling, and not at all to be compared to his former agony.   He uses no vegetables, his bowels are free, and he continues to take the medicine which you prescribed for him—Mist. aq. Calcis c. Mag., &c.

" There is no question that he has derived great benefit from your treatment, though he says ' he was near losing his teeth by it.'   He is much better than I had hoped ever to see him.   It will afford me much pleasure at any time to reply to any inquiries which you may be desirous to make concerning him.—I remain, dear sir, yours truly,

" ROBERT CANE."

In case of obstinate ulcers of the extremities, where no

suspicion of venereal taint could be entertained, I have occasionally at length effected a cure of the ulcers by administering mercury, so as to excite ptyalism, using still the same dressings which had been for a long time previously employed in vain.

I deem it unnecessary to mention here, what every practising surgeon must be familiar with, *viz.*, that mercurial fumigations, applied directly to ulcers, often prove a ready means of causing them to heal, although they may have previously resisted a great variety of topical remedies. I shall only beg leave to record my experience of the superiority for such purposes of the Sulphuretum Hydrargyri Rubri, and beg to refer the reader to page 75 of this work, where a most satisfactory and manageable mode of employing this article for fumigating any part is fully described.

The following case will prove most satisfactorily that mercury may be employed with the greatest benefit in some cases of ulceration which had long resisted almost every other treatment.

August, 1835. In the beginning of this month I was consulted by Mr. S. on account of an ulcerated tongue; his speech was by it rendered very indistinct; he had a flow of saliva from the mouth; he suffered but little pain, and was apparently in good general health. Besides the affection of his tongue, he also complained of epiphora of the right eye, with an obstruction in the right nostril, from which, by blowing the nose, hardened crusts, of amber colour, were occasionally thrown out. On the back of his neck was a pretty large ulcer, not very deep, and having callous edges. He informed me that he had laboured under ulceration of the throat, &c., for many years; and that, in April, 1832, I had seen him, his throat being ulcerated, and the ulceration having extended to the tongue; and reminded me that I had paid him two visits at the time the tongue was first attacked; that I had directed for him at that time pills of calomel, which he did not take for more than five or six days. On referring to my notes I found that I had seen him at the time he alluded to; that I had directed for him the mercurial pills, and that he did not seem satisfied when he learned the description of the medicine I ordered. Since that period he has been treated by different

surgeons with Corros. Sublim., and with almost every other
kind of medicine: topical remedies, without number, many
of them caustics, have been applied to the tongue. He
informed me also that, shortly prior to my first visits to him,
in 1832, his wife was delivered of a child, which has been
remarkably healthy. He has not had any eruption or pains
in the bones, but his throat was a long time severely attacked.
On looking into it I observe, on the right side, the velum
adhering to the back of the pharynx, and the cicatrix bears
evident marks of a former very severe ulceration of the fauces.
I ordered him Pil. Hyd. gr. v. sing. noct. Ung. Hyd. Fortis
Ɔi. om noct.

August 13th.   Mouth slightly but uniformly affected by the
mercury; the flow of saliva much reduced: the right nostril
more comfortable; epiphora less; the ulcer of the tongue has
made considerable advances towards healing; the swollen
edges of the ulcer less elevated; his general health is undisturbed by the action of the mercury.

August 19th.   His articulation is much improved; the
swelling and hardness of the tongue, in the vicinity of the ulcer,
much reduced. The ulcer now does not exceed the size of a
split pea. The eye and nostril much better; mouth pretty
sore. Repet. Pil. et Ung.

September 5th.   The tongue is quite healed, but it is in
some degree confined in its movements by the cicatrix, which
connects it to the floor of the mouth; the flow of saliva has
ceased; there is not the slightest weeping of the right eye;
he feels the nose better, but still some crust of hardened mucus
is blown down every day; the mouth not too much affected.

He did not desist from mercury until the beginning of
October, at which period I considered him as perfectly cured.

Few surgeons, extensively engaged in practice, have escaped
meeting with cases of this chronic or scrophulous ulceration
of the throat; and all know that this may exist singly, or it
may be combined with a similar affection of the mucous membrane of the nose, which is by no means a case of rare
occurrence. In more rare instances the ulceration spreads
forwards along the tongue.

The case above detailed is an instance of a combination of

all these affections at once, and exhibits the disease as continuing for a period of at least eight or nine years. When I first advised mercury for the cure of the throat, which had long been ulcerated, and of the tongue, which had only been recently attacked, I deemed the case much more favourable for its use than I found it on his last application when the disease had obviously spread to the nose ; for I need not say that, even in venereal ozæna, mercury is generally not capable of preventing most serious havoc being made in that organ ; but in the present case it has succeeded to the utmost of my wishes.

A few days ago I saw Mr. S., and found him in excellent health, and very much improved in flesh and complexion. He assured me that he then enjoyed a degree of good health to which he had for many years been a stranger. The tongue and throat remained perfectly healed, and not the slightest trace of any affection of the nose could be discovered, or even suspected.

Whether we consider or call this disease a scrophulous ulceration, or merely an obstinate ulcer, one thing is plain, that it was cured by a moderate ptyalism, although it had for eight or nine years previously resisted, I believe, every other mode of treatment, both local and constitutional. By adducing this case as an instance of the value of mercury in causing obstinate ulcers to heal, I do not mean to say or insinuate that it will in all cases of this ulcerated state of the throat or nose be equally efficacious; for I am well aware that, in a great proportion of such cases, we have to contend with a very morbid state of the general system,—one which is not likely to be remedied by mercury. But this much I will say, that in cases of this description, where ulceration is spreading to, and threatening the destruction of important parts, we may fall back upon mercury as our last resource. I think, too, that the local application of mercury directly to the ulcers may be of superior advantage, especially if used in the form of fumigations ; but on this subject I speak from a very limited experience.

I should, when enumerating the various remedies formerly employed in this case, have remarked, that he had repeatedly taken mercurials, especially Hyd. Corros. Sublim.

I am well aware that many cases of this chronic ulceration
of the throat have been treated, both by regular practitioners
and by empirics, with solution of corrosive sublimate; and
that, in some instances, the practice has been attended with
success—more frequently, I believe, when the doses of the
medicine have been very small. But, if I can judge by the
cases which have come under my observation, this treatment
has been much more frequently unavailing than successful.

If what I have said of the use of mercury in scrophulous
habits prove to be well founded, we shall then with some
degree of confidence resort to mercury, pushed to ptyalism in
obstinate cases of this affection of the throat, even when
accompanied, as it sometimes is, with enlargement of the
cervical lymphatic glands. I need not add that we should,
in such cases, strictly observe the rule to apportion the doses
of the medicine to the state of the general system of each
patient, avoiding as most dangerous the error of administering
it in doses too great for the constitution of the individual.

We not unfrequently meet with chronic tumours in the
abdomen, varying in size from that of an orange to that of a
foot-ball; these tumours are movable, and are generally
productive of some uneasy sensations, or of some slight derangements in the function of digestion. These tumours often
continue in an indolent state for years together; but at times
they may be affected with some degree of uneasiness, and even
of painfulness. When these tumours get into this state we are
called upon to use the most active means for removing it,
and restoring them to their original quiescent state. General
blood-letting, where it can be borne, and, in more delicate
habits, local bleeding, should precede all other measures.
Having premised this evacuation, the employment of mercury,
in such manner as to induce a moderate ptyalism in the
course of six or eight days, will be found of the most signal
service. As soon as the salivary system acknowledges the
action of the mercury, instantly does the pain begin to subside, and after a very few days ceases entirely. When I say
that if we do not quickly subdue this painful state of the
tumour, we shall see it speedily produce considerable enlargement of the tumour, and consequent increase of the preceding

distress, which, in most instances, are attended by dropsy, or by irritative fever, diarrhœa, and death ; I say, when we contemplate these results, we shall at once be able to appreciate the value of mercury used in such cases. Yet it must be admitted that mercury has no influence in reducing the size of the tumour while it remains in the indolent state. The same remarks apply to that painful condition which occasionally affects ovarian tumours. For here the use of mercury, preceded by general or local blood-letting, not only relieves the patient from present pain, but also prevents the very rapid increase of the tumour, which would otherwise immediately attend this painful state.

I have with success employed mercury, both as a local application and a constitutional remedy in some other diseases in which its use has not been generally adopted; but I decline to submit an account of these trials to my professional brethren, as the cases in each disease have been too few to justify us in founding on them any rule of practice.

PART II.

SELECTIONS FROM MINOR WORKS AND ESSAYS OF ABRAHAM COLLES.

THE following short essay on the operation of Lithotomy is selected from Colles' "Treatise on Surgical Anatomy," an unfinished work, the first part only of which was published in 1811. Although since then eclipsed by far superior works of the same kind, it, like the admirable work of Burns on the surgical anatomy of the head and neck, published in the same year, marks an epoch in the progress of surgery. Mr. Colles' treatise embraces a series of essays on the anatomy of inguinal, crural, and umbilical hernia, on the anatomy of the abdomen, the thorax, the neck and throat, the pelvis and the organs of generation, and the perinæum, concluding with observations on passing the catheter and the operation of Lithotomy.

ON THE OPERATION OF LITHOTOMY.

IN no other operation in Surgery is a knowledge of anatomy of more essential service to the surgeon than in that of Lithotomy. Let us now consider how it will guide him through the different steps of this hazardous operation.

The patient should be laid on a higher table than that ordinarily used for the purpose, as this will allow greater freedom of motion to the surgeon's hand, and lessen the dangers attendant on some of the most difficult steps in the operation. The staff should be of as full a size as the urethra can well admit—should have the handle made rough, which will enable the operator to hold it firmly without much difficulty. Having introduced the staff, now hold the handle of it firmly with the thumb and two forefingers of your left hand; this hold will enable you more sensibly to feel the point of the knife when it first enters the groove of the staff, and (which is of much consequence) will facilitate one of the most difficult and important steps of the operation, the lowering the handle of the staff. The staff thus held is to be drawn into the arch of the pubis, and then is to be made prominent in the perinæum. You are not, observe, to hold the handle of the instrument inclined over the right groin of the patient, as is generally directed. Let the staff be perpendicular to the horizon : let it at the same time be drawn up as closely as possible into the arch of the pubis, with its convexity bulging out in the perinæum.

If you take care to keep the staff well up into the pubis you will be secured against its slipping out of the bladder ; you will thereby save the rectum from being wounded, and you will avoid all risk of injuring the pudic artery. In a word, you will, by attending to this simple direction, render the different steps of the operation easy and certain. Then kneeling on your left knee hold the staff in this position : and, if you wish to render the right hand more steady, rest your right elbow upon the corresponding knee.

The staff being thus held, you now feel for the arch of the pubis with the forefinger of the right hand, and a little below this spot you commence your external incision, close to the left side of the raphe, and continue it obliquely, so as to pass midway between the tuberosity of the ischium and the anus. This first incision you will make, not with the point, but by laying the bellying edge of your knife fairly to the perinæum. The integuments will then fly asunder, their natural elasticity being aided by the divarication of the perinæum. In the same line with this you are to make your second incision, commencing it half an inch below the upper end of the first ; carry it deeply into the perinæum, by it you will divide the transversalis perinæi, a few fibres of the sphincter ani, a few fibres also of the transversalis perinæi alter (if present) a portion of the levator ani, and a portion of the ligamentous septum of the perinæum, taking care that your second shall be nearly equal in length to your first incision. Now enter your knife into the groove of the staff, which you will readily accomplish if you recollect that your incision is to open the membranous part of the urethra—that this part of the canal passes through the ligamentous septum of the perinæum, at the distance of one inch below the arch

of the pubis and two inches above the tuberosity of the
ischium.   Attention to these points will direct you to the
height in the perinæum, at which you are to enter into
the groove of the staff.   The depth at which this ligament
lies from the surface may be ascertained by attending to
the fullness of the perinæum in each individual, and recol-
lecting that it is attached to the rami of the pubis behind the
crura of the penis.

It is the more necessary for you to bear these anatomical
facts in your recollection, because the deeper parts of the
perinæum do not recede when divided.   On the contrary,
in corpulent subjects even the knife is concealed from your
view by the edges of the wound falling together as soon as it
has passed through them.

Therefore, holding the knife horizontally, you will push
it forwards and a little towards the right side of your patient,
taking care to enter it not at, but a little below, the upper
extremity of your external incision.   As you perceive the
point of the knife grating on the staff, move it from side
to side, that you may be sure of its being in the groove,
as you might be deceived were you to rely merely on the
rubbing of the knife against the staff; when thus assured
that your knife is fairly lodged in the groove you are to
bare it for about a quarter of an inch : this is to be done
while the staff continues to be held perpendicular, by moving
the knife in the same perpendicular direction.   If you do not
lay bare so much of the staff before you attempt the division
of the prostrate gland, you will have to encounter many and
most insuperable difficulties, of which we shall speak when we
come to describe the division of the prostate ; one, however,
may here be noticed, viz., the resistance given to the knife by
the levator ani and the triangular ligament, and by that
ligamentous structure which envelopes the membranous part
of the urethra.   Having thus divided a portion of the
membranous part of the urethra, you now proceed to the
most difficult part of the operation, viz., the division of
the prostate gland and the neck of the bladder on its left
side.   To effect this you must alter the position of your
instruments, because you know that the direction of the parts

now to be divided is very different from the direction of those through which you have already cut; for you are now to divide parts which lie behind the arch and symphysis pubis. Wherefore, while you hold your knife horizontally in the lower part of the incision in the urethra, you should now bring the handle of the staff down towards yourself by making it move on the point of your knife, as on a pivot, and at the same time keeping its concavity close up to the arch of the pubis; by this movement the back of the knife, instead of the point, comes to be lodged in the groove of the staff, and the beak of the latter is directed upwards. You will experience but little difficulty in running the back of your knife along the groove of the staff, if you but recollect the direction into which this last movement has thrown the staff, viz., that it has lodged it immediately behind the arch of the pubis, and, therefore, in order to give a corresponding direction to your knife, you must depress the handle of it, lowering your right wrist by throwing back your hand, and then pushing the knife on in the groove, taking especial care that you lower the wrist as you push on the knife.

I should have observed that before you begin to push your knife on along the groove you should incline its edge a little towards the left ischium, that you may divide the prostate gland on the left side. You will be sensible that your knife has entered the bladder by all resistance being removed and by the sudden flow of the urine.

I have said that it is a matter of the greatest importance to the successful, and indeed to the safe performance of this operation, that a considerable portion of the membranous part of the urethra should be divided before the staff is depressed, or that incision commenced by which the prostate and neck of the bladder are to be divided; for, if you have entered your knife into the urethra high up in the perinæum, and while the point of the knife is lodged there should depress the staff and attempt the division of the prostate, you will have to make it describe a portion of a circle at the time that it is dividing very resisting parts.

Nothing can be more unsatisfactory to the operator than the feel, when he attempts the division of the prostate, where he has entered the knife too high in the urethra; he feels as if the parts had not fully yielded, or indeed as if they had not yielded at all; and yet he is conscious that the degree of force which he uses cannot be continued without the danger of throwing the knife altogether out of the groove of the staff, and plunging it far forwards into the cellular substance between the bladder and rectum, or of sinking it into the rectum if the point should be at all depressed; so that the knife used in this manner is productive of all the dangers, and liable to many of those objections which apply to the gorget. Should the knife, on the contrary, be used in the manner here directed, you will not experience any resistance to the progress of it, except what you may naturally expect from the texture of the parts to be cut; and you have, in the complete absence of all resistance, the most satisfactory proof of the division being fully effected.

You may remark here that I have advised you to run the back of the knife, slightly lateralised, along the groove of the staff, and would wish you to have no other object in view when performing this movement; for if you accomplish this the prostate and neck of the bladder must at the same time be divided.

The breadth of your knife may in every instance be determined by this rule: use one of such breadth that when lodged in the groove of the staff it shall be nearly equal to the diameter of the canal of the urethra, and thickness of the prostate gland. A knife of this breadth, and with its cutting edge not above one inch long, will freely divide all those deep-seated parts which are to be divided, and from its dimensions and form cannot possibly divide the other parts in the neighbourhood, if used in the manner here directed.

Let us suppose the incision to be made into the neck of the bladder: the knife is now to be withdrawn by drawing its back a little way along the groove of the staff, and then by lowering the knife as you come out such of the external parts as have not been sufficiently divided can now be cut to the necessary extent. Here your knowledge of the wide flat pouch, formed by the rectum at this part, will prevent you from carrying your knife too low down before you have withdrawn it a little, lest you wound this intestine. Now examine the state of the incision with the index finger of your right hand, and if any bands of undivided cellular substance lie across the wound break them down with your finger. You now take the forceps with the handles lying on the same plane, and, by introducing them inclined from below upwards and forwards in the direction of the axis of the pelvis, you push them into the bladder, guiding them by the staff which you still hold, with its handle depressed.

This step of the operation requires a good deal of care, for were you to enter the forceps horizontally you run the risk, or rather you will scarcely avoid the danger, of pushing the instrument into the cellular substance between the rectum and bladder, however complete your division of the neck of the bladder may have been; for the edges of the wounded levator ani contracting expose this interspace, which now feels as a cavity, in consequence of the retraction of all that cellular substance which lies between these parts. The forceps being introduced you now withdraw the staff, and, standing up, you search for the stone. When you have laid hold on the stone proceed to extract it by withdrawing your instrument in the direction of the axis of the pelvis, viz., from above downward. If you attempt to withdraw the forceps in a horizontal direction the stone, if large, must injure the urethra by pressing it against the pubic ligament and the arch of the pubis. Nor can room be gained in this direction; a slight pressure towards the right side of the patient may gain some little room, but it is only in the direction above mentioned that you can gain any material room, and this too without inducing any contusion of the soft parts.

When you introduce the finger to try for a second stone be careful not to mistake, for the cavity of the bladder, the space interposed between it and the rectum.

If the stone has been unfortunately broken into small pieces by the forceps, you should endeavour to wash them out by throwing tepid water into the bladder with a large syringe, armed with a pipe three or four inches long. In performing this you must also be careful to pass the pipe completely into the bladder, and not to mistake for its cavity the space between it and the rectum, to which we have so often alluded. Some leave these fragments to be discharged with the urine, but this is objectionable, because although they may have fallen down towards the fundus of the bladder, yet they may be prevented from escaping by the inflammation and swelling of the lips of the wound; and while those fragments are allowed to remain the patient suffers to a considerable degree that series of distress for the removal of which he had submitted to the operation.

This, in point of the number of instruments employed, is the most simple mode of performing the lateral operation for lithotomy. It is the only mode that should be practised on children under six or eight years of age, because in them the urethra is too small to admit of the introduction of the instruments, hereafter to be described, without danger of lacerating this canal. At the same time we must admit that a more accurate knowledge of the anatomy of the parts, more dexterity in the use of the instruments, and more constant practice in this particular operation than fall to the lot of surgeons in general, will be required to enable the operator to execute it with confidence in himself and security to his patient.

By using two more instruments this operation can be performed with much greater facility, and with such security that few accidents have occurred during the operation, and still fewer instances of a fatal event, since this mode of operating has been generally adopted by the surgeons of this city.

The additional instruments required are a straight conductor and a knife, which is called the lithotome. These instruments had originally been invented by Mr. Daunt, an eminent surgeon in this city; they were improved by the late Mr. Dease, and owe their present perfect form to the ingenuity of Mr. Peile.

The first steps of the operation are the same as those above described. The position of the patient, the mode of holding the staff, of making the external incision, and of laying bare the groove of the staff, correspond in every particular, and therefore it is unnecessary to describe them here. The rules to be observed in the part of the operation to be performed by Mr. Peile's instruments are as follow :—The staff being laid bare, and the surgeon being assured, by moving the knife from side to side, that its point is lodged in the groove, must now bring down the handle of the staff towards himself, making it move on the point of the knife, as on a pivot; by this motion the back of the knife is sunk into the groove. You now divide the membranous part of the urethra and the anterior point of the prostate gland; this you effect by lowering the wrist while you move the knife onwards, taking especial care to make the back of the knife run in the groove, which can only be done by lowering the wrist in proportion as the knife is pushed forwards. The knife you now withdraw, retaining the staff in the present position. Next take up the conductor, catching a firm hold of it by applying your fore-finger along its stem, while the remaining fingers embrace its handle ; enter its beak into the groove of the staff,—you ascer-tain that it is fairly lodged by moving it from side to side,—and then, lowering the right wrist, run it along the staff, taking care to lower the wrist as you push the director forwards, until you have introduced it fairly into the bladder. The urine now flows along the groove of the conductor, assuring you of your success in this step. You now withdraw the staff by moving up the handle towards the abdomen of the patient, at the same time that you are drawing it out of the urethra, the conductor during this time being held immovable. Now, rising off your knee, stand between the legs of the patient, and, passing the two first fingers of your left hand into the ring, while your thumb is pushed against the handle of the instrument, raise it up as high as possible into the arch of the pubis. In this position you carefully hold it, as by this alone can a wound of the rectum be avoided. Now, holding the lithotome between the thumb and two fingers of the right hand, lay its beak on the lower edge of the groove, and, pushing it on until its point has got to the external incision, give it the necessary degree of obliquity or lateralisation, as it is termed, by turning the groove of the conductor more or less towards the arch of the pubis. Having determined on the degree of lateralisation which you judge necessary, now push on the knife, running it close and parallel to the conductor until it is stopped at the point of the conductor. Withdraw it cautiously by bringing it back again along the groove. By this means the division of the prostate is effected with the slightest possible force, for the operator is scarcely sensible of any resistance from prostate, and judges that it has been divided, not so much by his having overcome a certain degree of resistance, as by the knife having reached to the end of the groove.

The great advantages of this mode of operating are that any man who can lay open the urethra on the grooved staff, and has dexterity enough to introduce along it the straight conductor into the bladder, will certainly guard against dividing the rectum, will be enabled to give his knife the required lateralisation, which is secured without any further dexterity in making the incision, and therefore he will be able to avoid in every instance the division of the internal pudic artery.

Having withdrawn your lithotome, run your finger along the conductor into the bladder to satisfy yourself of the extent of the incision : but should you find that the prostate is not sufficiently divided, introduce the same lithotome again, now keeping the handle depressed below the stem of the conductor. The division of the gland will be increased in proportion as the handle of the knife is depressed, and therefore you can regulate the movement of the cutting part of the knife merely by observing the direction of its handle. Now introduce the forceps guided by the conductor, but passed from below upwards, or in a line corresponding with the axis of the pelvis, and conduct the remainder of the operation as already described.

The reader will perceive from the foregoing that Colles held the staff himself. It may well be doubted whether surgeons of the present day, who entrust to an assistant this important duty, have made any real improvement. The dexterous manipulation of the staff constitutes assuredly about the most important part of the operation. Up to the time when the lithotomy scarpel divides the prostatic portion of the urethra and enters the bladder, the nicety and success of the operation chiefly turn upon the rapidity, neatness, and precision with which the manœuvres of the staff are accomplished. It is not possible for any assistant, however skilled, to perform the necessary movements so well as does the left hand of him whose right hand guides the knife which seeks the groove in the staff.

In cases of fat perinæum, so deep that the finger can with great difficulty reach the neck of the bladder, the straight staff or director of Peile is a very valuable instrument.

ON THE OPERATION OF TYING THE SUBCLAVIAN ARTERY.

(Edin. Med. & Surg. Journal, 1813).

Soon after I had been raised to the office I now hold in the College of Surgeons, I turned my attention to the study of the subclavian artery. My first object was to discover the most effectual means of compressing the artery, and to ascertain the most eligible spot for this purpose ; my next, to investigate the practicability of tying this artery either before it reaches or after it has passed through the scaleni muscles. The anatomy of the parts satisfied me of the feasibility of the operation ; and Mr. John Bell's luminous statement and lively description of the anastomising arteries situated about the joint encouraged me to hope for its final success. Still, however, I did not feel myself justified in stating my opinion to the pupils of the college, for as yet the operation had not only been untried in practice, but the proposal wanted even the sanction of analogy. At length Mr. Abernethy gave to the world the first account of his operation of tying the external iliac artery. From this moment I felt myself authorised openly to express my sentiments : and every year since I have taken occasion, both in the anatomical and surgical course, not only to state to the class my sentiments generally on this subject, but to point out to them particularly the manner in which I conceived the operation might be performed.

A case of axillary aneurism having, in the year 1809, come under my care in the hospital, I proposed this operation, on which I had so long meditated, to the other surgeons in consultation. The proposal, however, was overruled, chiefly on the grounds of the operation never having been performed; and I confess that I yielded to the voice of the majority with the less reluctance, because I felt a secret apprehension of some terrible revolution being produced in the frame by tying an artery of such magnitude so close to the heart. This patient, in a few months, fell a victim to the disease; but neither by argument, nor influence, nor stratagem, could I obtain an opportunity of examining the body. Mr. Ramsden now published his account of the operation which he performed on this artery after its passage through the scaleni. Although the result was finally unsuccessful, the experiment served at least to prove to me one point incontestably, namely, that the fear of immediate danger from a general revulsion in the system was totally groundless. Emboldened by the establishment of this fact I again proposed the operation in the very next case that occurred, and it was unanimously agreed to in consultation. In this case I undertook to tie the artery before it had yet reached the scaleni. Several months elapsed before another case presented itself; but the patient in this instance positively refused to submit to the operation unless I would assure him of complete success. The unshaken fortitude and patient resignation with which he submitted to the wide-spreading ravages of this painful disease through all its successive stages proved that, in this determination, he was not influenced by a desire to avoid the comparatively trifling pain of an operation. Of this case I have related the appearances on dissection, as I think they suggest some useful practical inferences.

The third case was, of all, the most favourable for operation, and yet the result proved unfortunate. In this the artery was tied after it had passed through the scaleni.

Case 1.—Michael Cowel, a labourer, aged 33 years, was admitted into Dr. Steevens's hospital on 23rd September, 1811, for an aneurism of the right subclavian artery. He gave the following account of his disease :—

Thinks that he repeatedly hurt himself by endeavouring to

push on with his shoulder a loaded car, the wheels of which had sunk deep in passing over heavy ground. Soon after these exertions (perhaps in the course of a week) he began to complain at and near to the clavicle. At this time too he felt a tumour the size of a cherry, which throbbed with a continual pulsation. He points out a spot above the middle of the clavicle, and rather towards its humeral extremity, as having been the original seat of the disease. He did, in this early stage of the disease, observe a tumour or pulsation in a spot nearer to the sternum, but says that as the tumour increased in size it extended itself in that direction. It was not until the beginning of July he first felt a pain and numbness along the right arm and forearm.

At present the tumour extends from the sternal origin of the sterno-mastoid muscle along the clavicle, until it reaches a little beyond the arch of that bone. It rises in height nearly two inches above the clavicle, and is of a conical form, the apex of the cone being situated at the outer edge of the sterno-mastoid muscle. The finger can be passed with facility between the clavicle and tumour, but at no part can it be passed between the posterior surface of the tumour and the subjacent parts. The aneurismal pulsation is less strong, and the tumour more compressible between the heads of the sterno-mastoid muscle than at any other part. Pulse 74, and of nearly equal strength in both wrists. Of late he complains of slight numbness in the fingers of his right hand. He speaks of some uneasiness in the chest, which he terms "a draw on his chest." This has been much relieved since yesterday by blood-letting at the arm to the amount of fourteen ounces. His health in every other respect is good.

On Thursday, October 10th, the operation was performed, the patient being laid on his back upon a table, so placed that the direct light fell on the right side of his neck.

Two incisions were made through the integuments; the first running along the middle of the sterno-mastoid muscle, from the highest point of the aneurism to the top of the thorax; the second beginning at the middle height of the first, and ending at the sternal articulation of the clavicle. The integuments

being then separated from the subjacent parts, the lower
triangular flap of skin was laid down on the thorax, and the
upper portion turned upon the side of the throat.   The
mastoid muscle being thus exposed to view, its different
portions were next divided.   The sternal portion was first cut
through about an inch and a half above its origin.   This was
done with a sharp-pointed bistoury passed on a director which
had been introduced on the outer edge of the muscle.   The
clavicular portion was then divided to the breadth of about
half an inch, the director being passed on its inner edge.   In
both these cases the introduction of the director was much
facilitated by slightly scraping along the edge of the muscle.
The lower divisions of the muscle were then dissected from
the subjacent parts and turned down over the clavicle.   Now
the line of the carotid artery could be distinguished, and its
pulsations felt by the finger.   The sterno-thyroid and hyoid
muscles lying exposed, were divided in the same manner, and
at the same height from the clavicle as the sterno-mastoid had
been.   The division was carried only through so much of
their breadth as enabled us fully to expose the carotid artery.
We then proceeded to detach the lower divisions of these
muscles from the surrounding parts ; but here some delay was
occasioned by a large vein descending from the anterior part
of the throat into the internal jugular.   The vein, however,
being secured by two ligatures applied at a small distance from
each other, and then divided in the interval between them,
the inferior portions of the sterno-thyroid and hyoid muscles
were raised from their beds and turned down on the thorax.   Now
the sheath of the carotid artery was opened by pinching up a
small portion of it with the forceps and cutting the raised
part with the knife, carried horizontally.   The director,
introduced at this opening, was passed down towards the
thorax, and on it the sheath of this vessel, near to its root,
was divided.

During this and all the subsequent steps of the operation
the internal jugular vein was removed beyond the reach of the
knife by an assistant, who, with a flat silver instrument, not
unlike a powder-knife bended at the point, drew it outwards
and held it aside.

When the sheath of the carotid artery was divided the eighth pair of nerves became visible ; it lay deep, and to the outside, nearer to the carotid as they descended towards the thorax. An instrument, similar to that above described, but of larger dimensions and greater curvature, was now employed to protect this nerve, together with the internal jugular vein. The interval between the carotid and the eighth pair of nerves in the vicinity of the thorax was now scratched a good deal by cautious touches of the knife, for the purpose of obtaining a view of the origin and course of the subclavian artery. And now it was found that the aneurismal tumour had extended so close to the trunk of the carotid as to leave it uncertain whether any portion of the subclavian artery was free from the disease. Indeed, so strongly were the assistants impressed with the apprehension that the aneurismal tumour had extended to the aorta, or right common trunk, that the majority appeared disposed to abandon the operation altogether. Supported, however, by the concurrence of Mr. Wilmot, I determined to persevere in the attempt, at least till we could positively ascertain the actual state of the artery. I therefore proceeded to lay open that portion of the sheath which invests the very root of the carotid, in effecting which a small artery was cut close to the coats of the carotid, and yielded a smart hæmorrhage. Some attempts were made to secure it with a ligature, but these only served to impress us with the idea that some irregular and nameless branch of the carotid had been cut so close to its origin that it could not be tied. Our uneasiness, however, on this head was soon at an end, for the bleeding spontaneously ceased. The division of the sheath being completed, we could now see that only a small portion of the subclavian artery, lying between the aneurism and the forking of the arteria inominata, remained in a sound state. The length of this sound portion did not exceed a quarter of an inch, and on this we resolved to apply the ligature. For this purpose I endeavoured, with the point and nail of my fore-finger, and occasionally with the flat handle of a silver director, to separate the artery and adjoining portion of the aneurismal tumour from the parts in their vicinity. This was not accomplished without considerable difficulty and delay. But these disadvantages appeared to me of little consequence in comparison with the danger that might attend the use of the knife in this critical step of the operation. When the artery seemed to be sufficiently detached I took an aneurism needle of softened silver, blunt at the point and longer than those in common use, armed with a ligature of six silk threads, and, guiding it with my left fore-finger, passed it down on the inferior or thoracic side of the artery. But I could not push it upwards, so as to pass its point with safety beyond the cervical side of the vessel.

I then detached the artery still more completely from the surrounding parts, and again made repeated trials to pass the ligature, not only with the needle above described, but with the common aneurism-needle, and also with a loop of softened silver wire, but all without effect.

All these instruments had been introduced from below upwards with the view of guarding against the danger of wounding the pleura.

It was now proposed to try whether the artery might not be surrounded by passing the needle in the contrary direction. And with this advice I complied, rather indeed because I had been so long and so repeatedly foiled in my former attempts than because the danger of the proposed plan had escaped my recollection. With the same instruments, therefore, I made a few efforts to pass the ligature in the manner recommended, in some of which I had reason to fear that the pleura was wounded, both from the increased difficulty of respiration and from the bubbling of the air among the blood. I soon found it still more impracticable to pass the ligature in this than in the former direction. Different plans for passing the ligature were now suggested by the assistants. Among these, one was to tie one end of the ligature into a knot sufficiently large to prevent it from slipping through the eye, and then, with a sharp hook or pair of spring-forceps, to lay hold of the knot when carried in on the concave side of the needle. I made the attempt, but the eye was so long that it allowed the knot to remain behind the vessel when the needle was pushed forward, in consequence of which this expedient failed. After all these trials the ligature was at length passed with a common

silver aneurism-needle, introduced from the thorax upwards.
But it was necessary to use a good deal of force on the occasion,
so that the artery was raised up considerably out of its natural
situation.

I now put a single knot on the ligature, and ventured to
tighten it in the following manner :—I took a polypus-forceps,
the blades of which had been made fast, with an interval left
between them wide enough to receive the artery. I then
passed one end of the ligature through each eye of the forceps,
and, holding in each hand a blade of the forceps and an end
of the ligature, I pushed the points of the forceps down on
each side of the artery. By this means the ends of the
ligature, which crossed each other on the surface of the artery,
were drawn in opposite directions, and the knot was thus
gradually tightened.

The pulsation in the tumour and in the radial artery was
now considerably weakened, but not totally suppressed. Prior
to tightening the noose the breathing of the patient had
become more laboured, and he complained much of an
oppression at his heart. On tightening the ligature these
symptoms increased to such a pitch that every one present
apprehended his immediate death. He was raised up and
supported in a sitting posture, but without affording him any
relief; his countenance grew pale and indicative of instant
dissolution. Yet his pulse did not become intermitting, or in
any manner irregular, as was observed by a gentleman, who,
during these awful moments, kept his finger applied to the
right wrist. In this alarming state he continued for the
space of about a minute, during which some of the assistants
were so strongly impressed with the idea of his danger that
they quitted the room, lest he should expire before their eyes.
During these struggles the polypus-forceps were withdrawn,
and, as the ligature had been tied with only a single knot, it
is to be presumed that it yielded to the pulsations of the
artery. Yet the patient did not appear to be at all relieved.
Indeed, so great was his distress, and so strong my fears, that
I had actually passed the index finger of my left hand along
one side of the ligature, down to the noose, and was just
about to cut through it with a probe-pointed bistoury ; but,

perceiving at this moment that his distress was a little relieved, the idea occurred to me that the wound of the pleura had been the principal cause of his sufferings. I now closed the lips of the wound, and this procured him immediate and sensible relief. However, the patient was so exhausted by this painful and tedious operation, and had suffered so much on the tightening of the ligature, that all present agreed to leave the ligature in its place, and to wait the issue of a few hours or a few days. It was conceived that by thus allowing time for the wound in the pleura to be consolidated the operation might hereafter be completed, as in that case, the right lung being capable of performing its office, the heart would labour less when this ligature was again tightened. Accordingly the lips of the wound were brought into close contact by three stitches of interrupted suture passed through the integuments. The patient was laid in bed and supported in a sitting posture, being incapable of lying horizontally.

Thursday evening, 8 o'clock.—Pulse 96, full, regular, and equal in both arms. After the operation he had been unable to swallow any fluid, except by sucking a bit of linen wetted with his drink. Within the last two hours has been able to swallow better, but still only a small sup at a time. Complains only of his heart; says that it feels as if a knot were tied on it, yet points to the scrobiculus cordis as the seat of his distress. Countenance rather flushed; skin moist; temperature of both hands equal. Sixteen ounces of blood being now drawn from his arm procured him some relief.

Friday morning, 8 o'clock.—Did not sleep one moment during the night, yet says he passed it very well; his sister, who sat with him, says that he had a great many catchings at his heart. What she means by this term it is impossible for me to ascertain, as I have not seen him affected in the manner she attempts to describe. The blood drawn last night is very buffy, and much cupped: pulse this morning 120.

At 12 o'clock.—Pulse 116; distress rather less. It was determined that he should be bled again, but he was advised not to have the blood drawn till his distress should increase.

8 o'clock p.m.—He himself desired to be blooded, and sixteen ounces were taken from him at 9 o'clock. He declared himself much relieved by the bleeding; pulse only 96.

Saturday morning, 8 o'clock.—Has had some short slumbers during the night; lies much more flat and horizontal; swallows more freely, and takes large sups of his drink, swallowing it gradually; pulse 108, regular and equal in both wrists. He thinks himself much better, but still complains of his heart. No palpitation or fluttering of the heart can be felt when the hand is applied to the thorax.

Saturday evening, 9 o'clock.—Pulse 96, as he lies without speaking; when he speaks it increases to 110. He has eaten a small quantity of boiled bread and milk, and a little cake steeped in tea; thinks himself much better.

Sunday morning.—Pulse 108; has enjoyed better sleep last night; has lain on the left side for an hour, and now lies with his head less raised.

Sunday evening.—Has eaten boiled bread and milk in the day and some flummery in the evening; lies now with his head very little raised, and has lain on both sides at different times this day. Pulse 130 and small; complains of some pain in the wound; tongue is white; countenance rather flushed; no stool since the operation.—Enema purgans statim.

Monday, 12 o'clock.—He rested pretty well last night, and feels refreshed this morning; breathing free; he lies horizontally, and on his right side; pulse 114. The enema has procured a very copious evacuation; thirst has been relieved by taking lemon juice in his barley-water.

In the interval between our first seeing him and visiting the rest of the hospital he was seized with a shivering which lasted for ten minutes. During the shivering his skin did not feel cold to the hand, nor had he himself any sensation of chillness, but said that the shivering would go off directly. He got a little warm wine and fifteen drops of tincture of opium. His pulse and breathing seemed not at all affected by this rigour. In half an hour after the rigour had ceased we proceeded to dress him; the lips of the wound appeared nearly united: a very small quantity only of thin discharge

had flowed along the ligature. It was now deemed advisable
to cut out the two lowest sutures of the integuments, and
to separate the lips of the wound, for, in the present stage of
the wound, this could be easily effected, whereas, if it were
deferred for one or two days more, it would possibly require
the use of the knife. Accordingly the sutures were cut out,
and the lips of the wound separated by drawing the flat part
of an eyed probe along the line of the wound. The knot
of the ligature could now be felt at such a distance from the
artery that we thought that the point of the fore-finger could
just be insinuated a little between them. A consultation was
now held to decide on the propriety of immediately proceeding
to tie the ligature. Some proposed that it should be altogether
withdrawn ; others recommended that the attempt of tying it
should be made, but to postpone it for a day or two, till
the effects of the rigour should have shown themselves ; while
a third opinion urged the necessity of either instantly withdrawing the ligature, or tying it closely on the artery,—these
last assigning, as the grounds of their opinion, the danger
that the ligature, lying in contact with the artery, should
so far produce ulceration of its coats that they would
tear under the ligature if attempted to be tightened in one or
two days hence ; that the rigour had ceased, and the functions
of circulation and respiration were going on perfectly well at
this moment ; in short, that the present moment was as
favourable as any future period could be. The latter opinion
prevailed, and the third or uppermost suture being cut out,
and the wound thrown open, I proceeded to untie the remaining single knot. My object in doing this was to include
in the noose along with the artery a bit of flexible silver wire,
for the purpose of enabling us more readily to cut out the
ligature if we found, on tightening it, that alarming symptoms
would be induced ; another reason was that we might tie
on it the surgeon's knot, which would not slip, whereas with
the single knot the ligature could scarcely be tied tightly
at the bottom of such a very deep cavity. Without much
difficulty the knot was untied, a probe and aneurism-needle
having been successively used for the purpose of widening the
noose. The wire, very much bent, was included in the

new noose, on which was formed the surgeon's knot ; the ends
of the ligature were passed, each through the eye, on the end
of an iron wire ; these were thrust down, as I then fancied, to
the bottom of the wound, and attempts made to tighten the
ligature, but to very little purpose. Now, fearing that the
wire somehow prevented the ligature from being tightened,
one end of it was cut off close to the surface of the wound, and
the wire was then withdrawn by using a slight force. Now,
passing my finger down to ascertain the state of the ligature,
I discovered that we had not arrived at the bottom of the
wound ; the wires had descended only to the neighbourhood of
the anterior surface of the vessel. I now separated, with my
finger, the recent attachments of the artery to the surrounding
parts, and, passing each end of the ligature through the eye
of a polypus-forceps, the blades of which had been previously
taken asunder, I passed these down to the bottom of the
wound, and, giving the upper one to Mr. Wilmot to hold,
I held the lower one myself, and at the same time drew
it very tightly. This attempt reduced the strength of pulsation both in the tumour and at the wrists ; but we found that
this effect was owing, in a good measure, to the artery being
raised forward, and to the influence which pressing down the
blades of the forceps had on rendering it more tight. The
reason why I held the blade nearer to the thorax was, that I
feared if it had been directed downwards the pleura must
almost inevitably have been again lacerated. I next made the
attempt, having only the upper end of the ligature passed
through one blade of the forceps, while the lower, which, from
being short and wet, could not be held firm by the fingers, was
rolled round the other blade, while the fore-fingers of my
right hand was employed in giving a favourable direction
to the ligature, so as to prevent it raising the artery up
out of its situation. After one or two attempts in this way the
closing of the artery was effected, as was manifested by
the pulsation of the radial artery completely ceasing. Some
slight obscure pulsation, however, was still to be felt in
the tumour, which led some present to apprehend that the
ligature was not drawn sufficiently tight ; but on a moment's
reflection it occurred to us that this degree of pulsation

was communicated to the tumour by the pulsation of the sound part of the artery. A small quantity of florid blood flowed from a deep part of the wound, but not in a stream, just as the ligature was tied. A second knot was now made, and the wound closed by two points of interrupted suture.

During the tying of the ligature no alteration was observable in his countenance, respiration, or the actions of his heart. An assistant, who had held his hand on the left radial artery during the tying, could not discover the slightest alteration to have been induced by it; the patient did not exhibit the slightest symptom of pain; in fact, he seemed not to suffer in the slightest degree from this sudden revolution effected in the arterial system; he moved the arm as freely as ever.

I saw him an hour after the operation, at which time the temperature of both hands was equal, being in both 91 degrees. He continued free from any uneasiness. He had drunk very largely of whey during the operation, which lasted one hour: his thirst at this time continues unabated; he wishes for lemonade.

8 o'clock p.m., Monday.—Pulse 120; temperature of both arms, 91 degrees; thirst continues; has not had any sleep; tongue white and dry; skin moist, and feels of the natural warmth; he did not use any food since the operation this day.

Tuesday morning, 8 o'clock.—Pulse 130; in half an hour after, 120; passed the night without sleep, and has had very great thirst; yet he thinks himself better. He wishes for a few morsels of broiled meat, rather to indulge a medical theory than to gratify his appetite, for his stomach and bowels are extremely flatulent, which, he fancies, arises from want of solid food. Tongue is white; thirst is now less urgent: temperature of both arms equal.—Habeat haust. efferves, tertiis horis.

Tuesday evening, 9 o'clock.—He has just fallen asleep, and is talking almost incessantly, is tossing his limbs about, and would appear to suffer great distress; yet his sister says that this is not different from what occurs to him when in health; for that when he first falls asleep he talks over the

occurrences of the past day, and tosses himself about in the
bed, but that after he has been asleep for some time
his repose becomes quiet and profound. Pulse, 120; temperature of both arms apparently the same: has had some
sleep during the day, and thinks he might have had more had
he not wished to reserve it for the night; thirst is very
inconsiderable since morning; ate some flummery; tongue
more clean.

Wednesday morning, 8 o'clock.—He has been just disturbed
from a short sleep, and appears rather confused: has passed
a good night. but thinks himself weaker this morning. Pulse
120, smaller and weaker; thirst not more urgent; complains
of the fetor of the discharge from the wound. and wishes it to
be dressed. A considerable quantity of a yellow serous fluid
came from the lowest point of the wound; part of it flowed
spontaneously, and some escaped when pressure was applied
to the more remote parts of the wound. The skin did not
adhere to the subjacent parts; a small quantity of air seemed
to be interposed. No pulsation is felt in the tumour on the
humeral side of the longitudinal incision of the neck. He
appeared a good deal fatigued after the dressing. Tongue
white.—Habeat pil. purg. unam statim, et post horas tres,
enema purgans, nisi prius alvus soluta sit.

Wednesday, 3 o'clock p.m.—Has had some refreshing sleep,
and for a much greater length of time than heretofore: has
eaten five or six morsels of beef-steak, and, lately, a little
flummery; one copious stool after the injection. The nurse
reports that a small quantity of blood came off with the stool,
not more than was sufficient to streak it. He is now in a quiet
sleep, lying on his left side.

Wednesday evening, 9 o'clock.—Has got still more sleep;
temperature of the right hand, by the thermometer this day,
90; of the left hand, 92; but the right has been a good deal
exposed to the cold, being very seldom kept under the
bed-clothes; pulse 120. A few minutes before this visit
a shivering had come on, which still continues; yet he
does not experience the sensation of coldness, nor does his
skin feel cold to the hand. His breathing now appears a little
hurried, as he lies on the right side; yet, when desired to

make a full forced respiration, he does not experience any pain
from it, except what he refers to the heart, or scrobiculus
cordis, and this only in a trifling degree.

Thursday morning.—The rigour did not continue half an
hour ; he passed a good night ; had one sleep for three hours,
and during the rest of the night had a considerable share of
sleep.   Pulse 120, rather small ; tossing of the lower
extremities rather increased ; thirst much less ; discharge
from the wound copious, but has passed off; none of it
has lodged.   On pressing with the sponge an unusual
quantity of air escaped from under the skin ; no emphysema
noticed in the vicinity of the wound.   An obscure pulsation in the tumour was either felt or fancied by me ; certainly
none can be felt at the wrist ; size of the tumour a good deal
reduced.

Thursday evening.—Has slept well this day ; has eaten more
flummery than on any former day ; a slight emphysema
in the lower and anterior part of the neck.   At five o'clock
this evening some blood had been observed on his shirt
and on the bandages ; yet in every respect he appears as well
as usual.   At 7 o'clock p.m. the dressings were removed, and
the sutures cut out.   On raising the flap of skin a small
quantity of coagulated blood was found in the wound, though
not in its deepest part ; emphysema increased ; obscure
pulsation continues to be felt in the tumour.   Three small
pieces of sponge were laid in the wound ; a larger piece
was laid on the skin, and retained by adhesive plaster, as
pressure could not be applied by a bandage without confining
the thorax too much. Pulse 120; stomach moist; tongue white;
temperature of limb and state of breathing as this morning.

Friday morning.—At 8 o'clock a.m. I received, from one of
my pupils, the following note :— "About one o'clock last
night this poor fellow got a change for the worse ; appeared as
if strangling, with continued tossing of his legs and arms ;
complained of his heart, and said there was some one pulling
a cord round his neck.   He became quite delirious ; continued
in this state for nearly half an hour, after which he got comparatively easy ; and has been almost ever since insensible,
and is at present, I think, very near his end."

I visited him at 9 o'clock this morning. He lay insensible, rather pale, but not of that deadly paleness which so often attends great hæmorrhages; his breathing very laboured; in short, with every appearance of immediate death.

About 10 o'clock this morning he died.

In a few hours after his death we hastily removed the heart and large vessels, so low down as the clavicle, being prevented by the attendance of his friends from examining the parts *in situ*. No blood was effused into either cavity of the thorax. The lungs of both sides were sound; the upper extremity of the right lung had a slight adhesion with the parts in the vicinity of the ligature. The aorta was slit up from the heart to the forking of the arteria inominata: and thus we had an opportunity of observing the very diseased condition of the coats of this artery, as manifested by the wrinkled state of its lining membrane. A very remarkable pouch or recess, capable of receiving the point of the little finger, appeared in the wall of the aorta, on the right side of the artery, about one inch above the heart.

In the apex of this dilatation the coats of the artery were extremely thin, as if, the middle coat having been removed by absorption, the lining and outer coats had adhered together. The arteria inominata was dilated at the place of its forking. Between this dilatation and the aneurism, for which the operation had been performed, there was not above a quarter of an inch of the subclavian artery free from disease, and on this part the ligature had been applied. A pretty large opening had been made by ulceration through the coats of the artery, immediately above the ligature.

The next case of disease requiring the operation of tying the subclavian artery occurred in William Levanee, an invalided soldier, who was affected with aneurism of the axillary artery a little below the clavicle. The disease seemed to have arisen spontaneously. Tumour was the size of a pullet's egg at the time of his applying for relief.

In full consultation the operation was decided on; but this man refused to submit to it unless we assured him of complete success. He lingered for some months, and appeared to die, not from hæmorrhage, but rather from extensive suppuration and long-continued irritation.

Appearances on Dissecting the body of William Lerance, ætat 55 years.

26th April, 1813.—Very considerable swelling occupied the right side of the body, extending from the clavicle down to the eighth rib, and from the nipple of the breast across the arm. The arm and fore-arm were very much distended by an œdematous swelling. Three large ulcers a little below and before the axilla ; inferior costa of the scapula rough on its internal surface ; the inferior angle of this bone exposed and rough. On opening the thorax the viscera were found perfectly healthy, with the exception of slight adhesions of the right lung to the pleura costalis, and some water in that side of the chest. Heart rather small, but appeared perfectly healthy. Large arteries near the heart not at all dilated, and no other traces of disease in any of them than some very small opaque spots visible on their internal coats.

The pectoralis major was of a pale brown colour, and much thinner than natural; the pectoralis minor converted into a substance resembling a strong fascia. The parts above the clavicle were still in a healthy state : the different steps of the operation for tying the subclavian artery before it had reached the scaleni were performed on it. The trunk of this artery was traced of natural size and healthy texture, for an inch and a half below the origin of the vertebrals, and running over the front of the aneurismal tumour. At this place its coats became so thin that it was no longer possible to distinguish them from the coats of the tumour. Immediately before it became indistinct this part of the artery seemed dilated to the breadth of a quarter of an inch. The articulation of the humerus was opened and communicated with the aneurismal tumour. The branches of the subclavian artery did not appear at all enlarged. The basilic vein was thickened in its coats, but not at all enlarged in its diameter.

On the 19th June, 1813, I was consulted by the Rev. Mr. S., aged 48. He was a man of an uncommonly athletic frame ;

had constantly enjoyed the most robust health, and had always lived a regular and temperate life. The account he gave of his complaint was as follows :—One night before Christmas last, while lying in bed, his right arm was suddenly seized with a numbness, and the right hand felt as if instantaneously enlarged to twice its natural size. Soon after this the middle of the fore-arm and the insertion of the deltoid muscle were attacked with pains. He said that the physician whom he had consulted in the country had remarked, with some surprise, that he could not feel any pulse at the wrist of the arm affected. The disease passed for rheumatism until he came up to Dublin; but, on his arrival in town, having applied to Dr. Harty to be electrified, the doctor, during the process, observed a portion of the chain, which happened to lie on Mr. S.'s breast, to be moved up and down by a violent pulsation of some subjacent part. Struck with this occurrence, he proceeded to examine the seat and cause of the phenomenon, when he discovered an aneurismal tumour situated below the clavicle, and covered by the pectoral muscle.

On the day following I saw him, and found the aneurism nearly of the size of a goose egg. The tumour pulsated strongly, but seemed notwithstanding to be both deeply seated and thickly covered. The pulse of the right radial artery was as full and strong as that of the left. By pressing with my thumb upon the diseased artery, above the clavicle, along the outside of the scalenus muscle, I suppressed the pulsation both in the tumour and at the wrist. But a strong pressure on the tumour itself, even when continued for some time, excited no uneasy sensation whatever, either in the region of the heart or in any other part of the thorax; nor did it affect the powers of respiration in the slightest degree. In short, nothing preternatural could be discovered in the actions either of the heart or of the arteries, under any possible variation of circumstances.

On this very important case I requested a full consultation without loss of time; and accordingly, next day Messrs. Richards, Peile, and Wilmot, Dr. Harty, and I met in con- sultation. It was our unanimous opinion that, although the

present disease could not be clearly traced to an accidental
injury, yet, as not the slightest appearance of disease
could be discovered in any other part of the arterial system,
it was our duty to propose the operation of tying the subclavian
artery.

Mr. S., who was a man of uncommon fortitude of mind,
consented at once to submit to this formidable operation. He
wished, however, first to return to the country for some days,
in order to settle his affairs. To this delay we acceded,
advising him to avoid all bodily exertions, to live on very low
diet, and to debar himself of a full meal even of the coarsest
fare. We also prescribed some strong purging pills, to be
taken frequently, and directed that he should be blooded twice
at least before his return to Dublin.

On the 16th of July he returned, a good deal reduced in
flesh and colour. The aneurism had increased in size during
his absence, principally in the direction towards the clavicle.
Indeed, to judge by the eye only, the tumour seemed to have
extended to that bone ; but by the touch it could be ascertained
that the tumour and the clavicle were not yet in contact. He
now told us that some of his friends in the country had
reminded him of his horse having fallen and rolled over him
a few days before he felt the numbness of the right hand ; and
he can now distinctly recollect that he suffered considerable
pain about the shoulder at the time of the fall.

I commenced the operation by an incision beginning near
to the acromion, and terminating about midway up the side of
the neck. Another incision, commencing from the same
point, was carried along the clavicle, until it reached the
outer edge of the sterno-mastoid muscle ; the triangular
portion of integuments included between these incisions was
raised up from its apex, and laid back upon the throat. In
this first step of the operation some pains were taken to avoid
wounding one or two pretty large veins which ran very superficially along the side of the neck. When the integuments
were thus removed I detached the external jugular vein, at the
highest point where it lay uncovered, and, with a blunt
aneurism-needle, passed a ligature round it. I next tied up,
in the same way, a branch of this vein which entered the

trunk a little above the clavicle, in a direction from the acromion; and I then secured the lower extremity of the denuded trunk of the jugular vein ; but to effect this last was a task of more difficulty, for the vein here was not only covered by three or four layers of tough cellular substance and fat, but was besides, at times, suddenly puffed up to a very great size, especially whenever from pain the patient made a great exertion, accompanied with a deep and long-continued expiration. But notwithstanding all these ligatures the trunk of the vein was still freely supplied with blood from below, as appeared by its occasionally swelling in deep expirations. This obliged me to secure the trunk with a third ligature, about an inch above the clavicle, and, having done so, I cut the vein across between the two lowest ligatures, and laid the inferior portion of it down upon the clavicle. Now removing a little fat, I saw the omohyoid muscle covered with a fascia ; it was of an uncommonly large size, and lay very low, near to the clavicle ; it consequently became necessary to sacrifice it, in order to get at that part of the artery which lay near to the outer side of the scalenus muscle. For this purpose a very thin silver spatula was passed under the upper edge of the omohyoid muscle ; with this it was raised a little from its bed, and was cut across. On removing this muscle a thin fascia appeared covering a deep bed of fat. Having cut through the fascia, I now, with the flat end of a silver director, cleared my way among the fat, until I could feel, though not see, the artery. As the vessel was now lying at the bottom of a deep contracted cavity, I judged it necessary to convert this into a more open and less embarrassing form of wound. This I did by cutting across the edge of the trapezius muscle and the numerous layers of membrane which occupied the space included between the deep cavity and the acromial end of the external wound. In effecting this one small artery was divided and secured by ligature. When the wound was thus enlarged, the most anterior of the nerves going to the brachial plexus appeared to view. I now proceeded to lay bare the artery, holding my finger down upon it, and turning aside the superincumbent fat with the flat end of a silver director. But so great was the quantity of this substance to be removed,

that, although I cut away three or four large pieces of fat with
a scissors, and drew them out with a forceps, yet, to prevent
the remainder from falling down again into the cavity upon
the artery, I was obliged to have it held aside by three
long and broad spatulas of prince's-metal, bended at the end,
and applied one to each side of the wound.   This materially
facilitated the subsequent steps of the operation, by keeping
the artery uninterruptedly exposed to view.   My next object
was to detach from the surrounding parts that portion of
the artery which lay close to the scalenus muscle.   To effect
this I first made use of the flat end of the director, scraping
with it along the outer or acromial side of the artery.   I then
ventured to separate it more completely by introducing a
blunt aneurism-needle in the room of the director.   This
I moved upwards and downwards along the vessel until I was
enabled to pass it freely under the artery, and to receive
its point on the end of a narrow spatula, introduced between
the vessel and the scalenus.   Feeling the point of the needle
on the spatula I was satisfied that it had cleared the artery.
I now therefore depressed the handle.   On doing so I observed
that the point was covered by a dense white membranous
substance, closely resembling the coats of an artery.   I withdrew the needle, armed it with a double flat ligature (about
six threads thick), and, again introducing it on the acromial
side of the vessel, I quickly showed its point by the side of the
scalenus, covered with the membranous substance as before.
I hesitated for a moment, but every one present being satisfied
that this substance was not a part of the vessel, I readily
tore through it by scraping it with the flat end of the director.
Now, merely by depressing the handle of the needle, without
in the slightest degree raising the artery from its bed, I
was enabled to catch the ligature on the inner side of the
vessel.   Having got a secure hold of it I withdrew the needle,
leaving the ligature, doubled, under the artery.   I then
divided the ligature into two, and tied the lower with a
single knot.   Mr. Peile, who had applied one hand to the
tumour and the other to the right radial artery, announced,
as I tightened the ligature, that all pulsation had ceased
in both.

The assistants, being all anxious to satisfy themselves of these facts, I let go the ligature. Although I let it out of my hands but for the few moments I was necessarily occupied in these inquiries, yet, even during this short period, the pulsation perfectly returned; I therefore again drew the ligature tight, the pulse in the limb ceased, and I quickly made fast the ligature with a double knot. To enable us to tie this knot we had no occasion for any instrument; nothing more was necessary than to catch the ends of the ligature close to the vessel, and then to press on them with the points of the fore-fingers, held on a level with the trunk of the artery. No other precaution was necessary to save it from being pulled up from its bed, in drawing the ends of the ligature tight. The unanimous voice of all present directed that the second or upper ligature should be tied, which was done, close to the first, with a double knot. In this awful and critical step of the operation I kept my eye firmly fixed upon the patient. His countenance did not undergo the slightest perceptible change. No alteration was observable in the actions of his heart, or in his respiration; nor did he experience any pain or particular sensation in the limb on the tying of the artery.

It should be observed that a large vein, which, in the dead subject, I had often found running across the artery, in a direction from the scapula towards the thorax, and which, I had feared, might be productive of much inconvenience on the present occasion, was here found to be flat and empty, and to run so near to the thorax as not to interfere in the smallest degree with the passing of the ligature.

When the operation was finished three stitches of interrupted suture, aided by adhesive plaster, kept the lips of the wound in contact; on the flap was laid a piece of dry sponge, to bring into apposition the upper and under walls of the cavity, which had been left after the operation.

Shortly after he had been laid in bed he complained of cold, but had no rigour. In the course of an hour he felt himself growing too hot. In three hours after the operation his pulse was 96. In this short period he forced me to change his position two or three times, from his back to his sides, to try if he could by this means obtain any mitigation of what he termed a pain in the small of his back. He did not, however, complain of any pain or fulness in the head, or of any affection of the heart, or difficulty of breathing. His distress, though rather intense in degree, was of the same kind as I have frequently witnessed after amputation, or other great operations.

Mr. Wilmot, who remained with him from three o'clock this day until nine o'clock at night, informed me, on my return, that his pulse had got up so high that he had been almost tempted to bleed him; but a sweat breaking out, had reduced the pulse to its present state (120). For the remainder of the night I took charge of him, and during that time he was affected with a slight tossing of his lower limbs. He continued to complain of his back, and did not enjoy more than five or ten minutes' sleep at a time. He had some thirst, and in the course of the night his pulse rose to 134. He did not complain of the wound, except when he turned on that side or was awkwardly handled, and then he said it was sore. The sweat having at one time ceased for two hours, I observed that the right hand and arm became dry, as well as the left; and on the return of the sweat both arms were equally moist. The heat of both arms, too, appeared to be the same since the operation, and he has continued to enjoy the feelings and full power of the right hand.

Wednesday morning.—At six o'clock this morning I took twelve ounces of blood from his left arm; it flowed slowly, for he drew back his arm while I was making the incision. I gave him a draught, with two drachms of Aq. Ammon. Acetat.; and for a short time after this I fondly flattered myself that he became more easy and composed. These transient hopes, however, soon gave way to more gloomy apprehensions; and when I reflected on the quickness of his pulse, his restlessness and want of sleep, his forgetfulness and occasional wanderings of mind, I became greatly alarmed. I therefore summoned a consultation of all the gentlemen who had assisted me, for an earlier hour than that agreed upon the day before. We met at 11 o'clock a.m. His pulse was now 134, and the strokes ran so into one another that it was difficult to count them. The state of his pulse, we judged, would not be improved by a repetition of the blood-letting. It was therefore agreed that he should be first purged, then take tincture of digitalis, and, if his restlessness required it, take an opiate at night.

Three o'clock p.m.—At this hour the physic had only operated once, and then very sparingly. His pulse was now driving at a rapid rate. He complained of an occasional cough, which distressed him so much that he was forced to entreat the attendant to keep one hand on his head and to grasp his left hand with the other whenever he felt a fit of it coming on. He complained also of a soreness in his throat when he coughed or swallowed; and he pointed to the middle of his throat as the seat of this affection. Although I suspected that his fears made him magnify his distress in this particular (for he apprehended that the inflammation, spreading from the wound to the throat, must quickly prove fatal), yet, when I recollected that my former patient Cowell had complained of the same sensation, I deemed it prudent to meet it early, and therefore instantly commenced with the digitalis, although his bowels had not been sufficiently freed by the purging medicines.

At 4 o'clock p.m. he got the first dose of 10 drops.

At 8 o'clock the second dose of 15 drops.

At 10 o'clock, 25 drops.

At 2 o'clock a.m. Thursday, 30 drops; and at 6 o'clock the same repeated.

By this medicine his pulse was reduced, so that at 10 o'clock p.m. it beat 120, but still with the same indistinctness as before.

Thursday morning, 6 o'clock.—The longest sleep he has had in the course of the night has not been above half an hour; the tossing of the lower limbs continues; they are dry, and, though covered with woollen stockings, are inclined to be cold. Countenance is natural, but more reduced than when in health. Mental faculties more perfect than yesterday. Sweat continues on the head, trunk, and upper limbs. Has had many stools through the course of the night. State of tongue and thirst as yesterday. Pulse 120, and indistinct.

Moans at times to-day as he did the day before. Says that
he has an uneasiness at the small of his back, but that
it does not amount to pain.   Thinks his throat better.
There is some swelling, with a slight appearance of inflammation, across the top of the thorax.   He now entertains
better hopes of his recovery.   He begins, however, to complain of some uneasiness in his chest, but says that it is
relieved by sighing.   The bandage, which feels rather too
tight on the chest, is now cut across.   Heat of right hand,
measured by the thermometer, 94 ; that of the left hand, 96.

Thursday evening, 8 o'clock.—This day, at 1 o'clock p.m.,
he was suddenly seized with a rigour, which lasted for a
quarter of an hour.   This was succeeded by a profuse sweat.
During the cold fit he talked much and incoherently.   Soon
after the hot fit commenced he started up, jumped out of bed,
and became so violently outrageous that he required two
strong men to hold him.   In this state he continued for more
than an hour.   During this paroxysm he made the most
powerful exertions with the right as well as with the left hand.
After the paroxysm had subsided he enjoyed a sleep of half
an hour, during which his breathing was natural.   The cough,
which had distressed him to such a degree yesterday, is so
much relieved to-day as to cause only a slight and momentary
uneasiness in his chest.

Thursday night, 10 o'clock.—He had been raving a good
deal at 8 o'clock this evening, but at half-past nine he became
quite collected, and inquired very anxiously about his own
situation.   Pulse more distinct, though not less frequent.
Sweats continue.   Some wheezing in his chest.   Distress of
that part less than it had been early in the evening.   A large
blister is applied to his chest.

The inflammatory fulness and redness of the skin at the top
of the sternum and left side of the throat is considerably increased ; the slightest pressure on this swelling gives him
very sharp pain.   Mr. Wilmot and I, conceiving it possible
that his distress might be aggravated by some collection
of matter confined or pent up within the wound, removed the
dressings.   Only a small quantity of reddish water was
discharged.   I then cut out one of the sutures, and passed a

probe into the deepest part of the wound, conducting it along the ligature of the artery. We then made a pressure on the inflamed integuments, at the top of the sternum and left side of the thorax, but without obtaining the smallest additional discharge. At this time the aneurismal swelling was in appearance quite removed, and both sides of the thorax were to the sight and touch perfectly alike.

Friday morning.—He passed the night without any sleep; raved a good deal, but was not violent. Purging more frequent. Has taken two doses of a mixture, each dose containing ʒss. Oxymel Scillæ. Cough has not distressed him. He has expectorated some mucus. During the night he complained a good deal of the right fore-arm and hand,—at one time of numbness, and at another of pain. The hand is held with the fingers folded in, and rather stiffened. It now became too obvious that the storm, which had been collecting since the operation, had taken a fatal direction. The present state of the limb so strongly proved the near approach of mortification, that, however reluctantly, I was forced to relinquish every ray of hope, and to prepare my mind for the speedy termination of a case which I had watched with more fluctuations of hopes and fears than ever had agitated me on any former occasion.

Two o'clock p.m.—The fore-arm was distended at the insertion of the pronator teres, and also where the extensors of the thumb pass over the end of the radius; the hand and fore-arm cold; the fingers rigidly flexed. The usual applications were made to the limb, but without affording any relief. The mortification went on increasing; his delirium continued for the greater part of the night; and at four o'clock on Saturday morning he died.

Appearances on examining the body 26 hours after death.

Suppuration was established under the integuments on the fore-part of the throat, anteriorly to the sterno-mastoid muscles. On the thorax being laid open we discovered an appearance of suppuration commencing on some masses of fat lying in the anterior mediastinum. The upper lobe of the right lung was attached to the ribs by close and old adhesions.

Between three and four ounces of water found in the pericardium. The jugular veins, both the external and internal, were of a deeper colour than usual. Heart felt rather soft; right auricle full; right ventricle less full than usual. No disease in any of the valves on either side of the heart. The internal surface of the right auricle, especially on the septum of the auricle, more red than usual. The redness, though of a brighter tinge than that of the jugular veins, was not such as to indicate the existence of inflammation of this cavity. The venous blood was more fluid than usual. Coagula were found in the right and left ventricles, but not uncommon either in their size or other properties.

On endeavouring to separate the aneurismal sac from the walls of the thorax, it was found to have connected itself so intimately with the second rib as to have its cavity laid open by an attempt to disunite them. The form of that portion of the bone upon which the aneurism rested was changed from a flat surface to a cupped form. The extent of the altered part of the second rib was about half an inch.

The aorta, with the great arteries springing from its arch, and the humeral artery, being removed from the body, were examined next day.

The aneurismal tumour had extended more in length than in breadth, and had ascended so high as to have reached near to the spot on which the ligature had been applied. The aneurismal tumour was nearly as large as a pullet's egg.

The artery immediately above the ligature was seen thrown into folds and furrows, which would seem to be the inevitable effect of tying an artery of so great a size. On removing the ligature, and slitting up the artery, its internal coat was found ruptured by the ligature only at one part, to the extent of about one-fourth of its circumference. Through the remainder of the circle the internal surface of the artery presented a white line, obviously produced by the tightening of the ligature. Two very small portions of coagulable lymph were found immediately above the ligature. The trunk of the artery below the ligature was rather small. The branches of the subclavian artery were not at all enlarged.

I have been (perhaps tediously) minute in detailing the particulars of the foregoing cases, because we have, as yet, but one account published of the operation of tying the subclavian artery after it has passed through the scaleni muscles, and no instance, I believe, has hitherto been recorded in which this artery was tied before it had arrived at these muscles. Under such circumstances I would not presume to exercise my own judgment in omitting those particulars which to me did not seem important.

As I wished that every one who reads these accounts should be unbiassed in drawing his own inferences from the facts, I have studiously avoided interweaving any reflections of mine own. Such as have occurred to me in my attendance on these cases I shall now briefly state.

To lay bare the right subclavian artery before it reaches the scaleni will not be found difficult by any surgeon possessed of a steady hand and a competent knowledge of anatomy; but I fear that, with the utmost dexterity, much difficulty will be experienced in passing and tying the ligature, even in the most favourable cases. The instrument delineated in Mr. Ramsden's book appears to me well calculated for the purpose, and yet it is not free from objection. I should fear that the coats of the artery might be cut by the sharp edges of the flexible steel plate as it is passing round the vessel. Some means similar to those I have mentioned must be adopted to prevent the artery from being raised out of its bed while the surgeon is tying the knot.

This operation, difficult on the right, must be deemed absolutely impracticable on the left subclavian artery. For the great depth from the surface at which this vessel is placed,—the direct course which it runs in ascending to the top of the pleura,—the sudden descent which it makes from this to sink under the protection of the clavicle, and the danger of including in the same ligature the eighth pair of nerves, the internal jugular vein, or the carotid artery, which all run close to and nearly parallel with this artery; these all constitute such a combination of difficulties as must deter the most enterprising surgeon from undertaking this operation on the left side.

Even on the right side this operation will be very seldom
required ; for it can only be called for in injuries or diseases
of that small portion of this artery which lies between the
scalenus and the clavicle.   It is scarcely necessary to observe
here how frequently such a case will be complicated with
diseases of the great trunks of the arteries nearer to the
heart, or how difficult it must be to discover the existence of
those internal affections.   When examining any pulsating
tumour near the top of the thorax, the surgeon should bear in
recollection the remark made by Mr. Astley Cooper, as quoted
by Mr. A. Burns, in his ' Surgical Anatomy of the Head and
Neck,' " that aneurism of the aorta may assume the appearance of being seated in one of the arteries of the neck." The
quantity of blood lost by the haemorrhage, which yet we
presume was the immediate cause of death in this case, was
very inconsiderable, being infinitely less than I had ever before
known to prove fatal.

The operation of tying the subclavian artery on the acromial
side of the scaleni muscles is one which will be much more
frequently required, and which can be performed with equal
facility on the artery of either side.   The most striking and
most unfavourable circumstance of this operation is the great
length of time required to perform it, although the ultimate
and essential steps of it had been executed with so much
facility.

This delay is by no means unavoidable.   Indeed, any man
who will take the trouble of reflecting on the anatomy of the
parts concerned, or who will himself perform this operation
on the dead subject, must be convinced that it may be executed
in as short a time as the operation for femoral aneurism. The
causes of delay in this instance were, first, the almost unpre-
cedented nature of the operation ; the caution with which we
tread upon untried ground rendering every incision more slow,
and every step more deliberate.   Next, the great anxiety to
secure every vessel, even the smallest branch of an artery, or
a vein, lest our view should be obscured, or our progress
obstructed by any quantity of blood flowing into the wound,
was a principal source of this delay.   By referring to the
account of the operation the reader can readily estimate what

loss of time was occasioned by this object. I do not mean to say that we should pass through the various stages of the operation absolutely regardless of the flow of blood; but I am confident that we have no occasion to tie up any but the larger veins, and the suprascapular artery, if it should chance to be wounded. The curved spatulas, while they hold aside the lips of the wound, will, at the same time, serve to stop the bleeding. I cannot forbear to recommend, in the strongest manner, the use of these instruments, not merely in this, but in every other operation where the depth of the wound is considerable, compared with its length, and where it is of importance for the surgeon to distinguish the parts which lie at the bottom of such a cavity.

The necessity for removing any portions of fat can occur only in very corpulent subjects.

In passing the needle round the artery, it will be found absolutely necessary to have the scalenus muscle held back by one of these spatulas, otherwise it will be impossible to pass the ligature without including along with the artery some portion of the muscular fibres.

The facility with which the ligature was passed round the artery was in the highest degree gratifying to every one present. This inestimable advantage will be secured by an attention to the following points :—First, to extend the wound out towards the acromion, by which the form of it is changed from a deep cavity to a superficial wound; next, to introduce the needle on the outer or acromial side of the artery ; and, lastly, to select the most favourable part of the artery. This, on inspection, will be found to be where it has just passed through the scaleni. How necessary this selection is will appear by the perusal of the account of Mr. Ramsden's operation ; for he, by attempting to secure this vessel near to the first rib, or rather, as he says, at the lower edge of the first rib, found it almost impossible to turn the needle round the artery in the very narrow space between this bone and the clavicle.

Cases in which this operation may be necessary will not be very infrequent. In wounds of the axillary artery, either while it runs in front of the thorax, or while it lies along the humerus, this operation will be preferable to following the course of the wound by cutting through the pectoral muscles in the one case, or entangling ourselves in the brachial plexus, when the artery is wounded in the axilla. The pain and difficulties of the operation above described are trifling when compared to those which must occur in following the course of these wounds.

When an aneurism of the axillary artery shall require this operation, we may indulge a confident hope that the rest of the arterial system is free from disease, as it appears to have been in two of the foregoing cases, that of Levance and of the Rev. Mr. S.; one, where the disease arose spontaneously,—the other, where it could be traced to accidental injury. Although this operation has not yet proved ultimately successful, yet I think we should not despair. The history of surgery furnishes parallel instances of operations, now generally adopted, which, in the few first trials, failed of success.

PRACTICAL OBSERVATIONS UPON CERTAIN DISEASES OF THE ANUS AND RECTUM.

(Dublin Hospital Reports vol. 5.)

DISEASES of the anus and rectum have of late years much engaged the attention of the profession, as appears by the numerous publications on these subjects. Still, however, I think practitioners will allow that the diagnosis and treatment of these complaints admit of further improvement. With a wish to contribute to this desirable object, I beg leave to submit to the profession the following remarks on a few of these diseases :—

ORGANIC STRICTURE OF THE RECTUM.

This disease spares neither sex nor rank; it most frequently attacks those who are about the meridian of life ; sometimes, however, it afflicts children as early as the seventh or eighth year of their age. I have not met with any instance where it attacked a person at or beyond sixty years of age.

In some few cases the patient appears to be aware of the moment of the first attack of this disease, for he tells us that without any previous illness the bowels at a certain period suddenly became costive; that for the purpose of relieving them he took large and repeated doses of physic for three, four, or five consecutive days ; that at length his bowels suddenly gave way and a very severe purging took place, which, having continued for a day or two, was then succeeded by those symptoms which attend the disease when fully formed.

Many patients, however, cannot give any account of the first approach of this disease; they merely state that they have been for many weeks, months, or years subject to it ; that the symptoms from the commencement were pretty much the same as those they now labour under, but perhaps not quite so severe and urgent.

When organic stricture is fully formed (by questioning him), we learn from the patient that in the course of each day he has many and sudden calls to stool; that at each of these he is obliged to strain very much, and that the straining, which is not followed by any severe pain, produces a discharge of not more than a tablespoonful of mucus, which is sometimes streaked with blood, and very rarely mixed with a small quantity of fæces; that these evacuations are generally attended with a copious discharge of wind, and that as soon as the evacuation has taken place he feels free from pain or uneasiness; the remission, however, is of short duration, as he is soon again compelled to undergo the same unavailing distress. The number of these discharges are seldom less than from seven to twelve during each day and night; they do not take place at regular intervals, but generally a considerable number occur in quick succession, and then are followed by a pretty long interval of ease. The greater number of these evacuations are devoid of fæces; a feculent stool is passed perhaps once in two or three days; the fæces are then found passed in short pieces and of very reduced dimensions, not larger than a full-sized catheter, and in quantity not equal to what is passed in an ordinary evacuation by a person in health, yet after each feculent stool the patient feels much relieved.

There is not the slightest prolapsus ani with any of these evacuations.

The bladder is in some cases slightly affected; the patient then complains of a little difficulty or delay in passing his urine.

In some cases a fistulous opening forms in the nates or the perinæum, which will admit a probe to pass into the rectum; it yields a moderate quantity of healthy pus. The fistula undergoes little or no change from the time of its first appearance, even until the death of the patient. In some cases, especially in females, I have known the number of these fistulous openings to amount to twelve or twenty. The majority of cases of stricture of the rectum, however, are unattended by fistulæ.

Although this state of daily suffering proceeds with the

most unvarying regularity, not only for weeks and months, but even for many years, yet the constitution of the patient does not seem to sympathise in the slightest degree for a long time; not only do his colour and appearance proclaim the enjoyment of good general health, but even the most strict examination cannot discover in any of the functions the slightest deviation from health, except those above mentioned.

After a lapse of time, however, which is very different in different cases, a peculiar paleness of countenance and wasting of flesh announce an inroad on the constitution; those symptoms are soon followed by night-sweats, and now the patient complains of uneasiness about the sigmoid flexure of the colon, which soon extends along the left colon.

In this stage of the disease it too frequently happens in females that a communication is formed between the rectum and vagina, which causes the greater part of the entire of the fæces to pass through the latter. In men, but more rarely, a similar communication is established between the rectum and bladder.

As the disease draws to a close the patient begins to suffer much more serious distress from the state of the rectum, for, in addition to the frequent unavailing efforts to discharge the bowels, he is troubled on the slightest exertion—such as coughing, sneezing, or voiding urine—with an involuntary discharge of a thin brownish fluid of a muco-purulent nature. After some time this peculiar discharge comes away in the morning, the moment the patient rises from his bed. In the last stage of the disease this discharge comes away unceasingly without the consciousness of the patient, and the fæces will not pass unless in a perfectly liquid state.

This aggravation of local distress is accompanied by a corresponding decline in the general health; his appetite fails, thirst becomes very urgent, emaciation proceeds rapidly, hectic fever becomes fully formed, though more marked by profuse night-sweats than by mid-day or evening exacerbations; and now the patient, extenuated to the last degree, seems to be carried off as much by the exhausted state of his constitution as by the torments of the local disease.

Sometimes, in the advanced state of this disease, the patient is seized with symptoms of peritoneal inflammation, which puts a speedy termination to his sufferings, and he is suddenly carried off.   In such cases, on examination after death, we discover that the process of ulceration had opened the intestine immediately above the stricture, and that through this opening a portion of fæces had passed into the cavity of the abdomen.

In some very few patients the hectic fever seemed to have been arrested by the warmth of summer weather, but only to run the remainder of its course with unusual precipitancy on the approach of winter.

Among a considerable number of patients afflicted with this disease I have had an opportunity, in two instances only, of meeting with it in its incipient state.   In both of these the patient complained of different symptoms of irritation of the rectum, frequent stools, discharges mixed with mucus, and certain feelings of uneasiness.   On examination, by the finger, a thickening and slight projection of the gut was felt at a small spot on one side; this morbid alteration spread gradually round the entire of the canal, and extended along it only to a small distance; but, until the morbid derangement of structure had almost entirely performed the circle of the intestine, the patient did not exhibit those symptoms which I consider as the common and inseparable attendants on stricture of the rectum.   However constant in their attendance or unvarying in their course may be the symptoms of this disease, yet will the surgeon desire to be confirmed in his opinion by manual examination.   Proceeding to make this examination we often observe, at the orifice of the anus, the following appearance, which is indeed almost always present when the disease is seated near to the external sphincter, namely, at each side of the anus a small projection, which on its external surface appears as a mere elongation and thickening of the skin, but internally presents a moist surface, not exactly like the lining membrane of the gut, nor yet can we say that it is ulcerated; these two projections lie close together below and divaricate above, presenting a resemblance to the mouth of an ewer.   Whenever this external appearance exists I feel almost certain of finding a stricture of the rectum before the finger is pushed as far as the second joint into the gut. In some cases, however, this external mark has not been present.

When the stricture is situated pretty high up, the portion of the gut interposed between it and the anus is found to be in a perfectly healthy state; but when the finger arrives at the stricture it is arrested by the narrowness of the canal, which will barely admit the point of it; if now a slight degree of force, combined with a boring motion, be employed, the finger may be pushed through the thickened and indurated part, and will then (as well as its benumbed condition permits it to feel) find that the gut just above the stricture is in a very healthy state. The extent of the stricture, however, is very variable; sometimes it is little more than a mere ring, but at other times it extends along the canal as high as the finger can reach.

I have not yet met with any instance in which the intestine was strictured, only by means of bands thrown across its canal. Such cases, I presume, must be very rare.

In a few instances the stricture has been seated so high in the gut that it could be but barely touched with the point of the finger, until the patient was desired to "force down," and then a satisfactory examination of it could be made.

Cases of this disease examined after death present all the coats of the intestine very much thickened, except the peritoneal tunic, which, when closely inspected, is found to retain its healthy structure and appearance; the muscular, cellular. and mucous coats are much thickened; the latter is moreover hardened and raised into irregular ridges or folds, but without any ulceration.*

As this disease admits of examination by the touch, and produces symptoms so unchangeable in their course, and so uniformly alike in every case, it cannot, with ordinary

---

* In the fourth fasciculus of Dr. Baillie's Morb. Anat., and plate 4, we have an excellent specimen of the disease of organic stricture of the rectum; and in the same plate, fig. 2, we have a good specimen of cancer of is intestine.

attention, be confounded with any other; nor can it, when present, be overlooked by the practitioner. In consequence, however, of its presenting some symptoms in common with some other diseases, both of the rectum and the adjoining organs, I shall notice a few of the distinguishing characters between stricture of the rectum and these other affections.

1st. Cancer of the rectum has much in common with stricture, namely, the same narowing of the canal and hardness of its walls, the frequent straining stools and discharge of bloody purulent mucus; in the cancerous affection, however, the patient has a leaden sallow cast of countenance, and suffers severely from lancinating pains darting through the hips and pelvis into the groins and down the thighs and legs. Although cancer of the rectum, in its early stage, presents to the finger a feel very similar to stricture, yet, by repeating the examination after an interval of a few weeks, we discover that the cancerous ulceration has, in the interval, destroyed some portion of the hardened wall of the intestine, and has thus produced a condition of the parts very different from that found in cases of stricture of the same duration.

2nd. There is a rare form of scirrhus of the uterus and vagina, in which the latter passage is almost obliterated by a cancerous thickening of its walls; this gives rise to a train of symptoms not unlike those attending stricture of the rectum; examination by the finger, however, is alone required to remove all doubts.

3rd. The same test will apply to enlarged prostrate gland, if at any time the symptoms should simulate those of stricture of the rectum.

4th. Stricture of the rectum, in its early state, may be mistaken for an ulcer in this intestine; if, however, the ulcer be low down, it will become visible by expanding the anus, or by introducing a blunt polished gorget into the bowel, with its concavity towards the diseased side of the rectum; the finger, too, if steadily pressed against it, will be received into the cavity of the ulcer, although in a hasty examination this part feels as if it were a ridge; a severe pain, too, attends the evacuation of the bowels in cases of ulcer of the rectum, while stricture is free from such.

5th. A tumour in the pelvis may compress the rectum so as to narrow and materially obstruct the canal; in such cases, however, we shall find that the projection comes from one side only, and the coats of the intestine at this part retain their healthy structure and their natural softness.

And lastly. It may not be amiss to mention that we sometimes find in patients who are free from all symptoms of morbid condition of the rectum, that the finger in ano cannot discover any canal in the gut, the entire of its calibre above the sphincter being filled up with soft folds of the lining membrane; repeated observations, however, teach us that such a state is not morbid, as it in no way intercepts or disturbs the healthy functions of the intestine.

The treatment by bougie, usually recommended in this disease, appears to be well calculated to alleviate the sufferings of the patient. I feel confident, however, that a perfect cure of the organic stricture of the rectum has not been affected by any plan of treatment hitherto employed. I have paid great attention to the use of bougies, and yet I must candidly declare that hitherto I have not been so fortunate as to have effected a permanent cure in a single instance; nor have I had the good fortune to meet with any patient whom I *knew* to have been afflicted with this disease, who had been cured by another surgeon. No man can be more ready to proclaim and boast of our control over diseases, and therefore I trust that this declaration will be received by my brethren as it is intended, *viz.*, that it will cause them to consider the history and nature of this intractable disease, and to engage themselves earnestly in discovering some other plan of treatment which will gain the object of our anxious wishes.

I have not been contented with applying the bougie simply; I have often made it at the same time the means of conveying various applications to the seat of disease. For this purpose I have employed bougies with a deep groove running spirally their whole length, so that the ointment employed should not be rubbed off the instrument by the tightness of the anus and its sphincters.

I have nothing cheering to offer on the treatment of this disease; I have given the fullest and fairest trials to various internal medicines. Mercury I have employed in a great number of its preparations, and, although I have never yet met with a case of this disease which could be fairly traced to a venereal origin, yet I have pushed the mercury as far, and have kept up the ptyalism as long, as I have done for the most obstinate forms of venereal disease, but without any permanent benefit in a single instance.

Arsenic also I have administered, as far as was consistent with the safety of the patient, but without benefit; I must say the same of cicuta; neither has iron shown any power over this disease.

Large quantities of mucilage appear to give most relief. Blue pill, combined with a quantity of Dover's powder, has also occasionally afforded much temporary alleviation.

This is the only disease to which I think the title of stricture of the rectum should be applied: I would not give this name to the obstruction of the gut at the junction of the colon and rectum, nor can I admit the existence of what is termed the spasmodic stricture of the rectum. I am well aware that the greater number of modern authors not only speak confidently of spasmodic stricture of the rectum, but that even some describe various forms of this affection. Thus Mr. White speaks of, but certainly does not describe:—1st. Permanent spasmodic stricture; 2nd. Simple stricture; 3rd. Spasmodic constriction of the sphincter ani; and he declares the bougie to be the principal remedy for each of these. Even those authors who notice only one kind of spasmodic stricture of the rectum agree with him in extolling the bougie as the principal remedy.

I must, however, frankly declare that in the course of a pretty extensive practice for the last twenty years I have not been able to discover a single case of a disease corresponding to the description of the spasmodic stricture of authors. I therefore feel a considerable share of scepticism on this point. Indeed, I not only doubt the existence of any such disease, but I can recollect many cases in which this suspected condition of the rectum has yielded to the ordinary means for improving the state of the stomach and bowels, especially when combined with positive assurance to the patient that no such disease had existed; by this assurance alone we can sometimes remove all his anxiety and apprehension on this subject.

With respect to the obstruction and thickening of the colon, either at its junction with the rectum or higher up, I think that it is an affection very different in its symptoms and in its nature from that disease which I have just attempted to describe; for in cases of the former we find that in the advanced stages of the disease the patient cannot procure any evacuation without the aid of purgatives, nor is he (as in stricture of the rectum), harassed by frequent ineffectual strainings at stool; guided by his own sufferings, too, he can point to the spot where the disease is seated. On examination after death we find a very considerable thickening of the coats of the intestine in the circular direction, but of short extent longitudinally. The canal at the seat of the stricture is nearly obliterated, while the intestine immediately above is dilated into a wide pouch.

The symptoms which are stated by authors to indicate these diseases of spasmodic stricture, or of organic stricture seated beyond the reach of the finger, will, upon investigation, be found very fallacious; even supposing all those enumerated should be found combined in the same individual: thus we are told that in such cases we arrive at a knowledge of the existence of these diseases by examination with a soft bougie, by the very diminished diameter of the fæces, and by the admixture of blood or mucus with the stools.

The soft bougie, however, may, and very generally will, receive an impression from the projecting ridge of the sacrum; for it is by no means an easy matter to pass a rectum-bougie so that it shall not be arrested at this point, and consequently receive an impression from the projecting bone.

The diminished diameter of the fæces, too, may be produced by any irritation in the rectum, which will cause it to act frequently, and with increased contraction; this is often experienced even in perfect health.

The admixture of mucus or blood with the fæces is also frequently met with in other and very different affections of the bowels.

To prove the futility of these symptoms as a discriminating

test, and the danger of being misled by them, I shall briefly state the outlines of one case only.

Mr. ——, æt. ann. 36, who had lived rather fully and freely, applied to me six or seven years ago under the following circumstances :— He said that he had been, for the last two years, affected with stricture of the rectum; that in the preceding year he had an attack of dysentery, and that since that time he found the stricture worse; the diameter of the fæces diminished, calls to stool more frequent, and seldom passing without an admixture of blood and purulent mucus; he was losing flesh rapidly, his appetite and rest were very indifferent, and his mind was miserable. Having examined the rectum by the finger I expressed to him my hopes that he was not affected with such an intractable malady, by which I very nearly forfeited his confidence; for he told me that he had applied to a very eminent physician in the north of England, and then to another in London, both of whom assured him of the existence of this disease; and lastly, he had been under the care of the late Mr. White, of Bath, who furnished him with the bougies he was then using, and, immediately introducing one, he showed me the mark which the stricture impressed on it. My most urgent remonstrance could at this time only obtain from him a promise that he would not use the bougie as frequently, or for as long a period as usual. By small doses of blue pill, combined with compound powder of Ipecacuanha and an enema of Olei. Oliv. cum Subacet. Litharg. liquor., the irritation of the bowels was mitigated in the course of eight or ten days. Availing myself of this favourable change I again urged him to lay aside the bougie, and, with some difficulty, obtained a truce for ten days; within this period it fortunately happened that he passed one consistent motion, in which the fæces were of a large diameter; after this I had but little trouble in prevailing on my patient to lay aside altogether the use of the bougie. By persevering in the internal use of mild bitters and bark, and injecting every night an enema, consisting of Ol. Oliv. and Unguent. Supernitrat. Hydrargyr., into the rectum, his disease was finally cured, and since that time his bowels, though sometimes deranged by ordinary complaints, have never suffered from a return of the former affliction, and he has enjoyed very good health.

When we consider the irritable state of this patient's bowels, the wasting of his flesh, and the wretchedness of his mind, we may, I think, reasonably believe that a further perseverance in the use of the bougie would have rendered his disease eventually fatal; yet here were present all those symptoms which are said to indicate stricture high up in the rectum : the event has proved that no such disease did exist. And here let it be remembered that serious mischief has occasionally been committed by rude attempts to dilate a *supposed* stricture at the top of the rectum or termination of the colon, for I have known two cases where peritoneal inflammation and death speedily ensued; and I have heard, on the best authority, of two similar instances.

Spasmodic stricture of the sphincter ani must, I believe, be a very rare disease. Dr. Baillie has given one case of it in 5th vol. Trans. College of Physicians, and has, with his usual perspicuity, pointed out its characteristic symptoms. The only case which I had ever met with had been treated as stricture of the rectum ; it was in a young man of rank, who applied to me for relief from a stricture of the rectum. He told me that an eminent surgeon in a provincial town, another in Dublin, and another in London, had all assured him that such was his disease. He informed me that during the last two years he had suffered from it for two or three months at a time, and then he enjoyed immunity from it for four or five months. By minute inquiry into the symptoms, and by examination of the rectum, I satisfied myself that this disease was only a spasmodic condition of the sphincter. But, in order to convince my patient that his fears of stricture were unfounded, I passed a wooden globe, three and a half inches in circumference, mounted on a stalk of whalebone, ten inches up the rectum without having met with any obstruction.

VASCULAR TUMOURS OF THE RECTUM.

This disease is readily known by attention to the following symptoms :—Whenever the bowels are evacuated the patient feels a protrusion through the anus; the protruded part is observed to consist of a greater or smaller number of tumours of different sizes, soft and spongy to the feel, of a purplish colour, with a number of minute, yet very distinct, vessels on the surface of each. It will be absolutely necessary for the surgeon, in suspected cases, to examine the parts when protruded; for if, relying upon the accuracy of his touch, he merely examines by the finger introduced into the rectum, these tumours will, in all probability, elude his scrutiny.

In the early stage of the disease the protruded parts retire spontaneously, but in the advanced stages they require to be replaced by the hand. The evacuation of the bowels is followed by pain, which in that advanced stage of the disease does not cease for two or three hours. In some instances the patient informs us that the pain, which is pretty severe at the time of the evacuation, becomes almost insupportable in the course of half an hour or an hour after, and generally continues so for five or six hours. These last are cases of vascular tumours, complicated with fissure of the rectum, the very severe pain being caused by the latter disease.

There are some cases of vascular tumours in which the patient suffers, not so much from pain as from loss of blood at each evacuation; this I have known to reduce some of these unhappy sufferers to a state of exsanguine exhaustion. In this exhausted state of the patient we find the sphincter ani so very wide that it can easily admit all the fingers conjoined; yet this, in a few days after the operation, will resume its natural state of contraction. If we watch this disease for some months together we shall find that when the vascular tumours are combined with fissure there are periods of exacerbation during which the patient, after an evacuation, writhes in severe agony, and cannot allow the parts to be returned. At this time the tumours feel hard and seem to be inflamed.

I had an opportunity of examining the structure of these tumours in a patient who died of another disease. On slitting up the rectum I saw three blood-vessels, each as large as a crow-quill, running for some way down the intestine, and then dividing into a number of branches; these vessels ramified very profusely, and each seemed, by interweaving of its branches, to form one of these tumours. The trunks and branches were covered only by the lining membrane of the intestine. This examination shows us how inapplicable to this disease are the terms " varicose tumours," " hæmorrhoidal excrescences."

The opinions of the profession are divided as to the best mode of removing these tumours, some preferring the ligature, others excision. I am decidedly in favour of the latter, for this reason, that Tetanus, the consequence which is to be dreaded in the former mode of proceeding, cannot by any means be foreseen or guarded against; whereas hæmorrhage, which alone is to be feared in the operation by excision, may be prevented, and can most certainly be suppressed.

The following mode of operating I have found to be uniformly and permanently successful; and it is considerably less severe than that generally recommended :—

The tumours having been made to protrude by means of a purgative injection, I direct my assistant to pass a hook or common tenaculum through one or two of the largest, while I seize another lengthwise with a polypus-forceps; then, drawing the tumour a little towards the axis of the gut with a large pair of scissors passed behind the forceps, I cut off all that portion which is engaged between its blades. I then proceed in the same manner to remove those tumours which the assistant holds transfixed by the hook. By fastening and drawing out the tumour with the forceps we much facilitate its removal by the scissors. Proceeding in this way I guard against these tumours being drawn up within the sphincter as soon as the first had been removed. I do not think that any case will require the removal of more than three of these tumours, and not unfrequently the cure will be ensured by cutting off only two of them. When the operation is finished the protruded parts generally retire within the sphincter. Should any part remain out, it must be completely pushed in with the finger.

In order to guard against the danger of hæmorrhage, I take care not to prolong my incision higher on the bowel than what I conceive will, when replaced, lie within the circle of the sphincter; for if we cut the gut higher up this part, when returned, may bleed freely from not having any surface closely opposed to it. Besides, we know that by cutting higher up we are in danger of cutting the trunk of the vessel, instead of confining our incision to the tumour, which is composed solely by the convolutions of its very minute branches.

I have said that if hæmorrhage should follow this operation it can most certainly be arrested. This I have done satisfactorily in a case where the patient had suffered for nearly twelve hours by bleeding into the gut, from which it was at intervals shot forth *involuntarily*, with forcing and straining. Many and unavailing were the methods employed by three or four surgeons and myself, until at length I bethought me of Petit's method of stopping an hæmorrhage from the rectum. In this case I made use of sponge instead of the lint recommended by that author.

I should be afraid to adopt Mr. Hey's method of cutting away all the protruding tumours, together with the skin at the verge of the anus, lest the patient should afterwards experience that distress which a too contracted state of this outlet must occasion; for in one case where, for the purpose of extirpating warts, a ring of skin at the verge of the anus had been cut away along with these excrescences, the condition of the patient was rendered truly miserable.

It will be satisfactory to timid patients to learn that this disease of vascular tumours may be rendered very mild, if not ultimately cured, without the necessity of resorting to an operation. By causing the patient to inject four or five ounces of an injection, consisting of two grains of sulphate of zinc to each ounce of water, every night going to bed, and making him retain it during the night, I have known the disease so much alleviated that patients have flattered themselves they had obtained a perfect cure. I have known some of them, though not living in the most regular manner, yet enabled to enjoy all the sports of the field without inconvenience. I fancy that this plan of treatment has succeeded best when begun immediately after one of the painful fits the disease. I have known many of these patients much benefitted by smearing, after each evacuation, the protruded parts with a liniment of Ol. Oliv. ʒii. and Plumbi Subacetat. liquor., ʒi.

ULCER OF THE RECTUM.

We sometimes find within the rectum, at a short distance above the anus, an ulcer unconnected with any other disease. In such cases the patient complains that he observes his linen stained with a purulent discharge, which often flows when he is not at stool; on examination this will prove different from healthy pus, frequently containing an admixture of thin bloody fluid; at times the quantity of discharge is much lessened, and then the sufferings of the patient are aggravated, but on the flowing off of a larger quantity he experiences great relief; he suffers sharp pain on going to stool, and this continues for an hour or two. On examination the finger soon discovers the seat of the disease, which at first feels raised and rather tough, but, by pressing the finger firmly on this spot, the point sinks into a small hollow cup of an ulcer, the edges of which are found in some degree hardened. We may obtain a satisfactory view of the ulcer by passing upon the finger a blunt polished gorget, the cavity of which is to look towards the seat of the disease; then, by everting the anus as much as we can, we shall obtain a full view of the ulcer by the light reflected from the gorget.

The unhealthy nature of the discharge, the degree of pain and hardness of the edges of this ulcer, might make us apprehend some carcinomatous affection; but in any doubtful case, by temporising and watching the course of the disease for a few weeks, we shall have all doubts removed; for if it be cancerous the ulcer will, in this period, have changed in form and extended in size; besides, on strict examination, whilst the point of the finger lies in the cup of the ulcer, it will be found that the hardness is confined to the edges, whilst the bottom of the ulcer is soft.

The remedy for this disease is to introduce into the rectum a convex-edged scalpel, and make an incision through the entire length of the ulcer, continuing it through the sphincter, and dividing the verge of the anus; as soon as this wound has got into a state of suppuration we should dress it and the ulcer with some stimulating ointment, introduced on a dossil of lint.   The cure goes on without interruption, although it is rather tedious and slow of healing.   I need hardly add that the final cicatrisation will be promoted by the occasional application of nitrate of silver.

FATAL CONSEQUENCES RESULTING FROM SLIGHT WOUNDS RECEIVED IN DISSECTION.

(Dublin Hospital Reports, vols. 3 & 4.)

CASE 1.—On Monday, December 1st, 1818, Mr. William Hutchinson, one of my pupils, of rather a delicate constitution, received a slight scratch of the knife on the outer side of the first phalanx of the right thumb on opening the body of a man who had died of cynanche laryngea in Dr. Steeven's Hospital. The cellular membrane on the external surface of the larynx and pharynx contained that amber-coloured fluid, resembling melted jelly, which is so often met with in such as have been carried off by this disease. The wound was not more than one-sixth of an inch in length, and so superficial that he took scarcely any notice of it at that time.

In the evening of this day he was drowsy, and retired to bed earlier than usual; the next morning he complained of headache, sick stomach, and of a most acute pain in the right shoulder and axilla. In the course of the day these symptoms became so severe that he was desired to take an emetic, which, after a short interval, was followed by a purgative medicine.

On the third day the pain had increased in an extraordinary degree; it was confined to the shoulder joint; there was no discoloration of the integuments, but there was some swelling about the joint and above the clavicle; he suffered no uneasiness in the elbow joint; no inflamed lymphatic vessels could be traced along the arm, nor could any enlargement of the glands be discovered either in the axilla or above the clavicle, but he suffered exquisite pain in both these places from the slight pressure necessary in making the examination. The scratch on his thumb was at this time quite free from inflammation, the cuticle was raised into a small flattened vesicle, and was about half-filled with a very white milky fluid.

In this state he continued for three or four days, suffering the most agonising pain, and labouring under a violent fever, accompanied with the greatest dejection of spirits.

Fomentations were applied, but without relief. Leeches in the axilla and cold washes were also tried, but with no better success. Large doses of opium failed to procure the slightest mitigation of pain, and the fever was not in the smallest degree controlled by the usual remedies.

He now at length had some relief from the severe pain in the shoulder, but this was not accompanied by a corresponding remission of the fever. In the course however of a day or two he for the first time complained of a pain along the right side of the thorax. A diffused erysipelatous redness was seen extending from the axilla nearly half way down to the ileum : and, although a fulness appeared on the ribs, yet nothing like a phlegmon could be discovered. In a day or two more this redness had extended down that side as low as the great trochanter, and could be traced across the abdominal muscles to the groin ; the skin had a doughy feel, and easily received the impression of the finger. In some places the surface presented to the eye the appearance of distinct vesicles, yet these parts, when examined by the finger, felt perfectly solid ; in fact, they resembled accurately those elevations which are observed in the site of leech-bites long healed, when the skin to which the leeches had been applied becomes swollen by inflammation. The fever continued unabated, and his strength seemed nearly exhausted, and was supported only by large quantities of wine. In this extremity, on the 15th of December, an incision was made over the fourth and fifth ribs, in the hope of finding some lymph or matter diffused through the cellular substance, but none was discovered. Neither the local inflammation nor the febrile symptoms seemed at all influenced by this incision. In a few days more the inflammation ceased to extend, the cuticle began to desquamate, and the pain to diminish ; the skin which had been affected still retained its morbid feel. At the termination of three weeks after the accident he was suffering no pain in his side, the shoulder had long since ceased to trouble him, the cicatrix in the thumb could scarcely be observed, and his general health was gradually recovering ;

yet at this time he began to experience pain in the arm, along the inner edge of the biceps muscle: this was but slight for a day or two; it soon, however, became very violent. There was a considerable degree of induration along the biceps, which extended even to the pectoralis and latissimus dorsi; the arm became red, swollen, and œdematous; the swelling extended below the elbow; the pain, which at all times was exceedingly great, was rendered almost insupportable by any movement of the arm. Fomentations and poultices were applied, and a fluctuation being obscurely felt along the inner edge of the biceps, an opening was made on the 1st of January, which gave exit to a small quantity of matter. In a few days after the abscess in the arm was opened a fresh attack of inflammation took place in his side, which ended in the formation of another abscess, situate over the sixth rib, and not very far from the spine. After this period he rapidly improved, so that in the latter end of January he was able to go to the country.

On the 14th of February, when he had been three weeks in the country, the abscesses were healed, the hardness along the biceps was greatly diminished; but even then he could not make any use of the arm; the integuments along the side had regained, in a considerable degree, their natural feel. His general health was wonderfully improved. He did not regain the full use of the arm for some weeks after this; the induration of the integuments of the side was also very slow in dispersing.

From the commencement of the disease his spirits were sunk; he frequently raved; his pulse throughout was never less than 120, and sometimes it was 130. His stomach was at the commencement very irritable, and got better; but it again became alarmingly irritable about the beginning of January. This might have been partly caused by the large quantity of opium he used, for he could not procure any relief, except by a large dose, amounting to a drachm or four scruples of the tincture daily.

Case 2.—On Saturday, February 13th, 1819, Mr. Dease, late Professor of Anatomy and Surgery to the Royal College

2 c

of Surgeons in Ireland, lectured on the cervical nerves and brachial plexus. The subject (which was dissected for him), a female about forty years of age, had died in one of the hospitals of a chronic pulmonary affection ; the body had not been buried, and was nearly fresh, for she had not been dead above forty-eight hours.

On the morning of Friday, when she had been just brought into the College from the Hospital, I saw the thorax opened by Mr. Shekelton, who called my attention to a brown thick fluid contained in the pericardium. The lecture was delivered at one o'clock on Saturday by Mr. Dease, and on Sunday morning early he awoke with a violent shivering and sickness of stomach ; the former was very severe, and the latter continued for upwards of two hours. He threw up his dinner, consisting of fish, of which he had eaten very freely, and afterwards he vomited a large quantity of bile. Even at this time he complained of acute pain in his left shoulder. His friend, Dr. Sheridan, saw him early this morning.

He sent for me at half-past eleven o'clock, and earnestly besought me to bleed him for this pain of the shoulder, which he described as very severe, and as aggravated by the slightest attempt to move the arm. I found him at this time labouring, as I supposed, under high symptoms of the prevailing fever, and, conceiving that his complaints of the arm were in some measure the effects of impatience, I declined bleeding him unless by the desire of his physician. At 3 o'clock p.m. I again saw him, and although I still was of opinion that he was affected with common fever, attended with derangement of the stomach and liver, greater than ordinary, yet I could no longer resist his importunate solicitation to be blooded, and I took away nearly twenty ounces of blood from his arm by a large orifice ; the blood flowed freely. He thought himself relieved of the pain in the shoulder while the first cup was filling, but this was probably ideal, as he did not express relief towards the conclusion of the operation, and as I found his pain not at all mitigated at 9 o'clock p.m., when the fever was as high as it had been at any time of the day. The blood was neither buffed nor cupped, and the proportion of serum was considerable. I now observed a slight fulness above the clavicle, along the left side of the neck, in the space between the sterno-mastoid and trapezius muscles, and, being in doubt whether this apparent fulness might not be owing to the position of the head, which was held rather towards this side, I wished to satisfy myself by the touch, but on applying my finger, even with the slightest pressure, he complained of exquisite pain.

Monday morning, February 15th.—I was called upon at 8 o'clock this morning to see him, and found that he had spent a very restless night, owing to the pain of his shoulder; and when I went into the room he had the entire of the joint covered with leeches, to the amount perhaps of one hundred. We advised a draught with Elect. Scam. fomentations, and opiate-liniment. At 5 o'clock p.m. we met again, Dr. Sheridan, Dr. Brooke, Mr. Richards, and myself. We learned that the draught was rejected almost as soon as it had reached the stomach, and, although the pills had brought away some liquid stools, yet the fulness of the abdomen remained unreduced. No relief has been derived from the fomentations and liniment. Repeat the pills, and assist their operation by enemata.

Tuesday morning, 16th.—Bowels have been more satisfactorily freed; symptoms, however, remain as yesterday. We felt at a loss to account for the pain of the shoulder, which however was less severe. We advised him to persevere in the use of purgatives, giving him the Inf. Sennæ et Tamarind, with Tinct. Jalapæ.

9 o'clock p.m.—On visiting him this evening he accidentally mentioned an uneasiness which he felt in his left side. On examination I discovered a colourless swelling on the side of the thorax, a little behind and below the posterior border of the axilla; upon which the recollection of Hutchinson's case at once occurred to my mind. On my suggesting to him my suspicions of the cause of his sufferings Mr. D. denied his having received any cut, of which he was so positive as almost to refuse to let me examine his hand. I discovered on the dorsum, rather towards the ulnar side of the second joint of the thumb, the mark of a slight scratch, not one fourth of an inch long; this formed the diameter of a vesicle,

which was almost half filled with a fluid of a milky whiteness and consistence, and about this size ⬭

I should have observed that the pain of his shoulder was better on Monday night and Tuesday morning, so that he was not (as on Sunday or Monday morning) fixed to one spot, but he could now lie with his body slightly turned towards the right side, and could raise himself in the bed by pulling, with the left hand, a cloth tied to the foot-post. The skin above the clavicle at this time bore pressure very well; the skin at the swollen part of the side was not discoloured, but was possessed of the most painful sensibility to the touch. It should be observed that every evening, about six o'clock, he had an exacerbation of restlessness and depression of spirits.

Wednesday, February 17th.—This morning we resolved to give a bolus of calomel, with a liquid purgative, and at night a draught with twenty-five drops of laudanum.

Thursday, February 18th.—The discharges from the bowels in the course of yesterday were very free and natural. He slept for four hours after taking the draught, and awoke cheerful and refreshed. He was directed to take a bolus of carbonate of ammonia, together with some diaphoretic medicine; also to foment the side, and rub it gently with a liniment of camphorated oil and aqua ammoniæ. Draught to be repeated at night with forty drops of laudanum.

Friday, February 19th.—Had a very bad night; some delirium this morning, but is now, at half-past eleven o'clock, more composed. Face with a yellow tinge; countenance sharp, yet not indicative of much pain or inward distress; pulse smaller. He desired me to look at the right arm, which had been blooded. I found the incision inflamed in the ordinary way; but I remarked on the forearm, about two inches below the incision, a small vesicle containing a fluid like that produced by the original wound on the thumb.

N.B.—This vesicle did not increase in size, or alter in any way, until the time of Mr. Dease's death.

Saturday, February 20th, 1819.—At our visit this morning (11 o'clock) we observed his manner quick, and bordering on delirium. Pulse 126 and smaller. The entire side, from a very little below the axilla down to the hip, was swelled.

This day we observed the swollen part studded pretty thickly with small elevations, to the eye like vesicles, yet hard to the touch. . They bore a resemblance to the elevations which arise in the cicatrices of a part which had been scarified, when it is affected with swelling.

An erysipelatous blush, which had been first observed on Thursday, and had rather increased on Friday, was now more strong, but occupied only a small portion in the middle of the swelled side. Tongue covered with a white coating; countenance less yellow than yesterday, but still contracted and small; abdomen full.

Saturday evening, 9 o'clock.—Delirium set in soon after our visit, and has continued high; perspiration warm and rather general; however, the left side of the trunk of the body, where the clothes lightly cover him, is quite dry though warm; pulse weaker; the swelling has passed more from the abdomen towards the back; erysipelatous redness more extended and more strong, occupying a considerable portion of the side; bowels free. Although the perspiration is warm, while he remains covered, yet if his hand remain uncovered for a few minutes it feels cold and clammy.

Sunday morning, February 21st, 9 o'clock a.m.—At ten o'clock last night he took Tinct. Opii gutt. xl., but the delirium continued for three hours after it; he insisted on being moved to an adjoining bed, and there he lay for nearly three hours without any clothes. On being again laid in his own bed he was stupid, and his left leg was stiff; countenance still sharp, but not yellow; lips of a good red; he has clammy sweats on the head and upper parts of the body; frequent deep inspirations, like sighing; pulse 126 and weak. Wine negus, with nutmeg.

12 o'clock.—General state as at 9. Inflammation of the side extends up to the axilla; and on the posterior edge of the axilla appears as if there was an abscess, but is without fluctuation; colour of the inflamed part darker; he has passed no urine since nine o'clock last night, although he had an enema this morning. We advised a poultice to the

side, a bolus of scammony, calomel, and jalap, and cordial draughts every two hours. We now observed a swelling on the anterior part of the right arm, occupying about a hand's-breadth of the belly of the flexor muscles, beginning about an inch and a half below the orifice made in bleeding him. The vesicle on this arm remains as when first observed : a poultice to this swelling.

5 o'clock p.m.—We agreed to puncture this tumour, although he appeared approaching fast to dissolution. A quantity of serous fluid, nearly a teaspoonful, flowed from the opening, but did not in the slightest degree reduce the swelling.

Sunday evening, 9 o'clock.—Pulse in right wrist not to be felt; heat of limbs not reduced; breathing quick and laboured. He passed urine at five o'clock this evening, and at ten o'clock he died.

Monday morning.—I was called early this morning to examine his body, as one of his pupils, who remained in the room, fancied that he heard him breathing. I now observed

that two or three vesicles of this size ⬭ had formed

on his back; the swelling had extended down the thigh; the left arm was swelled, and rather hardened from the elbow nearly up to the shoulder; this swelling was chiefly along its anterior surface, but it could also be felt all around; yet there was no redness or vesication on the limb.

Along the left side of the trunk the raised hard spots still continued as in life.

It is not irrelevant to mention here the outlines of Mr. Egan's case. Mr. Egan had, on Saturday, 13th February, after Mr. Dease's lecture, employed himself in dissecting a part of the same subject. On Sunday evening he had a rigour, followed by febrile paroxysm. On Monday he felt himself not well, yet he went into company, and danced in the evening. On Tuesday he complained of hoarseness; inflammation attacked the metacarpal joint of the thumb, accompanied with pain and erysipelatous redness around it, the pains passing up along the back of the forearm. This was attended with a smart degree of fever. On Friday he complained of tenderness under the border of the pectoral muscle, and a gland could be felt enlarged under the edge of this muscle. He had no perspiration in this febrile state, and his chest appeared to be seriously affected. This affection commenced on Tuesday.

On Sunday, February 28th, I saw him. Fever very high; pulmonary affection and distress very severe. An abscess, which had been collecting for some days in the axilla, now appeared to me fit to be opened, although there was no redness of the skin, no pointing, no surrounding hardness; the matter discharged was purulent, of an unusually thick consistence, and the cavity proved to be very extensive, passing across from the pectoral to the latissimus dorsi muscle.

N.B.—This night he had retention of urine, probably owing to an opiate draught; he had no recurrence of this distress. He gradually recovered.

REFLECTIONS.

I have been thus minute in relating the circumstances of these two cases, because I consider them as presenting us with the character of a disease both formidable and new.

The severe pain felt on the point of the shoulder; the peculiar colourless swelling or fulness above the clavicle within a few hours after the infliction of the wound; the wound itself at this time free from inflammation or pain; no red streaks, no pain, nor even tenderness on pressure through the whole course of the limb. In short, no trace at the time of any morbid connection between the wound and the seat of the pain, were the most remarkable features of this disease.

Another remarkable feature was the very peculiar appearance of the pustules (unlike to any I had ever witnessed), and the striking resemblance of the pustules to each other. Add to this the appearance of a similar pustule on the forearm of Mr. Dease, observed on Friday the 19th. The character of this pustule could not have been influenced by the wound of the lancet, for that wound had only the ordinary appearances observed in cases where a slight inflammation supervenes on blood-letting; besides, this pustule was seated lower down on the limb, and the skin intervening between it and the wound remained perfectly natural, free from discoloration, swelling, or pain. The elevations of the cuticle, to the eye resembling vesications, while actually solid ; and the swelling on the forearm, which was opened on Sunday, only a few hours before Mr. Dease's death, present characters which tend still further to remove the resemblance of this to any other disease usually consequent on wounds.

The description of the swelling, passing from the axilla along the side down to the hip, and, being accompanied with redness, may induce some to consider these two cases merely as instances of erysipelatous inflammation. But to such I would beg leave to observe that the redness was of a bright healthy colour, did not appear until the fourth day, and at no period did the redness occupy more than one-sixth of the swollen part. So very unlike were the characters of this disease to those of erysipelas that the late Mr. Richards could not be persuaded to believe that there existed any connection between the wound and the disease, until I brought with me Mr. Kinchela (a very intelligent pupil, who had kindly attended Mr. Hutchinson), and who, on seeing the peculiar swelling along the side of the thorax, instantly pronounced this to be the same disease as Hutchinson's. It should be observed that erysipelatous redness, though more considerable in Mr. Hutchinson's case, never occupied more than one-half of the swollen surface. The inflammation seizing on the arm in three weeks after the receipt of the wound in Mr. Hutchinson's case, and in eight days in the case of Professor Dease, presents another striking point of resemblance.

In whatever way the question of the nosological arrangement of this disease is disposed of, this much is obvious, that neither habits of dissection the most lengthened, nor a state of the subject the most free from putrefaction, can secure the anatomist from the danger of this formidable disease. For upwards of twenty years was Mr. Dease, during the winter months, in the habit of dissecting human bodies. If my recollection serve me, I have found most of the instances of inflammation and fever following wounds received in dissection among those pupils who have arrived at the third season of their anatomical pursuits.

It is an opinion universally held by the public, and pretty generally received by professional persons, that the danger from wounds in dissection is in proportion to the putrefaction of the subject, or to the contagious nature of the disease which had caused death. On the contrary, according to my observation, unpleasant consequences have been so rare, where the subject was far advanced in putrefaction, as to induce me to think that putrefaction rather gives protection to the anatomist; nor could I trace any connection between the disorder in question and the contagious nature of the disease which had destroyed the subject of dissection, and surely during the winters of 1818—19, when fever raged so extensively in Dublin, we should have had some striking instances of such a connection had it existed. Perhaps the majority of the affections in question have occurred in cases of examination of bodies to ascertain the cause of death, and where the body had not even been interred.

According to the paper contained in Hist. de l'Academie Royal des Sciences de Paris, Année 1776, p. 53, the infection from animal matter, under peculiar circumstances, is more dangerous when communicated from other animals than from human bodies, for in the former it did not require the presence of a wound to produce effects the most strange and violent. And this may serve to impress those who pursue comparative anatomy with due caution.

I have never witnessed any of those dangerous constitutional affections which are mentioned by some authors as attending the dissection of bodies in a state of putrefaction. I have no doubt often seen persons suffer a momentary sickness or faintness from the fœtid vapours which issued as soon as the cavities of the thorax or abdomen were opened; but never did I observe fever supervening. Some of our dissecting pupils every season suffer from fever, but these are cases of the ordinary type, and are to be traced to ordinary causes of fever; for instance, irregularities in their mode of living, excessive fatigue, sitting in wet clothes, great anxiety of mind, or contagion. Affections of the lungs more frequently than any constitutional disease can be traced to the occupation of the anatomical student, but these arise from the cold and damp of the dissecting rooms, and not merely from hanging over the dead and putrefying body.

In cases of the severe consequences following wounds received in dissecting, does an infection enter by the wound and cause all the mischief? Or are the symptoms to be accounted for by some peculiarity in the constitution of the individual? The peculiarity of the symptoms and the similarity between the cases of Professor Dease and Mr. Hutchinson is so strong as to lead to the opinion of both having arisen from infection; besides, I think the number of severe cases arising from slight wounds to anatomists is much greater than what we find in those classes whose occupations expose them to similar slight wounds and punctures of the fingers while handling dead animal matter; yet, on the other hand, it must be admitted that in Mr. Egan's case the symptoms were very different from those under which Professor Dease suffered, although both received their wounds in a few hours from each other, and while dissecting the same subject.

The opportunities of treating the cases of Professor Dease and Mr. Hutchinson have not enabled me to discover any plan or remedy of service in this formidable disease. It is therefore the more incumbent on anatomists to attend to the means of prevention.

The advantages of the immediate application to such wounds of caustics in a solid or liquid form is too obvious to require any comment; they are the only means upon which we can rely with absolute certainty.

But those who are acquainted with the zeal of anatomists must doubt whether any practice which interferes with the prosecution of their pursuits will be generally adopted; the pain and the delay which must be occasioned by applying caustic to every little scratch will prevent this plan from being as generally followed as it should be, or as the safety of individuals requires.

I would therefore recommend that each dissecting-table be furnished with a cup of oleum terebinthinæ, into which the anatomist should plunge his finger the moment it is wounded. The smart which this produces is so inconsiderable that it will not cause any serious delay, while the irritation may counteract the power of infection, or alter the mode of inflammation in the wound.

On Tuesday, May 18th, 1824, Mr. Shekelton was engaged in examining the body of a man who had died of peritoneal inflammation, consequent on the operation of lithotomy. The examination took place in a very few hours after death, the body still retaining its heat. I observed that soon after the abdomen was opened, Mr. Shekelton pricked himself with the point of the knife, which called forth the usual involuntary expression of pain, but was not further attended to. He proceeded to take out the contents of the pelvis, so that his hands were necessarily immersed in that cavity for a considerable time.

On the following evening (19th) he felt himself unwell; on Thursday (20th) he gave an anatomical demonstration as usual, and immediately after it he went home very ill, but owing to the natural reserve of his disposition he did not apply to any of his medical friends. On Friday evening (21st) Mr. Cusack accidentally saw him, and found some of the glands swelled in the left axilla, from which he was apprehensive that the disease was produced by the wound received on Tuesday. When Mr. Cusack again visited him, on Saturday (22nd), the condition of these glands was so much improved as to induce him to look upon the case as one of simple fever brought on by cold and fatigue, to which Mr. S. had lately been much exposed.

On Sunday morning (23rd) I first saw Mr. S. He told me, in a most emphatic manner, that he had passed a *most wretched night;* yet the febrile symptoms were not very severe. Pulse only 84, though his skin was hot, and there was an indescribable anxiety and distress very perceptible: he said that he felt as if his stomach was too full. The glands in the left axilla were slightly enlarged, and tender to the touch; but there was neither swelling nor redness along the arm. He complained of pain along the course of the ulnar nerve, and down the left side of the thorax. There was no appearance of inflammation in the ring finger, which had been wounded;

indeed we could scarcely discover any trace of the wound. Were it not that I recollected to have seen him wound the finger, and that his present anxiety and distress very far exceeded in severity the other febrile symptoms, I should have supposed him affected with common fever only.

By his own desire leeches were applied to the axillary glands, but he suffered so much from the bites of each of these, that, after the fourth had fixed, he refused to let any more be applied. He took an emetic, which operated well, but did not produce any marked impression on the symptoms.

Monday (24th).— He complained of uneasiness in the stomach and bowels, and felt as if he would be much relieved by having a free stool. His tongue was covered, but not very thickly, with a white mucous coating; skin of the face assuming a yellow colour; features rather sharp; no uneasiness in the course of the lymphatics of the arm; swelling and pain of the axillary glands had ceased; so that some of his medical friends still indulged the hope that the case would prove to be but a severe instance of common fever.

Tuesday (25th).—Constitutional distress increased, and the real nature of the disease was too obvious to all his medical attendants. We agreed that he should have a full opiate at night.

Tuesday evening.—This afternoon a red spot, the size of a shilling, appeared on the right patella. For two hours this evening he suffered such severe torture in the knee that he declared he would much rather endure the pain of having every limb amputated in succession than again undergo the pain he had suffered in that knee; this pain he referred to the cavity of the joint.

Wednesday (26th).—At ten o'clock last night he got sixty drops of laudanum, and in the course of two hours twenty more: these produced very profuse and general perspiration, but no alleviation of symptoms followed. He dozed a good deal, and his attendants reported that he had passed a good night; he himself declared that it was a most uncomfortable night.

During the night his brother, Dr. Robert Shekelton, observed a large red and swollen patch over the right tibialis anticus muscle. This, and the spot on the patella, now had a solid feel, and were not very tender to the touch.

Early this morning (at two o'clock) the right arm, from the shoulder down to the elbow, was observed to be swelled, but without discoloration; the right thigh also was swelled, but not discoloured. A red patch appeared on the dorsum of the left foot, and he complained of a pain in the left scapula and shoulder. Tongue covered with a thick coating of white mucous; countenance of a deeply yellow tinge.

Thursday (27th).—He now complained of pain in the left arm, with considerable swelling of it and of the forearm. Yellowness of skin was now universal, and more deep than before; adnatæ of the eyes very yellow; countenance more sharp and contracted; tongue as yesterday.

At six o'clock, this evening, his weakness became extreme, his features very sharp and contracted, yet his pulse was regular, and not very quick.

Friday (28th).—Twice in the course of last night he was supposed to be dying; yet he soon after called for solid nourishment, ate a teacupful of panada, and took the yolk of an egg beat up with brandy: this he did from some notion that he now only required solid food in his stomach to complete his recovery. He was obviously not free from delirium.

During the night he passed a large quantity of urine of a peculiarly dark brown colour.

At five o'clock this morning he sent for me, and requested that I would make incisions into the left arm and forearm, which were much swollen, but free from redness; indeed, the only place where any redness could be seen was on the left scapula. I could not refuse to comply with his wishes, although I knew how unavailing this measure must be; for the hand was livid, and cold as a gangrened hand: he had the power of moving the thumb only; all his extremities were cold; this was the coldest; the right arm and forearm were less swelled, and soft; the spot on the middle of the right leg, which always had a solid feel, was now more extended, and less raised.

I made an incision into the red patch on the back of the

left shoulder, on the inner side of the arm, on the forearm, and along the outer edge of the biceps at the bend of the arm. All these incisions were carried through the fascia of the limb. Nothing, except a small quantity of blood, issued from any of these incisions; in none of them did the skin or cellular substance exhibit any diseased appearance, except in that at the bend of the arm, where the cellular membrane had a deep yellow colour. His brother, in the course of the night, had punctured the bursa on the olecranon of the left arm, which was very tense, red, and painful; from this a large quantity of red watery fluid still continued to flow. During the night he complained that half a glass of claret felt as strong and hot in his stomach as if it had been ardent spirits.

He died at nine o'clock this morning.

In this case we recognise the same train of symptoms as in the cases of Mr. Hutchinson and Mr. Dease. The present case tends to strengthen the opinion I advanced, "that slight wounds received in dissecting fresh bodies sometimes give rise to a peculiar disease, perfectly distinct from every other disease consequent on similar wounds." In addition to the other symptoms there described as being characteristic of this disease, we may mention this striking one, viz., that previously to the disease terminating either in death or in recovery, swelling and inflammation seize upon the portion of the limb interposed between the original wound and the first seat of pain. We see that this took place in Mr. Hutchinson's case at so late a period as three weeks; the fever, although it had remitted, not having ceased until this inflammation had occurred. In the fatal cases of Mr. Dease and Mr. Shekelton it appeared, in the former on the ninth, in the latter on the tenth day.

I must entreat the attention of the reader to this fact, that the redness which is seen on the swollen parts is very unlike to that of erysipelas, for the colour is that of a peach-blossom, is of very small extent compared with the extent of the swelling, is seen for a few days, perhaps for a few hours only, on the same spot, and next is observed in some very distant part, possibly in the opposite limb; besides, this peculiar redness vanishing quickly from a part does not leave any vesication or desquamation after it, as is seen in cases of erysipelas.

If any proof be wanting to establish an essential difference between this disease and phlegmonoid erysipelas, it will be found in the state of the swollen parts when cut into. For, although four incisions were made into different parts of the left arm of Mr. Shekelton, yet no discharge, except a small quantity of blood, issued, nor was any change of structure visible except that slight one which has been noticed at the bend of the arm.

This sudden shifting of the swelling and redness from one to another, and very distant part, is not to be confounded with what we sometimes observe in chronic diseases, viz., an abscess suddenly appearing in a part remote from the original seat of the disease. For in such cases the abscess quickly forms, and with this peculiarity, that we feel the fluctuation of pus although we have scarcely any marks of preceding or accompanying inflammation : the skin is not reddened until the ulcerative process is about to give exit to the fluid. In such cases, although we may have a succession of abscesses, yet we have not any instance of a sudden swelling accompanied by a light blush of redness, and of its equally sudden disappearance, leaving after it only a slight degree of swelling, without any other symptom or trace of inflammation.

In short, this peculiar disease, the effect of slight wounds received in dissection, presents much less of inflammation of the wound or its vicinity than occurs in the various other diseases to which slight injuries more frequently give rise. Here it seems to produce mischief by exciting a fever, which, in its turn, induces a swelling and redness of very peculiar characters, although at length (if the patient chance to survive), it will end in inflammation and suppuration of the wounded limb.

Whatever difference of opinion may be entertained as to the nature of this affection, it will be allowed on all hands that, although some few have escaped, yet the plans of treatment hitherto pursued have all proved quite unequal to contend with so formidable a disease.

The plan which I would suggest as most likely to succeed (for as yet it is untried), is to administer calomel with the view of speedily exciting ptyalism. Many experienced practitioners I know are in the habit of combining this medicine with opium, when their object is to excite ptyalism very quickly. But I should prefer giving it in an uncombined form, and in doses of three grains every three or four hours. When administered in this manner it seldom fails to produce salivation in thirty-six or forty-eight hours, provided that the two or three first doses affect the bowels. In cases where the bowels are not moved by the first doses of this medicine it will be necessary, in order to ensure its effects, either to combine the calomel with some purgative, or occasionally to interpose a purgative. When thus administered it will very seldom disappoint our expectations. Mercury administered in other forms of fever has so often succeeded in effecting a cure that I think we may, with some confidence, anticipate its good effects when administered in the fever consequent on wounds received in the dissection of very fresh human bodies.

POSTSCRIPT.—Since the preceding pages were written a very valuable essay on the subject of Diffuse Inflammation has appeared in the first volume of the Transactions of the Medico-Chirurgical Society of Edinburgh, from the pen of Dr. Duncan, jun. The extensive collection of cases contained in that essay demonstrates the dangers which too frequently arise from slight wounds ; but I cannot agree with the author in considering all these cases as examples of diffuse inflammation. Some few of them obviously are instances of the same disease as that of Mr. Shekelton. It is remarkable how exactly the morbid appearances in the thorax of the female subject, whose dissection gave rise to cases 8 and 9, corresponded with those of the subject which Mr. Dease was dissecting when he received his fatal wound.

As so few histories of the dissection of these fatal cases have been published, I shall make no apology for drawing the attention of the reader to this sentence in the report of morbid appearances in the 9th case :—" The cellular substance of the arm was everywhere healthy, and there was not the slightest vestige of disease in the forearm, nor could any connection be traced between the abrasion on the finger and the morbid parts."

I shall only add that every reliance may be placed on the accuracy of this account, for the dissection was made by Mr. Lizars.

Mr. Travers, of London, in his "Inquiry concerning Constitutional Irritation," a work rich in valuable materials, and replete with ingenious reasoning, considers the effect of slight wounds received in dissection as a distinct disease; and he seems inclined to ascribe the mischief to the introduction of a poison, which he thinks is generated by the first stage of decomposition of the human body.

I shall not pretend to decide whether the disease be owing to this cause, or rather (to use Mr. Travers's words), " to the condition of the remaining animalisation, which has hitherto resisted the operation of external agents"; but I must say that I cannot admit the general conclusion that " this disease, bearing a specific character, may be derived from absorption of the fluids of both fresh and stale bodies." No doubt it has arisen when the body under dissection has been cold, and nearly twenty-four hours dead, as certainly as where the subject was so recently dead as still to retain some of its warmth. I do not think, however, that we have on record a single well-marked case of this disease having arisen from the dissection of a body in which any of the obvious signs of putrefaction were present.

In No. 88 of the Edinburgh Medical and Surgical Journal is a paper by Dr. James Dease, Surgeon to the Forces, the second case of which is clearly an instance of this disease, and one in which calomel had been given pretty largely; but it does not appear that it was given with the view of inducing salivation. Ptyalism was not excited, and the case terminated fatally.

OBSERVATIONS ON SOME MORBID AFFECTIONS OF THE NAIL OF THE GREAT TOE.

(The Dublin Medical Journal, 1843.)

THE morbid affections of this part, which are attended by ulceration, do not appear to me to have received that attention from writers on practical surgery to which they are entitled, both on account of their troublesome and painful character as also on account of the very severe operations which have been suggested for their removal. Authors allude to these affections as if they were only different stages of the same disease, viz., onychia, and for their cure have recommended the evulsion of the entire nail and the excision of all the ulcerated parts, together with the matrix of the nail. I hope, however, to show in the following observations not only that these diseases differ from one another in many essential points, but that also in all of them those very severe operations to which I have alluded may be dispensed with, and that the cure of these troublesome affections can be effected by means of a bloodless operation, and by external applications of a mild nature.

I shall first allude to that form of fungous ulceration which is described by writers under the title of "*the nail growing into the flesh.*" In this disease we observe, at the angle of junction between the anterior and external edges of the nail, an ulcerated fungus, into which this angle, as also a portion of the outer edge of the nail, are more or less sunken. The colour of the fungus is rather florid, its surface is smooth, the discharge is purulent, in small quantity and tolerably healthy, unless the parts have been irritated by too much exercise of the limb or by some external application or local injury; there is little or no surrounding inflammation, no enlargement of the toe, and the pain is in general very trifling, unless during exercise, when the weight of the body on the limb causes the nail to press into the soft substance of the fungus, which thus often induces considerable uneasiness and lameness.

This disease does not appear to me to have any tendency to spread to, or to involve the adjoining parts, as I have seen cases in which it has remained stationary for some months, and in one for two years, at the end of which period the symptoms were in no way more severe than at the commencement, although most writers assert that it generally passes into malignant onychia. The origin of this troublesome affection is usually attributed to the effects of a tight boot or shoe, or to some accident in cutting or breaking off the end of the nail; in many instances, however, no cause can be recollected or assigned for its occurrence.

The unaided powers of the constitution do not appear capable of curing this disease; hence various methods of treatment have been employed by different surgeons, such, for example, as (according to Dessault) introducing a small plate of tin under the edge of the depressed nail, and insinuating it between the latter and the fungus, with the twofold intention of compressing the fungus and of elevating and everting the edge of the nail, so as to make the latter completely overlap the former. This plan of treatment has not stood the test of experience, and Dessault himself admits that it is very tedious, and often very painful for several days, that the fungus does not entirely disappear before two months, and that even after this the metal plate must be occasionally worn to guard against a relapse.

Caustics of different sorts have been employed by others, and Mr. Wardrop, in his paper in the 5th vol. of the ‘Medico-Chirurgical Transactions,’ particularly recommends the Argentum Nitratum; I have employed this remedy myself, and have seen it also used by others, but, I regret to say, not with that degree of success which Mr. Wardrop's recommendation would lead us to expect. I have, however, found some benefit from this application; under it the fungus has become reduced in size, also less irritable, but these improvements have continued only for a few days, after which the parts returned to their former condition.

Sir Astley Cooper, in his lectures, and Dupuytren, in his Leçons Orales, recommend an operation for the cure of this disease, the directions for performing which are essentially the same by each of these distinguished surgeons, viz., pass one blade of a scissors beneath the nail, from its anterior edge up to the root, then cut through the entire length of the nail, next seize the outer segment with a strong forceps, and by means of these tear it off the toe.

This operation inflicts a great degree of suffering, because in this disease the nail is not, as in onychia, separated from the vascular and highly sensitive matrix, except only through a small extent of space, not more in any direction than a quarter of an inch at its external angle, and, therefore, the scissors pushed upwards between the nail and the adherent matrix, and the forcible evulsion of the former by the forceps, must cause exquisite pain, which, though of short duration, can be regarded as nothing short of actual torture. A similar operation is also recommended by Mr. Liston in his recently published work on Surgery. Notwithstanding these authorities in favour of this plan of treatment, I am by no means an advocate for this peculiarly painful and distressing operation, but on the contrary I believe we may be relieved from the necessity of performing it, and that we can in all instances effect a permanent cure by a very simple operation, and one comparatively free from suffering, namely, by confining the excision of the nail to so much only as is already detached from the matrix : all of this portion, as well as that imbedded in the fungus, must be removed.

To effect this I proceed as follows :—While an assistant, with a spatula, presses down the fungus, I seize with a forceps (with strong flat blades, like those of the torsion forceps) the edge of that portion of the nail which is to be removed. I then pass a probe with a thin flat extremity beneath the nail, close to the fungus, as high as it can go, taking care to direct it towards the outer edge of the nail; this enables me to judge how far the nail is detached. I then take a strong crooked scissors of a large size, with one sharp-pointed blade; this I introduce beneath the nail as far as the probe has directed me; with one stroke of the scissors I then cut off all this detached portion of the nail, while, by means of the forceps, I draw it away with moderate force. Should this attempt fail to remove it, I then re-examine the part with the probe,

and again introduce the scissors as high as they can be
pushed; a second cut will then complete the separation, and
admit of the easy removal of this part of the nail; this
second attempt is sometimes attended with a sharp momentary
pain, as the point of the scissors often enters a short distance
into the sensitive matrix.   The portion of detached nail
presents no change of texture whatsoever; but a few drops of
blood follow this operation.   The only dressing required is a
small bit of dry lint, which is to be pressed firmly with the
probe between the fungus and the edge of the nail.   In a few
hours the toe is free from pain, and the patient can walk
without any lameness or uneasiness in three or four days after
the operation.   The dressing continues perfectly dry, and
need not be changed until the fourth day; at this time the
fungus will be found much reduced in size, perfectly dry, and
of a firmer consistence.   A small bit of lint is to be reapplied
as before; it should not, however, be pressed down so firmly
as at the first dressing.   In the course of ten or fifteen days
the fungus will have entirely disappeared, and the parts be
restored to a healthy state.   I have never found it necessary
to apply the olive-shaped cautery to destroy the fungus,
as described by Dupuytren.   In the course of this treatment I
also advise the patient (in compliance with the rules laid down
by authors) to scrape the upper and outer surface of the nail
with a sharp penknife, or with a bit of glass, an instruction,
however, which the patient neglects as soon as he feels himself
relieved from pain.

It affords me much gratification to be enabled to state that
in no instance have I met with a case of relapse after this
operation has been efficiently performed.   But the result of the
operation itself is not in all cases so successful as I have
hitherto represented, for in some instances, four or five days
after the operation, the patient will complain of some uneasiness
in the toe, when we shall find, on examination, that the
dressing is moistened with a little discharge, and that a
small portion of a whitish substance, like soft and swollen
leather, is rising up through the fungus.   This substance,
may be, is regarded as a sort of accessory ungual filament,
arising close to the original nail from the anterior and outer

border of its matrix, and which is now altered in texture and direction : this filament is so soft that it breaks and tears, if caught by the common dissecting forceps ; in order, therefore, to remove it (which it is necessary to do), we must seize it with the torsion forceps, and excise it with one cut of the scissors, passed well and fully beneath it; the lint dressing is to be reapplied in the manner before mentioned, and the case will proceed without further interruption to a perfect cure.

Before I dismiss this subject of " the nail growing into the flesh " I may add that I have seen many cases of this disease remaining unchanged for months, and even for a year and upwards ; and I have never yet met with an instance in which the ulceration has extended so as to induce that more severe disease of " onychia maligna," although some writers of high reputation have described this latter as an increase or extension of the former.

There is another morbid affection which occasionally engages the anterior and inner angle of the great toe-nail, and which causes considerable lameness and uneasiness, particularly on pressure ; this affection is often mistaken for an attack of gout, particularly in those persons where such an attack may be expected, or even desired. In this disease there is no swelling or redness, but pain, on pressure, at the anterior and internal angle of the nail. On a close examination of this spot we find that this angle rests on a hard white mass of laminated, horny cuticle, which we can easily remove in bran-like scales, when we shall see a small cup-like cavity without ulceration or disease. The ungual angle appears thick and bulbous opposite this point, and the pain is caused by its pressing against this mass. This affection is easily cured by scraping away all this substance and exciting the bulbous angle of the nail, and then interposing a little lint. Attention is required for some little time to remove any unhealthy growth of the cuticle or nail, and to secure the patient from any further uneasiness. Finally, I may remark, I have never seen this disease engage the outer angle, neither have I seen that last described engage the inner angle of the toe-nail.

ONYCHIA MALIGNA OF THE GREAT TOE.

This painful disease is often the result of some local injury, such as a heavy weight falling upon or rolling over this part; it occasionally occurs without any assignable cause. In some instances the nail is found very much altered from the healthy state, being of a dirty brown or black colour, and adhering to its matrix in detached spots only, while it is separated from it both at the root and to a considerable extent of its under surface. In other cases it has been altogether thrown off, and a foul ulcer occupies its site : this ulcer presents jagged edges and unhealthy surface, with an ichorous discharge, and extends along the toe above or beyond the matrix of the nail; both phalanges are swollen, so that the circumference of the toe is increased by at least a quarter of an inch, the last phalanx presenting a very peculiar bulbous shape. When the original nail has been cast off, we usually see projecting from the sides and tarsal border of the ulcer a narrow plate of a white substance, not unlike white leather soaked in water; this sometimes forms one continuous shelf all round the ulcerated border, projecting in a peculiar prominent manner, that is, rather at an angle to, instead of being on a plane parallel to, the dorsal surface of the phalanx. In some cases this white substance (which is the result of an abortive attempt to produce a true nail) appears only in detached spots or flakes, the intermediate parts of the ulcerated margin being devoid of any such growths; they are most frequently seen at the posterior and anterior angles of the nail, but occasionally in other parts of the circumference. The surrounding integument is discoloured, being often of a livid or purplish tint; it is also indurated, and exudes a copious perspiration of a peculiar heavy odour. This ulceration sometimes induces caries of the bone, and even extends to the phalangeal articulation.

All modern authors concur in the opinion that onychia depends upon the morbid condition of the secreting matrix of the nail, and that it can be cured only by removing, by excision, this diseased organised structure ; and no writer that I can recollect encourages the idea of a permanent improvement being possibly effected by any other local treatment.

I admit that the operation of the complete removal of the entire of this deceased matrix does effect the cure in a very short space of time, provided the bone or joint is not diseased (in which case amputation is inevitable), and that subsequently rest and simple dressing will alone accomplish the healing process, the place of the nail being supplied by a dense hard skin. But still this operation is not without objection; it is always attended with severe pain, and which sometimes continues for many hours; it also too frequently happens that the disease returns in some one spot or other, owing to the matrix not having been wholly eradicated, which indeed it is often extremely difficult to do; for the shape of the toe is so bulbous and so deformed, the textures so changed and so condensed by chronic inflammation, and the edges of the ulcer are so raised over the part to be removed, that even an anatomist cannot easily recognise the relations of the several tissues involved in the disease, or ascertain the exact extent of the substance to be excised; during the operation also the flow of blood is so profuse as to prevent the operator distinguishing one texture from another; accordingly, it occasionally happens that the patient, having enjoyed an immunity from pain for some days after the operation, becomes alarmed by feeling a slight return of his former uneasiness on any exercise of the limb, or on any pressure on some particular spot, generally on one of the angles of the original ulcer; and the surgeon, on a careful examination of this spot, finds there is still a little ulceration and a fresh production of that ungual growth already described as abnormal both in texture and in direction, and which indicates the persistence of some of the diseased matrix. This will again act as a foreign body, exciting constant irritation, and will soon lead to a renewal of all his former sufferings, and for the removal of which the excision of this morbid tissue must be repeated. Any surgeon who has experienced this disappointment can, I am confident, bear testimony to the horror with which the patient has contemplated, and the reluctance with which he has submitted to, a repetition of this peculiarly painful operation. I have known it necessary to repeat this even a third time; and therefore, whenever these secondary operations become necessary, I should remind the surgeon that to insure success he must cut much more deeply and extensively than he might at first suppose to be necessary.

Notwithstanding these objections to this operation of excising the diseased matrix, I yet admit that when it is properly performed it will be followed by a speedy and a perfect cure; however, I am of the opinion, founded on some experience, that this disease may be cured, without resorting to such a painful measure, by external applications alone, which are unattended with pain, and which, in a few days, will induce a considerable amendment, and even a perfect cure in the course of three or four weeks. The plan of treatment which I have pursued for some years, and as yet with invariable success, is as follows:—I confine the patient to bed, and direct a poultice to the toe for two or three days; I then cleanse the ulcer carefully by directing on it, from some height, a small stream of tepid water from a sponge; I next cut away as much of the loose nail as I can without paining or irritating the sensitive surface around, and then I fumigate the part by means of a mercurial candle, containing ʒi. of the Hydr. Sulphuretum Rubrum (olim cinnabar), to two ounces of wax.* This fumigation is to be applied night and morning, and after each the toe should be gently enveloped in lint or linen, lightly spread with Ung. Spermaceti. In four or five days the patient will express himself as considerably relieved; the discharge from the ulcer will be found of a healthy purulent character, and the appearance of the whole part much more favourable. The fumigation is still to be persevered in, and all projecting portions of nail to be closely cut; I consider this latter direction as very essential, as thereby the mercurial fumes can have more free access to the surface of the ulcer. In proportion as the ulcer improves, it is interesting to observe so does the condition of the growing nail; it acquires not only its natural firm and horny consistence, but also assumes its proper horizontal direction. For some time after the general surface of the ulcer has been healed there still remain small spots of ulceration, generally at the angles, around some white germs of new nail. Against these

* See Colles's Practical Observations on the Venereal Disease, p. 76.

points the full force of the mercurial vapour should be directed; this can be effected by adding a small conical ivory tube to the funnel. I attribute much of the success of this treatment to the use of the mercurial candle in preference to fumigation in the ordinary mode. I must observe that during this course of treatment the patient must absolutely abstain from walking, or even standing on the affected limb; exercise but for a single day will counterbalance all the amendment produced by a week's rest and fumigation. In the cases which I have thus treated I have had no recourse to any constitutional treatment, or to any peculiar regimen. No doubt cases must occur in which there will be derangement of the general health, and which will require suitable constitutional remedies before we can expect to cure the local disease by mere local applications. I make no doubt that this ulceration may sometimes arise from, or be so intimately connected with, some constitutional derangement as to be incurable without the aid of constitutional treatment. Mr. Wardrop, in his paper above alluded to, has recommended a cautious and judicious use of mercury, and has recorded some instances of its success.

The uniform success which has attended this plan of treating onychia by mercurial fumigation—during the last few years I have adopted it—leads me to hope that the surgical operation I have already alluded to may in future be dispensed with; but even should more extended experience prove that I have been too sanguine as to this expectation, yet I feel persuaded that this practice cannot be injurious, and that it will never fail to improve the condition of the ulcer and of the surrounding parts, reducing the enlargement and removing the deformity of the toe; so that, even should the operation be ultimately necessary, the surgeon will be better enabled to ascertain the exact situation and extent of the disease, and thus by a single operation to remove the whole of the matrix so effectually as to secure his patient from any relapse; therefore the time which has been occupied in the fumigating process cannot be considered as lost or misapplied.

ON THE FRACTURE OF THE CARPAL EXTREMITY OF THE RADIUS.

(Edinburgh Medical and Surgical Journal, Vol. 10, 1814.)

THE injury to which I wish to direct the attention of surgeons has not, as far as I know, been described by any author; indeed the form of the carpal extremity of the radius would rather incline us to question its being liable to fracture. The absence of crepitus, and of the other common symptoms of fracture, together with the swelling which instantly arises in this, as in other injuries of the wrist, render the difficulty of ascertaining the real nature of the case very considerable.

This fracture takes place at about an inch and a half above the carpal extremity of the radius, and exhibits the following appearances :—

The posterior surface of the limb presents a considerable deformity; for a depression is seen in the fore-arm, about an inch and a half above the end of this bone, while a considerable swelling occupies the wrist and metacarpus. Indeed, the carpus and base of metacarpus appear to be thrown backward so much as on first view to excite a suspicion that the carpus has been dislocated forward.

On viewing the anterior surface of the limb we observe a considerable fullness, as if caused by the flexor tendons being thrown forwards. This fulness extends upwards to about one-third of the length of the fore-arm, and terminates below at the upper edge of the annular ligament of the wrist. The extremity of the ulna is seen projecting towards the palm and inner edge of the limb; the degree, however, in which this projection takes place is different in different instances.

If the surgeon proceed to investigate the nature of this injury he will find that the end of the ulna admits of being readily moved backwards and forwards.

On the posterior surface he will discover, by the touch, that the swelling on the wrist and metacarpus is not caused entirely by an effusion among the softer parts; he will perceive that the ends of the metacarpal and second row of carpal bones form no small part of it. This, strengthening the suspicion which the first view of the case had excited, leads him to examine, in a more particular manner, the anterior part of the joint; but the want of that solid resistance which a dislocation of the carpus forward must occasion forces him to abandon this notion, and leaves him in a state of perplexing uncertainty as to the real nature of the injury. He will, therefore, endeavour to gain some information by examining the bones of the fore-arm. The facility with which (as was before noticed) the ulna can be moved backward and forward does not furnish him with any useful hint. When he moves his fingers along the anterior surface of the radius he finds it more full and prominent than is natural; a similar examination of the posterior surface of this bone induces him to think that a depression is felt about an inch and a half above its carpal extremity. He now expects to find satisfactory proofs of a fracture of the radius at this spot. For this purpose he attempts to move the broken pieces of the bone in opposite directions; but, although the patient is by this examination subjected to considerable pain, yet neither crepitus nor a yielding of the bone at the seat of fracture, nor any other positive evidence of the existence of such an injury, is thereby obtained. The patient complains of severe pain as often as an attempt is made to give to the limb the motions of pronation and supination.

If the surgeon lock his hand in that of the patient and make extension, even with a moderate force, he restores the limb to its natural form, but the distortion of the limb instantly returns on the extension being removed. Should the facility with which a moderate extension restores the limb to its form induce the practitioner to treat this as a case of sprain, he will find, after a lapse of time sufficient for the removal of similar swellings, the deformity undiminished. Or, should he mistake the case for a dislocation of the wrist, and attempt to retain the parts *in situ* by tight bandages and splints, the pain caused by the pressure on the back of the wrist will force him to unbind them in a few hours; and if they be applied more loosely, he will find, at the expiration of a few weeks, that the deformity still exists in its fullest extent, and that it is now no longer to be removed by making extension of the limb. By such mistakes the patient is doomed to endure for many months considerable lameness and stiffness of the limb, accompanied by severe pains on attempting to bend the hand and fingers. One consolation only remains, that the limb will at some remote period again enjoy perfect freedom in all its motions, and be completely exempt from pain; the deformity, however, will remain undiminished through life.

The unfavourable result of some of the first cases of this description which came under my care forced me to investigate with peculiar anxiety the nature of the injury. But while the absence of crepitus and of the other usual symptoms of fracture rendered the diagnosis extremely difficult, a recollection of the superior strength and thickness of this part of the radius, joined to the mobility of its articulation with the carpus and ulna, rather inclined me to question the possibility of a fracture taking place at this part of the bone. At last, after many unsuccessful trials, I hit upon the following simple method of examination, by which I was enabled to ascertain that the symptoms above enumerated actually arose from a fracture seated about an inch and a half above the carpal extremity of the radius.

Let the surgeon apply the fingers of one hand to the seat of the suspected fracture, and, locking the other hand in that of the patient, make a moderate extension until he observes the limb restored to its natural form. As soon as this is effected let him move the patient's hand backward and forward, and he will, at every such attempt, be sensible of a yielding of the fractured ends of the bone, and this to such a degree as must remove all doubt from his mind.

The nature of this injury once ascertained, it will be a very easy matter to explain the different phenomena attendant on it, and to point out a method of treatment which will prove completely successful. The hard swelling which appears on the back of the hand is caused by the carpal surface of the radius being directed slightly backwards instead of looking directly downwards. The carpus and metacarpus, retaining their connections with this bone, must follow it in its derangements, and cause the convexity above alluded to. This change of direction in the articulating surface of the radius is caused by the tendons of the extensor muscles of the thumb, which pass along the posterior surface of the radius in sheaths firmly connected with the inferior extremity of this bone. The broken extremity of the radius being thus drawn backwards causes the ulna to appear prominent toward the palmar surface, while it is possibly thrown more towards the inner or ulnar side of the limb by the upper end of the fragment of the radius pressing against it in that direction. The separation of these two bones from each other is facilitated by a previous rupture of their capsular ligament, an event which may readily be occasioned by the violence of the injury. An effusion into the sheaths of the flexor tendons will account for that swelling which occupies the limb anteriorly.

It is obvious that in the treatment of this fracture our attention should be principally directed to guard against the carpal end of the radius being drawn backwards. For this purpose, while assistants hold the limb in a middle state between pronation and supination, let a thick and firm compress be applied transversely on the anterior surface of the limb, at the seat of fracture, taking care that it shall not press on the ulna; let this be bound on firmly with a roller, and then let a tin splint, formed to the shape of the arm, be applied to both its anterior and posterior surfaces. In cases where the end of the ulna has appeared much displaced, I have laid a very narrow wooden splint along the naked side of this bone. This latter splint, I now think, should be used in every instance, as, by pressing the extremity of the ulna against the side of the radius, it will tend to oppose the displacement of the fractured end of this bone. It is scarcely necessary to observe that the two principal splints should be much more narrow at the wrist than those in general use, and should also extend to the roots of the fingers, spreading out so as to give a firm support to the hand. The cases treated on this plan have all recovered without the smallest defect or deformity of the limb, in the ordinary time for the cure of fractures.

I cannot conclude these observations without remarking that were my opinion to be drawn from those cases only which have occurred to me, I should consider this as by far the most common injury to which the wrist or carpal extremities of the radius and ulna are exposed. During the last three years I have not met with a single instance of Dessault's dislocation of the inferior end of the radius, while I have had opportunities of seeing a vast number of the fracture of the lower end of this bone.

THE foregoing short but accurate description of the fracture, now so very generally spoken of as "Colles's Fracture," is, without doubt, of all his writings that by which Mr. Colles's name is most universally known. If we except the observations of Petit and Pouteau, there are no earlier accounts of the injury, and the characteristic deformity accompanying it, worth being referred to.

Since the appearance of Mr. Colles's memoir few subjects in surgery have been more carefully studied: innumerable essays and lectures have been published about it, and its literature has attained large proportions. Much additional light has been thrown on the exact pathology of the injury in question, and many excellent suggestions have been offered as to its treatment. Yet 1 venture to think that few more accurate accounts have been given of the symptoms and appearances by which the surgeon may recognise this fracture in the vicinity of the wrist-joint than that contained in Mr. Colles's paper. As Professor R. W. Smith says, the most important diagnostic signs are given by Colles; and when we recollect that at the time the account was published the author had had no opportunity of investigating *post mortem* the nature of the accident, it must be admitted that he conjectured its anatomical characters with tolerable accuracy. Sir Astley Cooper says, " I have seen this accident frequently, and at first did not understand the nature of the injury; indeed, dissection alone taught me its real character;" and yet, although furnished with the information derived from *post mortem* examination, he observes, " that there is an evident projection of the radius and ulna on the dorsal surface, and of the carpus on the palmar surface of the forearm."* Surely, adds Professor Smith,† dissection could not have

----

* Treatise on Dislocations and Fractures. Edited by Mr. Bransby Cooper, p. 494.

† A Treatise on Fractures in the vicinity of Joints. Dublin, 1847.

taught him this, for the carpus is displaced backwards with the lower fragment of the radius, and the inferior extremity of the ulna projects in front and internally.

The fracture is usually the result of a fall upon the palm of the hand, and is liable to happen whenever a person, in the act of falling forwards, throws out before him his arms and hands in a state of extension, which he does, as it were, instinctively, to save the head and face from injury. Under these circumstances (if luxation of the bones of the forearm at the elbow-joint does not occur), from the influence, upon the one part, of the weight of the body and impulse of the fall, and, upon the other, of the resistance given to the hand by the ground, the radius, which receives almost the whole force of the shock, breaks at its weakest part, that is, its lower extremity, for it is here that the cellular structure is most abundant, and the compact tissue thinnest ; the carpus escapes uninjured, owing to the number of its articulations, which, as it were, divide and decompose the shock, and it is further protected by the palmar fascia, and by the numerous tendons which traverse the front of the carpus.

Severe pain is immediately felt at the seat of the fracture; the patient is sensible of having sustained some severe injury, and is in general conscious of something having given way in the limb ; he finds that the hand is powerless, that he cannot, without material aggravation of suffering, allow it to hang unsupported ; and therefore the limb is usually presented to us for examination resting by its ulnar margin on the palm of the other hand, and usually in a middle state between pronation and supination. If we desire the patient to supinate the hand, he attempts to do so, and generally succeeds in the effort ; but if we watch closely the manœuvre by which he accomplishes the movement we find that the motion takes place not in any part of the forearm, but in the shoulder-joint ; the patient rolls the head of the humerus outwards, and frequently is obliged to incline the entire shoulder towards the side of the injured limb before he can accomplish such an amount of rotation as will be sufficient to supinate the hand. He finds it extremely difficult and painful to maintain the forearm and hand in the horizontal posture by the unaided efforts of the muscles ; the fingers are usually flexed, and the neighbourhood of the wrist presents a singularly distorted appearance ; the lower fragment of the radius and the carpus incline to the side of supination, while the shaft of the radius tends to the side of pronation ; one fragment, the superior, is slightly drawn forwards, while the lower undergoes a considerable displacement backwards, and causes a remarkable prominence upon the dorsal surface of the forearm ; immediately above this projection is seen a well-marked sulcus, the direction of which is generally somewhat oblique from above, downwards and inwards, towards the lower end of the ulna.

This obliquity is seen even in cases in which the line of fracture is accurately transverse, and is owing to the double displacement which the lower fragment undergoes.

The dorsal prominence extends across the entire breadth of the fore-arm, but, as might be expected, is most striking towards its radial border, and gradually sinks as it approaches the ulnar margin of the forearm. Upon the palmar surface of the limb there is likewise seen a tumour, which is less prominent than that upon the dorsal aspect, but more extensive, reaching up a considerable distance along the forearm ; it ceases below at the annular ligament of the carpus, where a deep and narrow transverse sulcus is seen, and which remains evident, even though a considerable amount of swelling has set in : these two prominences are not placed upon the same level, and, generally speaking, the dorsal is more evident and striking than the palmar, for in the latter direction the swelling is usually greater, and the form of the tumour not so circumscribed. This want of correspondence between the anterior and posterior projections is one cause of the appearance of obliquity which so many of these cases present.

The lower extremity of the forearm assumes a rounded form ; its antero-posterior diameter is increased, but though there may be some, there is by no means a corresponding diminution in its transverse measurement ; the alteration in the latter direction is much more marked when the fracture engages the body of the bone about two or three inches above its lower extremity ; it depends upon the displacement, towards the ulna, of the upper end of the lower fragment, the fracture being in general above the pronator quadratus muscle, the influence of which is then, of course, exerted upon the inferior fragment alone. The inter-osseous interval, however, in the situation of Colles's fracture is so narrow that the deformity arising from this cause is very slight, and the diminution in the transverse diameter of the forearm scarcely perceptible.

The convexity of the ulnar edge of the forearm is slightly increased, but when the eye is cast along the radial margin of the limb the latter is observed to present a double concavity, one (supposing the arm to hang by the side) directed forwards, and projects at the ulnar border of the carpus.

The most usual seat of the fracture is from three-quarters of an inch to one inch above the radio-carpal articulation : sometimes it is only a quarter of an inch above the joint, but I have never seen it higher than one inch ; it always *appears* to be higher than it really is ; but should the lesion of the bone take place at two inches or two inches and a half above the radio-carpal articulation, the injury no longer presents the peculiar and remarkable characters which distinguish Colles's fracture of the radius.

Professor R. W. Smith,[*] in his admirable treatise, next turns to the pathology of the injury. I am not acquainted with any abler or more accurate criticism of the view of those who have maintained that fractures

---

[*] A Treatise on Fractures in the vicinity of Joints, &c. By Robert William Smith. Dublin, 1847.

of the lower end of the radius, within an inch of its carpal surface, are at least in general, if not always, examples of impacted fractures. This view has been advocated with much ingenuity by Voillemier; Nelaton has, in fact, adopted the theory in his 'Elemens de Pathologie Chirurgicale, t. i., p. 742, and the late Mr. Callender produced in support of it actual recent pathological specimens. That "*fractures par pénétration*" are, however, those most usual when a breach of continuity occurs within an inch of the carpal end of the radius can hardly any longer be maintained in face of the evidence accumulated by Professor Bennett, to which I shall presently refer.

"In a fall upon the palm of the hand," writes Monsieur Voillemier,[*] "where the radius rests chiefly upon the dorsal surface of the upper range of carpal bones, the whole force of the shock is transmitted to the radius, tending to approximate its extremities; it breaks, therefore, at its weakest part, which is a short distance above the wrist-joint, where the compact structure of the shaft is thinnest, and the cancellated tissue most abundant; the compact tissue of the upper portion is, by the continuance of the force after the fracture has taken place, driven into the cancellated tissue of the lower fragment. This may occur in several ways: if the end of the bone is large; if the shock has been transmitted in the direction of its axis; and if the bone has yielded throughout its whole circumference at the same moment; then the whole extremity of the upper fragment penetrates into the lower, and thus the two fragments are locked together. But if the force of the shock has been very considerable, then the upper fragment continues to descend, and the lower, compressed between it and the carpus, is split into several pieces; the styloid process is broken off, and the lower end of the bone crushed. Dupuytren has alluded to cases of this description, and has termed them 'Fractures par écrasement.' This mode of penetration is, however, as might be supposed, very rare.

"But there is another mode of penetration, and the more common; it also happens from a fall upon the hand, the force of the shock coming obliquely on the radius, so that the posterior surface of the bone has to support the greater part of it; the lower fragment is at the same time carried backwards, and thus the posterior wall of the compact tissue of the upper fragment begins to penetrate into the spongy structure of the bone, while the anterior wall glides upon the palmar aspect of the carpal fragment.

"In all the specimens of fracture of the lower end of the radius, examined long after the occurrence of the accident which broke the bone, a line of compact tissue is found, covered over with cancellated structure, directed vertically, and descending to within a greater or less distance of the articular surface of the radius, as the case may be. Free below, it is manifestly continuous above with the body of the bone, the compact wall of which becomes all at once thicker where it is joined by this line.

---

[*] Archives generales de Medicine. May, 1842.

This line is never double; it is continuous with the posterior wall of the bone when the fracture is attended with displacement of the lower fragment backwards, and with the anterior when the inferior fragment is thrown forwards.

"The explanation of this fact, not hitherto noticed, becomes sufficiently simple if we admit the doctrine of fracture by penetration. Let us take, for example, a case of fracture with displacement of the lower fragment backwards—the more common species : in consequence of the peculiar nature of the displacement (*renversement*) the penetration can only happen posteriorly, for in front "*ce véritable mouvement de bascule*" will tend rather to separate the two anterior edges of the fracture, or will leave them merely in contact. Thus the line of compact tissue can only exist posteriorly, for it is only the compact tissue of the posterior surface of the upper fragment that penetrates the cancellated tissue of the carpal end of the bone.

"In the case of reciprocal penetration the posterior border of the upper penetrates the lower fragment, and the anterior wall of the carpal end of the bone ascends into the superior fragment. In this case there should be two lines of compact tissue, when the bone is divided from before backwards; one only, however, is found. The explanation of this fact offers itself when we reflect upon the extreme thinness and delicacy of the layer of compact tissue which invests the front of the lower fragment, in consequence of which it is broken and destroyed at the moment of penetration into the upper fragment, but posteriorly it is the thick resisting compact structure of the shaft that penetrates the carpal fragment. We cannot judge of the extent to which this penetration has taken place by the length of the line of compact tissue, for this always appears longer than it really is, for the periosteum, in passing from one fragment to the other, fills up in part the depression produced at the level of the fracture; there is also a remarkable thickness of the compact structure of the lower end of the bone, in old fractures, of which it is difficult to give an explanation."

Monsieur Voillemier further supposes that the circumstance of the fracture being impacted explains the difficulty of detecting crepitus in cases of fracture of the lower end of the radius.

This theory of M. Voillemier is admitted by Professor Smith to be ingenious; nay more, he admits that the appearances are quite correctly described. In every instance of Colles's fracture which he had an opportunity of examining, he found, upon making a section of the bone from before backwards, a line of compact tissue continuous with the posterior wall of the shaft, and extending into the reticular structure of the lower fragment.

But, he adds, do the appearances above described warrant us in concluding that these inquiries are examples of fracture with penetration ? This question, after a careful and exhaustive investigation, Professor Smith answers in the negative. According to him the compact structure of the shaft of the radius *appears*, in old cases where union has long since taken place, to have penetrated the lower fragment in consequence of its having become encased in osseous matter deposited for the union of the fracture.

The impaction theory is liable to the following objections:—1. The distance between the line of compact tissue and the posterior wall of the lower fragment is but the measure of the amount of displacement backwards of that fragment. 2. This interspace is considerable even in those cases in which the fragments are found to be on the same plane in front. 3. Were the theory correct the amount of shortening of the posterior surface of the bone should be much greater than it ever is. 4. There is no correspondence between the length of the line of compact tissue and the amount of shortening of the back of the radius. 5. The possibility of either fragment being driven into the other is difficult to be conceived as long as the ulna remains entire and the ligaments uninjured. 6. Were it possible to separate the united fragments, and draw down the inferior so as to extricate it from the apparent impaction, we would not thereby succeed in restoring the normal form of the lower end of the radius.

In addition to these arguments Professor Smith adds that the appearances disclosed by the examination of recent specimens are opposed to the impaction theory. This, however, we now know is not altogether true. The examples of recent fracture dissected by the late Mr. Callender and others prove beyond doubt that penetration may take place, and that, too, without any fracture of the ulna.

Professor Bennett, by a careful examination of a large number of specimens, seems to have placed himself in a good position to decide between the relative merits of those who have advocated the impaction or the non-impaction theory, and of whom Voillemier and R. W. Smith are the respective champions. His memoir* appears to me so valuable a contribution to the pathology of this subject that I do not hesitate to quote from it at some length the part referring to this topic:—

" Since their views have been published, we find authorities ranging themselves on opposite sides, according as they adopt the non-impaction theory of Smith or the opposing doctrine of Voillemier, that all these injuries are fractures by penetration. Of late the number of recent dissections have forced most writers to admit the occurrence of each form. We have ample proof that the simple transverse fracture without impaction, the impacted, and the fracture *par écrasement* with a shattered lower fragment, are all possibilities. Intimately connected with the theory of impaction is the explanation of the variability of the symptom

* Remarks on Colles's fracture, &c. By Ed. H. Bennett, M.D., M.Ch., Professor of Surgery in the University of Dublin, &c. Read in the Section of Surgery at the Annual Meeting of the British Medical Association in Cork, August, 1879.

so much relied on in the diagnosis: crepitus. Yet anyone who will take the trouble to investigate the statements contained in the accounts of recent cases which have been submitted to dissection may find that this symptom has been absent in each form of the injury. We cannot, therefore, rely on its absence as a proof of the existence of impaction in any given case, nor are we for the same reason to abstain from attempting reduction of the deformity for fear of undoing the impaction, or again rely on it as a safeguard against displacement, and so adopt a lax treatment. I will notice here that, valuable as recent dissections are in guiding our examination of united specimens, too high an importance has been attached to them by some. Certainly I cannot agree with Mr. Callender that the dissection of three cases where the fracture was caused by exceptionally great violence are sufficient 'to remove all doubt on this subject.' I cannot agree with him even when he makes the following remark :—' There can be no question but that impaction is the cause of the displacement.' This surely is to mistake the effect for the cause. The following passage from Voillemier shows that, however fully impressed he was in favour of the theory of impaction, he avoided the error of committing himself to its absolute adoption :— ' Mais il faut être prévenu que dans certains cas de pénétration peu marquée, de fractures par arrachement, il peut n'y avoir aucun changement dans les axes du membre, pas de raccourcissement, pas de mobilité des fragments, pas de crépitation et seulement de la douleur du gonflement autour du poignet, enfin tous les symptômes d'une entorse, et ces cas ne sont pas très-rares.'

"Here the author admits a group of injuries sufficiently common, in which impaction is absent or inappreciable; in which, too, crepitus and deformity are absent. If such cases be overlooked, as they too commonly are, the most extreme deformity establishes itself in a short time, as I have more than once observed. These facts are certainly inconsistent with the theory that deformity, the result of displacement, is solely due to impaction, and that muscular action takes no part in its production.

"In opposing Voillemier's views Smith falls into the opposite error, in almost absolutely denying the occurrence of impaction. He says:—'I am inclined to believe that the doctrine of fracture with penetration is untenable.' He says, indeed, in another passage, ' Until, therefore, the result of the examination of recent specimens can be adduced in support of the theory, I shall be inclined to believe that impaction is only apparent.' I have the good fortune to be able to exhibit here the specimen, the recent dissection of which seems to have so strongly influenced his opinion, a specimen familiar to the readers of his book, which the College of Surgeons have kindly exhibited here. This specimen, with many others since recorded, obtained by *post-mortem* examination and by experiment, establish the existence of the simple fracture, no matter how many cases of the impacted variety may be found. The circumstance which seems to have confirmed him in his opinion is expressed in the following passage in that part of his paper in which he discusses the opinion of Voillemier on the value of the constancy of the third line of compact tissue seen imbedded in the cancelli of the lower fragment of united specimens :—' If it indicated the forcible impaction of the upper into the lower fragment, we should expect to find the latter split into two or more pieces, but this form of fracture of the lower end of the radius is comparatively rare.'

" I can prove from the collection before me, a collection made without any reference to this opinion, and certainly representative of the general features of the injury, that the opinion that comminuted fractures of the ower fragment are comparatively rare, is incorrect.

" I think, too, that I can show proof that Voillemier's opinion as to the relative frequency of the forms of impaction is not correct. He describes four possible varieties of impaction :—1. The penetration of both walls of the upper fragment into the tissue of the lower without its being burst asunder. 2. The same mode of penetration, with crushing of the lower fragment,—Dupuytren's fracture *par ecrasement*. 3. The bending or hinging, so to speak, of the anterior walls, with impaction of the upper into the lower fragment posteriorly. 4. The reciprocal penetration of the fragments, the upper into the lower behind, and the lower into the upper in front. Of these possible forms of impaction he believes the third to be the most common, founding his opinion apparently on the observation ' that the impacted line is never double '; an observation in which Smith agrees.

" I am able to show, among the specimens before the meeting, that exceptions occur to this rule ; further, that comminution is more common than Smith has estimated it to be, and that the paths of the fractures which occur in the lower fragment present a remarkable constancy in position. If I rightly interpret these appearances the conclusion must be arrived at that, in comminuted fractures, reciprocal impaction is the most common form of the accident.

" The number of specimens which I submit is in all sixty-two. Of these fifty-four are true Colles's fractures, transverse in the general direction of the fracture, the lower fragment displaced backwards, with more or less distortion in the direction of the inferior articular surface, the most constant element of which is dependent on the elevation of the styloid process. Four specimens are examples of the rare injury in which the lower fragment is displaced forwards ; three are epiphysary separations ; and in one a fracture has occurred, detaching the styloid process of the radius, with a small portion of the articular surface adjoining it.

" In estimating the frequency of comminution of the lower fragment I have included under this head only such specimens as exhibit distinct evidence of fracture having passed completely through the carpal articular surface ; many specimens present evidences of splintering of the non-articular surface of the lower fragment, and some of the upper also, but I think such should be excluded in the present discussion. Applying this test we find, of the specimens before us, thirty-one are simple, twenty-three comminuted, a fact that is inconsistent with Smith's statement 'that this form of fracture is comparatively rare.'

"Before stating the remaining facts exhibited by these comminuted specimens, I may delay a moment to consider the mechanism of the injury. Arguing *à priori*, as the path of Colles's fracture is generally transverse with regard to the long axis of the radius, and the displacement of the hand tends constantly in the direction of abduction, we should expect that the upper fragment, acting as a wedge, if this be the cause of the comminution, should split the lower first and most constantly at that border of the lower fragment next the ulna, against which the force is directed, and where the lower fragment is thinnest. This is exactly what the specimens before us prove to be the case. Among those of the simple fractures which border on the comminuted in their characters, we find a crack first appearing on the ulnar facette, reaching just to the carpal surface, placed at the centre of the curve of the facette.

"In ten of the twenty-three comminuted, a fissure, starting from the same point in the ulnar facette, runs into the carpal articular cartilage along its posterior edge, breaking out into the dorsal surface of the bone, either, in the least extensive injuries, at the outer side of the common extensor groove, or in others running along as far as the groove for the radial extensors, and, in a few, breaking out at each of these points. The frequency of this position of fracture has been already noticed by Packard.

"Its frequency appears to have led Rhea Barton into the error that there was a form of fracture limited to this region of the bone requiring to be distinguished in practice from Colles's fracture. His idea of the mechanism of the injury is quite the reverse of what we see in action here; for he supposed that pressure of the carpus in over-extension detached the fragment. In the eleventh specimen of the series a branch is traced passing off from the first fissure towards the anterior depression in the scaphoid facette; this fissure is seen in the bone, but, as we often find in such, has left the cartilage intact. In the twelfth specimen the fissure, starting as before, passes along the route indicated in the preceding, but without any posterior branch. In the next five specimens the primary fissure exists with its anterior and posterior branches. In three of the four remaining specimens the features of the comminutions are the same, but chronic rheumatic arthritis has masked their exact details, so that it is difficult to follow them. In one specimen only are the features of the lines of the fracture entirely different; in it an oblique line seems to have detached the styloid process, passing from in front backwards and inwards across the scaphoid facette.

"There appears then sufficient ground for the assertion that the mode of comminution is constant, and that it results from the impaction of the upper into the lower fragment, taking effect first and chiefly on the side of the fragment next the ulna.

" A further examination of these specimens shows that, where the fractures remain sufficiently distinct in outline to trace their details exactly, the impaction of the fragments is reciprocal ; that while the posterior surface of the upper pierces the lower, the anterior of the lower penetrates the upper, it may even lead to its splintering. This conclusion is at variance with the opinion of Voillemier ; but I believe the explanation of the discrepancy between us is, that hitherto the examination of comminuted specimens has not been made with sufficient care. I believe that in many simple fractures no impaction whatever occurs, and again, that in many injuries, Voillemier's third variety, the bending of the anterior wall with posterior impaction, occurs. I cannot resist the conclusion to which this collection leads, that in comminuted fractures, which occur in proportion to simple injuries almost in the ratio 2-3, the characters are constant, and the result of reciprocal impaction of the fragments."

I feel that the foregoing observations on Colles's fracture of the radius would be incomplete unless supplemented by some reference to the methods of treatment now usually approved of by surgeons. Upon this subject also much has been written. It would be tedious to enter at large into the merits of the various methods of treatment; yet I should not do justice to the subject without putting the reader in possession of the views of two surgeons who have devised splints for the treatment of this accident, and explained the rationale of its treatment in the most practical and lucid manner. I allude to Doctors Henry Bond and Alexander Gordon.

Dr. Bond read his original paper before the College of Physicians, Philadelphia, in December, 1851. It was published in their Transactions. The account here given is a full abstract derived from the 'American Journal of Medical Sciences,' vol. xxiii., 1852 :—

" The usual mode of treating these fractures, as is well known, consists in the use of two long straight splints, with compresses or cushions, and bandages. The palmar splint extends from the elbow down to the extremities of the fingers. Some, however, do not allow this to extend below the second joints of the fingers. The dorsal splint extends sometimes only to the extremity of the metacarpe. When this dressing is applied the longitudinal axis of the forearm will be continuous, or parallel, with that of the hand.

" There are several objections to this mode of dressing the fracture, which I will attempt to point out. In the first place, it violates what ought to be regarded as a surgical canon in the treatment of fractures, viz., to adopt such a position as will put all the muscles acting on the part as much in repose, as free from tension, as possible ; so that the least counteracting force will be required. 2. The constrained position of the hand demands tighter bandaging, in order to prevent derangement of the fragments by paralyzing or subduing the muscles that are rendered tense by the position assumed. 3. This constrained position and tight bandaging greatly increase the danger of that protracted or permanent rigidity of the hand and fingers which is a too frequent result of those injuries. 4. This mode of dressing, by long straight splints, not only increases the danger that it will result in rigidity, but that, when it does occur, the hand will be left unsightly, inconvenient, or useless. 5. There is another objection to it, which will be regarded by the surgeon as of more or less importance, according as he is actuated more or less by the feelings of humanity. I refer to the distress or discomfort which must result from a constrained position and the force applied to maintain it.

"The muscles that act on the hand are least tense, or most in repose, when the hand is inclined backwards, so that the metacarpe forms a considerable angle with the forearm,* when it is also inclined inwards towards the ulnar side of the arm, and when the fingers are moderately flexed. In this case it will be perceived that the longitudinal axis of the forearm, if prolonged, would not correspond with that of the hand, but would pass through or very near the point where the thumb and index finger most easily and naturally meet. Thus, in the innumerable manipulations with the thumb and fingers (as with a pen, pencil, button, needle, money, &c.), their points most easily and naturally meet in this axis of the forearm. This will be found to be the position of the hand when it hangs by the side with all the muscles relaxed.

"When the hand is placed in the position above described, so as to take off tension from all the muscles, there will be so little tendency to displacement of the fragments that a very gentle pressure of compresses and bandages will be adequate to maintain them in their proper relation to each other. The dressing may be removed earlier, so as to give motion to the hand and fingers without danger of producing derangement of the fragments, and the gentle pressure of the dressing will be less likely to deprive the tendons and sheaths of their lubricity, and thus to cause permanent adhesions.

"There are cases, as before observed, where such violence is done to the bones and soft parts, especially in elderly persons of a rheumatic or gouty diathesis, that it may be impossible to avoid permanent adhesions and rigidity. In such cases, if the usual authorized mode of treatment be adopted, the result will be a most awkward, unsightly, useless member.

"But if the hand can be placed and retained in the unconstrained natural position above mentioned (to say nothing of the better chance of escaping permanent stiffness), *in the first place* the unsightly deformity will be avoided, and, in the *next place*, the hand will not entirely have lost its uses. For the hand, thumb, and fingers, being placed very nearly in the position of their most frequent uses, the interossei, the lumbricales,

---

* Malgaigne calls this *la flexion habituelle de la main en arrière.*

and the several short muscles of the thumb will, by causing only a very limited motion, enable the hand to perform very many of its useful functions.

"I can say with confidence, not only from *a priori* reasoning, but from some experience within the last few years, that the dressing of a limb on the principles here inculcated will very materially conduce to the comfort of a patient. I shall here make no comment upon what is said about paralyzing the muscles by tight bandages, nor upon the power of the body to accommodate itself, by a very painful discipline, to very distressing necessities.

"The importance of the position of the hand in the treatment of fractures of the radius has been fully recognised for a long time by eminent surgeons. In these cases Mr. Cline did not allow the splints or the sling to extend below the wrist. His object was to let the hand, by its own weight and without any impediment, incline towards the ulnar edge of the forearm; and, while the ulna acted as a counter-extending force, this inclination of the hand would prevent the fragments from over-riding or overlapping each other, and make it very easy to keep them in apposition. He understood well the mechanism of this accident· When the radius alone is broken the ulna affords all requisite counter-extension, and in proportion as the hand is inclined towards the ulna will the lower fragments be drawn down, so that there will be hardly a chance for one fragment to overlap the other; certainly there will be little difficulty in keeping the fragments in apposition with very gentle means. But Mr. Cline's method of dressing, in order to accomplish the indication, was too indeterminate; he could not depend upon maintaining steadily the same degree of inclination of the hand, and one might suppose that there would be danger of producing artificial joints. Nevertheless, I am persuaded that with Mr. Cline's method of treatment with short splints there would be fewer cases terminating in deformity and loss of the use of the hand than when the arm and hand are tightly swathed in long straight splints.

"Sir Charles Bell long ago inculcated the importance of the inclination of the hand in the treatment of fractures of the forearm, and he has given a plate illustrating his opinion. Boyer is very explicit on this point. He says 'the extension should be made by inclining the hand towards the ulnar edge of the forearm.' Yet this obvious principle and this explicit direction are wholly disregarded in the present usual mode of dressing with long straight splints and tight bandages. Dupuytren, whose lectures on this subject should be studied by every surgeon, devised a splint—his *attelle cubitale*—with the special object of maintaining this inclination of the hand towards the ulna. Notwithstanding this great man devoted such deep attention to this subject, there were serious defects in his apparatus, which have been pointed out by subsequent French writers.

"I have attempted to devise a mode of dressing these fractures, having reference to the principles advanced in this paper, and that will meet the following indications :—

"1. To maintain such an inclination of the hand upon the forearm as shall most effectually relieve the muscles from tension or put them in repose.

"2. To maintain the hand and fingers in a position that, if rigidity should result, the member shall be as little an incumbrance and retain as many of its uses as possible.

"3. To make it easy of application, requiring no extraordinary skill or dexterity, and little liable to be deranged or displaced.

"4. To make the dressing easy and comfortable to the patient, while it does not lack efficiency.

"My own experience of its use within the last three years convinces me that I have to some extent accomplished these indications. How far this shall be corroborated by others can be known only when others shall have had time, opportunity, and disposition to test it.

"To enable others to test the principles herein maintained, in the mode of treating these fractures, I offer the following directions for preparing the dressing, with some explanations as to its application :—

"1. With a light board of proper thickness for a splint take a profile of the forearm and hand of the patient, placing the hand in its habitual inclination towards the ulnar side of the arm, and extending the profile from the elbow downwards, so that it shall reach the second joint of the fingers on the inside, when these are moderately flexed—as much flexed as they are when the points of the thumb and fingers are brought into contact. The lower end of the board must be cut off obliquely (at an angle of fifteen or eighteen degrees) in a direction corresponding with that of a body grasped in the hand, when the hand is inclined to the ulna as above indicated.

"2. Cover the board thus prepared with sheeting or other strong fabric. This may be done by winding around it, from end to end, a narrow rolling bandage, covering all of it as nearly as may be with few or no duplications. This is the most expeditious method. A neater one is to cut a piece of sheeting, of the general form of the board, but extending beyond the board on every side, and fastening it upon the board either by a few stitches drawing towards each other the overlapping edges, or gluing down those edges upon that side of the board which is to be towards the arm, and which edges are to be covered with the pasteboard.

"3. Prepare a block of soft light wood, from seven-eighths to eleven-eighths of an inch thick, and from two to two and a half inches wide, according to the size of the patient's hand, and of a length corresponding with the width of the board in the palm of the hand. This block is to be carved and rounded, so as to adapt it to the form of the hand and make it easy for the thumb, and in the grasp of the hand when it is placed on the board. It is to be fastened there by screws or nails, so that the remote edge of it shall correspond exactly with the lower oblique end of the board.

" 4. Upon that part of the board not covered by the palm-block fasten, by means of small carpet tacks, a piece of bookbinder's pasteboard, extending on each side beyond the edges of the board about an inch. If the pasteboard be very thick and stiff make a slight incision in it along the edge of the board, in order to bend more easily the two projecting portions of it, thereby making a kind of box for the lodgment of the arm.

" It seems to me that this splint, or one constructed on the same principles, will meet the above-mentioned indications in the following manner—1st. The form given to the board retains the hand in its habitual inclination towards the ulnar edge of the arm, accomplishing the object aimed at by Dupuytren's *attelle cubitale* with as much certainty, with more simplicity, and more comfort to the patient. 2nd. The palmar block retains the hand in its habitual inclination backwards, and it gives the fingers that moderate flexion which most relieves the muscles from tension, and likewise that position which, if stiffness should result, will not only save the hand from a most inconvenient, ungraceful deformity, but will reserve to it the power of performing very many of its most frequent and useful functions. In addition to these advantages this block contributes much to the comfort of the patient. 3rd. The object in covering the board with a strong fabric, as above described, is to retain the bandage with certainty in its place without applying it with a dangerous tightness; for, by fastening the roller to this covering with pins, the surgeon need never have his patience tried by finding his dressing deranged at his next visit. I can speak with confidence on this point from having used it repeatedly in cases where this quality was fully tested. 4th. The pasteboard is not an essentially necessary part of the splint, but it will be found to contribute to the comfort of the patient and the convenience of the surgeon.

" The requisites for dressing with this splint are flannel or other soft fabric, to cover or line the inside of the splint; two compresses; a roller; sometimes, but not always, a dorsal splint.

" The flannel or other fabric with which the splint is lined should extend a little beyond the edges of the pasteboard, and the same piece may be extended over the palmar block; but it will be better to cover this block with a separate piece. For this purpose take a piece of flannel large enough, when it is doubled, to cover the block. Through the doubled edge, with a proper needle, carry a small string (such as ligature-twine), and tie this around the splint immediately above the block. The covering of the block thus applied may be conveniently changed without removing the arm from its bed.

" Two compresses will generally be required : the anterior or palmar, and the posterior or dorsal. The proper construction and application of the former of these are a most important point in this dressing, and certainly not less so when long, straight splints are employed; and deformity of the radius or wrist will most frequently result from negligence or want of skill in its use. If the compress be deficient in thickness, and the bandage be applied with its usual tightness, there will not fail to be either a curvature forwards or a sigmoid flexure, which are the usual deformities. If the thickness of this compress be excessive there may be a curvature backwards, which I think seldom occurs; but there will be such undue pressure by such a compress as to increase the danger of adhesions, and to aggravate the discomfort of the patient.

"These preparations being made, the fragments, if deranged, are to be reduced, and the splint applied. I have very seldom, if ever, found any difficulty in replacing the fragments. Let an assistant, grasping the hand (not the fingers), incline it towards the ulnar side of the arm, according to the direction of Boyer, making steady, but not very vigorous, tension, and, with his thumbs and fingers, the surgeon will easily press the projecting fragments into their proper relation to each other. In making this extension Dr. Dorsay directs the assistant to 'grasp the hand.' Dr. Barton says an assistant ' makes extension from the fingers.' It seems to me that Dr. Barton errs on this point. The objections are, first, that by grasping and pulling the fingers the flexor muscles are brought into so much more tension that they will offer more resistance to the extending force. In the next place, by pulling at the fingers the hand will be brought more nearly into a direct line with the forearm, instead of being inclined towards the ulna, the utility of which inclination in such cases is, I think, so obvious that it would be superfluous to cite authority or to attempt a demonstration.

"After the forearm is laid into the splint apply the dorsal compress. This compress is seldom essentially necessary in these cases, but it may always be advisable to use it. Its thickness is comparatively unimportant, especially when a dorsal splint is not employed. It may be made of folds of a bandage of about the width of the wrist, and so long as to cover the lower fragment of the radius and the wrist, but not extend upon the hand. After adjusting this compress apply a roller, beginning upon the lower fragment of the radius, carrying it down over the wrist, the metacarpus, and the first joints of the fingers, leaving the thumb free; then returning with the bandage to the upper end of the splint, and attaching it in several places by pins to the woven covering of the splint. If the compresses have been properly made and adjusted it is unnecessary, with this splint, to apply the bandage with anything like the tension ordinarily employed in dressings with the long straight splints; and those accustomed to the use of these splints will be liable to err on this point."

No account of the treatment of Colles's fracture would be complete without some allusion to the ingenious apparatus devised by Dr. Alexander Gordon.* His "improved splint" is not readily described in words: it

---

* A Treatise on the Fractures of the Lower End of the Radius, &c. By Alexander Gordon, M.D. London: Churchill, 1875.

is figured in most of our surgical handbooks; but to see it and use it is the only satisfactory method of ascertaining its form, mode of application, and practical merits. It consists of two portions; one an anterior splint which goes on the front of the forearm, and has a comfortable support for the ulnar side of the hand, which is maintained in a partially flexed position; the second a curved splint, which, with a firm pad, is so applied on the back of the forearm as to press forward the carpus, and with it the lower fragment of the fractured radius.

"In consequence," says Dr. Gordon, "of the displacement, outwards and backwards, of the lower upon the upper fragment the lower end of the ulna becomes unnaturally prominent in front and internally. This is observed immediately after the accident; but if the primary deformity persists, the prominence of the lower end of the ulna will become more marked, its posterior surface more curved, its inner border more convex, and its head twisted so as to look outwards. The displacement backwards of the carpus on the carpal surface causes absorption of the posterior half of that surface, increasing its obliquity, and giving to it an aspect looking more backwards. This is one cause of the shortening of the posterior surface of the radius.

"In consequence of the accident the fractured surfaces have suffered more or less contusion, they inflame and soften. The backward pressure of the carpus, acting continuously on the fractured surfaces behind, causes interstitial absorption of a large quantity of the cancellous structure of the lower end of the bone.

"The outward and upward pressure produces effects even more decided. In one of twenty-seven specimens the styloid process has become almost horizontal and very much thinned, in five it is much absorbed, and in twenty-one it presents various degrees of absorption and displacement backwards; but the most marked displacement is that of the cancellated tissue outwards, to the extent of three-fourths of an inch in the worst instances. Remodelling of the whole of the cancellated tissue posteriorly and externally, with vertical shortening and transverse elongation, are the effects of the altered bearing of the carpus on the lower end of the radius. The compact tissue in front is not shortened, but it has become more or less convex. Behind, at the seat of fracture, the radius has become concave from the displacement backwards of the lower end of the lower fragment: there a new formation of cancellated bone has taken place, strengthening the bone at the broken part, helping it to resist the upward and backward pressure, and masking the exact course of the fracture behind.

"This new disposition of bone behind, with the shortening of the radius externally and behind, have led many surgeons to suppose that these alterations must necessarily be the result of impaction; an opinion which I hold to be now untenable, and which the production of the fracture artificially, controverts in the most decided manner. Colles's fracture is not, nor can it be, an impacted fracture; its mechanism declares impaction to be a mere phantom of the imagination, resulting from the erroneous interpretation of pathological facts."

It is impossible to concur in the last assertion. Impaction cannot be called a phantom of the imagination in face of the specimens accumulated by Professor Bennett, and the actual dissections of recent cases made by Callender, and published in Bartholomew Hospital Reports.

WEST, NEWMAN AND CO., PRINTERS, 54, HATTON GARDEN, LONDON.

www.ingramcontent.com/pod-product-compliance
Lightning Source LLC
Chambersburg PA
CBHW021343210326
41599CB00011B/739